Multichannel Optical Networks

Network Theory and Applications

Volume 1

Multichannel Optical Networks

by

Peng-Jun Wan

Department of Computer Science,
Illinois Institute of Technology,
Chicago, IL, U.S.A.

SPRINGER-SCIENCE+BUSINESS MEDIA, B.V.

Library of Congress Cataloging-in-Publication Data

Wan, Peng-Jun, 1970-
Multichannel optical networks / by Peng-Jun Wan.
p. cm. -- (Network theory and applications ; v. 1)

1. Optical communications. 2. Multichannel communication.
3. System analysis. I. Title. II. Series.
TK5103.59.W35 1999
621.382'7--dc21 99-23435

ISBN 978-1-4613-7426-8 ISBN 978-1-4615-5317-5 (eBook)
DOI 10.1007/978-1-4615-5317-5

Printed on acid-free paper

TO MY WIFE
XIAOFENG FOR HER
PATIENCE AND
UNDERSTANDING

Contents

List of Figures

List of Tables

Preface

The huge bandwidth of optical fiber was recognized back in the 1970s during the early development of fiber optic technology. For the last two decades, the capacity of experimental and deployed systems has been increasing at a rate of 100-fold each decade–a rate exceeding the increase of integrated circuit speeds. Today, optical communication in the public communication networks has developed from the status of a curiosity into being the dominant technology. Various great challenges arising from the deployment of the wavelength division multiplexing (WDM) have attracted a lot of efforts from many researchers. Indeed, the optical networking has been a fertile ground for both theoretical researches and experimental studies. This monograph presents the contribution from my past and ongoing research in the optical networking area. The works presented in this book focus more on graph-theoretical and algorithmic aspects of optical networks. Although this book is limited to the works by myself and my coauthors, there are many outstanding achievements made by other individuals, which will be cited in many places in this book. Without the inspiration from their efforts, this book would have never been possible.

This monograph is divided into four parts:

- *Multichannel Optical Networking Architectures,*
- *Broadcast-and-Select Passive Optical Networks,*
- *Wavelength-Switched Optical Networks,*
- *SONET/WDM Optical Networks.*

The first part consists of the first three chapters. Chapter 1 provides a brief survey on the networking architectures of optical transport networks, optical access networks and optical premise networks. It also introduces various networking design issues that will be studied in this book, together with their background and motivations. Chapter 2 describes the exciting application of the emerging optical integrated circuit (OIC) technology in a number of optical routing devices including wavelength router, optical add-drop multiplexers (OADMs), optical cross-connects (OXCs), passive star couplers. As a fundamental building block, arrayed-waveguide grating (AWG) exhibits elegant optical routing properties. These properties are then exploited to construct optical routing devices of large-scale, high performance and low cost. Chapter 3 investigates how to build an optical network with large number of working wavelengths with cheap optical transceivers of small tunable ranges. In general, the problem to find the minimum tunability in order to achieve full concurrence is NP-hard. Efficient approximation algorithms for waveband/wavelength assignment are then obtained for general network topology. For many regular topologies, optimal waveband/wavelength assignments are presented.

The second part includes the next three chapters. Chapter 4 determines the subnetwork structures of multi-transceiver broadcast-and-select optical networks embedded with various regular interconnection topologies. The subnetwork structures play critical roles in many networking designs such as the selection of the number of transceivers, the transmission scheduling, and the optimal routing. In particular, its application in optimal routing is studied in Chapter 5. Traditionally, the routing follows the routing algorithms of the embedded regular interconnection topology. But a wavelength assignment to embed a topology actually creates a logical topology which is a supergraph of the embedded topology. Thus an optimal routing in the logical topology can potentially be better than that in the embedded topology. Chapter 5 identifies the logical topologies of passive optical networks embedded with hybercubes, star graphs and rotator digraphs. An optimal routing is found if the embedded topology is hypercube. When the topology is a star or a rotator, only suboptimal routings are presented and the optimal ones remains open. Finally Chapter 6 investigates the minimal disjoint channel-set assignment in multi-star cluster interconnection networks, which was proposed in the literature to achieve great scalability. Depending on the network configurations, the minimal disjoint channel set assignment problem is formulated into two versions of vertex coloring. Optimal or near optimal vertex coloring schemes are then obtained in various networks.

The third part consists of three chapters. Chapter 7 gives approximation algorithms for minimal wavelength assignment to lightpaths in rings and trees of rings. These algorithms are simpler and more efficient than those in the literature. In addition, tighter relations among the wavelength number, the load and the clique number of a set of paths in these two network topologies are obtained. Chapter 8 studies how to support maximal number of connection requests subject to limited available wavelengths in a wavelength-switched optical network without wavelength conversion. This maximal throughput problem is NP-hard even in many special network topologies. Polynomial-time approximation algorithms with provable performance are obtained in various networks. Chapter 9, on the other hand, considers wavelength-switched optical networks with wavelength conversion. It investigates the placement of minimal number of wavelength converters in a wavelength-switched optical network such that the number of wavelengths required by any set of lightpaths is reduced down to the natural congestion bound. This optimization problem has been shown to be NP-hard in general in the literature. Chapter 9 shows that for most practical networks it is solvable in polynomial time. In particular, when the underlying topology of the network is a tree, a tree-connected rings or a tree of rings, efficient algorithms are obtained.

The fourth part consists of three chapters, which explore the optical bypass capability of OADMs to minimize SONET add-drop multiplexers in WDM self-healing rings. Chapter 10 studies the grooming of arbitrary traffic in WDM BLSRs. Because of hardness of this problem, the traffic grooming is decomposed into two subproblems: generation of primitive rings, and grooming of primitive rings. Both subproblems are in general NP-hard. Various approximation algorithms are developed for both subproblems. Their approximation ratios are also analyzed. Chapter 11 addresses the optimal grooming of all-to-all uniform traffic in WDM BLSRs. By applying classical combinatorial designs such as Steiner triple systems, partial triple systems and graph decompositions, elegant and efficient grooming algorithms are obtained. Finally, Chapter 12 is devoted to single-hub WDM self-healing rings. If all wavelength channels have the same line-speed, optimal grooming of uniform traffic and near-optimal grooming of non-uniform traffic are provided. If two line-speeds are supported, optimal traffic partition and grooming is obtained. The optimal traffic partition and grooming of non-uniform traffic is also addressed as well.

It should be emphasized that while this book provides many completed results, it also leaves many open research problems that are either partially solved or completely unsolved. Most of these problems are graph theory and algorithm oriented. I hope this book can provide some applied research topics to the theoretical computer scientists. As the book covers both fundamental network architectures and advanced networking design issues in sufficient detail and depth, and most importantly and uniquely, presents many open problems, it is ideal to be used a graduate textbook or reference book in optical networks. Finally I also expect that industry professionals will find this book useful as a well-rounded reference to optical networks.

I am very grateful to many colleagues, friends, and students for all kinds of contributions. Among them are Prof. Charlie Colbourn at University of Vermont, Prof. Gruia Calinescu, Prof. Ophir Frieder, Mr. Liwuliu at Illinois Institute of Technology, and Mr. Xiangyang Li at University of Illinois. I would particularly thank to Prof. Ding-Zhu Du of University of Minnesota for his encouragement and guidance. My editor at Kluwer, John Martindale, was generous beyond belief with my many demands on their time and patience.

PENG-JUN WAN

I

MULTICHANNEL OPTICAL NETWORKING ARCHITECTURES

Chapter 1

WDM NETWORK ARCHITECTURES

Wavelength Division multiplexing (WDM) increases fiber transmission capacity by sending data over multiple wavelength channels simultaneously in a single fiber. These wavelength channels are spaced sufficiently apart to keep neighboring signal spectra from overlapping. WDM has been in existence since 1980s. Early WDM systems, known as coarse WDM (CWDM), were restricted to transmit at two widely separated channels: the 1310-nm band where conventional single-mode fiber has zero dispersion, and the 1550-nm band where conventional single-mode fiber has their lowest attenuation. As a two-wavelength technology, the CWDM system could *only* double the transmission capacity. Furthermore, these two wavelength channels are inherently conflicting, making span engineering more difficult [133]. The underpinning of the fiber optical technology shifted in the early 1990s. As these technologies advanced, dense WDM (DWDM) systems that can support 16, 32, 40, 80, 96, 128, and now 160 wavelengths have been commercially available since late 1995. At the same time, the capacity of each WDM channel is rising, slowly from 2.5 Gbps to 10 Gbps. As the result of these progresses, the realizable transmission capacity of a single fiber has soared to hundreds of gigabits per second and may even rise to terabits per second in the near future.

The deployment of WDM have been evolving for a wide range of networking applications. Due to the unique combination of technology and business forces, WDM has the most immediate and extensive pay-offs on *transport networks*–the underlying high-capacity infrastructure for core inter-office, metro inter-office, and broadband business access networks. Driven by inexhaustible demands for bandwidth, WDM is slowly but steadily migrating from the transport networks toward residential and enterprise access networks. In addition to the public telecommunications networks, WDM is also entering into the world of premise networks such as local area networks (LANs) and parallel interconnects. The networking architectures for these applications should balance a web of trade-offs among a set of critical factors, such as service transparency, maintenance and survivability, and also the ability to successfully meet associated technical and networking challenges. Implicit in the balancing exercise is the ubiquitous cost-functionality trade-off [9]. Rather than a single solution for all applications, a range of optical networking architectures will arise as each market segment applies its unique priorities to make fundamental architecture trade-offs. In this chapter, we will provide a brief survey on a number of these network architectures and their associated design issues. More detailed discussion can be found in [9, 58].

The remaining of this chapter is organized as follows. In Section 1. we will introduce *Synchronous Optical Network/Synchronous Digital Hierarchy* (SONET/SDH) networks, the current transport infrastructure based on *time division multiplexing* (TDM). They are worth of introduction here as the optical transport networks will highly resemble to SONET/SDH networks in many architectural aspects, including network topologies, types of network elements and provisioning of survivability. We will then present a practical vision of optical transport networks in Section 2.. After that we will describe various network architectures for optical access networks in Section 3.. Finally, we will explore different network architectures that can be used for optical premise networks in Section 4..

1. SONET/SDH NETWORKS

1.1 FROM PDH TO SONET/SDH

Early digital transmission systems have been built around *plesiochronous digital hierarchy* (PDH) [30]. Two plesiochronous hierarchies, with different bit rates but following the same principle, have been defined by ITU-T through the years: the North American and European hierarchies. Both are based on bit-interleaved asynchronous TDM. The multiplex signals of the i-th level are made by multiplexing a given number of signals of the $(i-1)$-th level. Tables 1.1 and 1.2 summarizes the North American and European PDH bit rates. It is worthwhile noticing that the T-4 signal has not been standardized by ITU-T. The digital transmission systems implementing the PDH standard in North America and European are known as *T-Carrier* systems and *E-Carrier* system respectively. In North America, T-carrier systems are mostly T-1 and T-3 systems. The T1 line typically consists of two twisted pairs of copper wire are often used to connect a switch to a remote (digital loop carrier) terminal, which in turn connects copper pairs to the individual users in a carrier serving area. T-3 carriers are mostly provided on fiber while some on coaxial, fiber, microwave, or infrared. They were deployed to interconnect service-provider nodes or to connect those nodes with inter-exchange carrier locations.

DS Level	Norminal Bit Rate (Mbps)	Number of Voice Channels
T-1	1.544	24
T-2	6.312	$4 \times 24 = 96$
T-3	44.736	$7 \times 96 = 672$
T-4	274.176	$6 \times 672 = 4032$

Table 1.1. North American PDH.

DS Level	Norminal Bit Rate (Mbps)	Number of Voice Channels
E-1	2.048	30
E-2	8.448	$4 \times 30 = 120$
E-3	34.368	$4 \times 120 = 480$
E-4	139.264	$4 \times 480 = 1920$
E-5	564.992	$4 \times 1920 = 7680$

Table 1.2. European PDH.

The PDH multiplexing is quite awkward and costly due to the asynchronous timing structure (each terminal in the network runs on its own clock without being locked onto a common clock). To accommodate the large variations occurring in the clock rate and thus the signal bit rate, extra stuffing bits are added to synchronize the various low-speed signals to a common rate before multiplexing, and are combined with other framing bits to form a high-speed stream. Due to these stuffing bits, the low-speed signals are neither visible nor accessible within the high-speed digital signal frame. Indeed, the traffic must be completely demultiplexed at the multiplexers and or switches to make the payload accessible for further processing. Thus a transmission often requires expensive *back-to-back multiplexing* and/or the cumbersome *back-hauling* to reduce the expense of repeated multiplexing and demultiplexing. The situation gets even worse if multi-stage multiplexing is present. In addition, the PDH systems are very limited in their management operations, and provide very little support for end-user control for the provisioning of service. Because of these drawbacks, they were progressively replaced by the SONET/SDH standard since late 1980s [15] [16] [132].

SONET standardization efforts have been underway since 1985. Following ANSI's development of the SONET standard, the CCITT undertook to define a synchronization standard that would address interworking between the European and North American transmission hierarchies. That effort culminated in 1989 with CCITT's publication of the SDH standard. SDH is a world standard, and as such, SONET can be considered a subset of SDH. The frames and signals of the SONET hierarchical levels are byte-interleaved synchronous multiplexing, named *synchronous transport signal* of level n (STS-n), for $n = 1, 3, 12, 48, 192$. On the other hand, those of SDH are named *synchronous transport module* of level n (STM-n), for $n = 1, 4, 16, 64$. For reference, the standard hierarchical levels of SONET and SDH and their bit rates are summarized in Table 1.3. A STS-n (STM-n) signal is obtained by byte-interleaving of n STS-1 (STM-1) signals in a single stage, instead of the recursive multiplexing in PDH. Because all network terminals can be traced back to a highly stable reference supply, neither stuffing bits nor framing bits are used. Thus the bit rate of the STS-n (STM-n) signal is exactly n times of the bit rate of the STS-1 (STM-1) signal. Individual STS-1 (STM-1) signals can be extracted without demultiplexing the entire (STM-n) signal [30].

SONET Level	SDH Level	Line Rate (Mbps)
STS-1		51.84
STS-3	STM-1	155.52
STS-12	STM-4	622.08
STS-48	STM-16	2488.32
STS-192	STM-48	9953.28

Table 1.3. Summary of SONET/SDH.

The SONET/SDH system can accommodate applications at various speeds. In the following, for the sake of brevity, we will simply focus on the SONET. The STS-1 signal is specifically intended to provide transport for a T-3 signal. Lower-speed signals such as T-1 and T-2 are supported by designating portion of the STS-1 frame as a *virtual tributary* (VT). Table 1.4 lists the three sizes of VTs. To accommodate different mixes of VTs in an efficient manner, the STS-1 payload is divided into seven VT groups. A VT group may contain one VT-6, three VT-2s, or four VT-1.5s, and must contain only one size of VTs. However, different types of VT groups may be mixed into one STS-1 payload. Virtual tributaries can further be concatenated to VT-6 in increments of VT-1.5, VT-2, or VT-6. The VTs are individually and independently accessible within STS-1 frame through pointer information. A pointer is located in a determined position within the STS-1 frame and identifies the position of the first byte of the pointed VT within the STS-1 frame. In order to compensate the clock frequency offsets present in the clocks of PDH systems, VTs are allowed to shift independently in the STS-1 frame. However, their position is always tracked by the respective pointers. This improved accessibility improves switching and grooming at the finer VT levels. SONET also offers the flexibility of concatenating STS-1s to provide the necessary bandwidth for single high-speed inputs such as is required for DBDQ, ATM, or FDDI. Typically, the concatenation is done in multiples of STS-3c or STS-12c.

Type	Transport for	VT Rate Mbps
VT-1.5	1 T-1 (1.544)	1.728
VT-2	1 E-1 (2.048)	2.304
VT-6	1 T-2 (6.312)	6.912

Table 1.4. SONET virtual tributaries.

The high-speed transmission capacity of SONET/SDH system is achieved through the optical fiber technology, which provides superior performance versus the older microwave and cable systems. The SONET/-SDH terminating equipment converts an electrical STS-n signal (or the corresponding STM signal) into an optical carrier OC-n signal at the same signalling rate. The OC-n signal is transmitted over single mode fiber at either 1310 nm or 1550 nm. It can travel up to approximately 40 Km, without requiring any regeneration or amplification. An end-to-end fiber connection between two SONET terminals is decomposed into three hierarchical layers: the path layer, the line layer and the section layer.

The SONET/SDH provides extensive measures for monitoring the integrity of digital signals that impinge on network elements. Although the monitoring mechanism is intrusive and requires OEO conversion, it quickly detects and isolates faults. In addition, during loss of signal (LOS), loss of frame, or any other path failure, SONET/SDH provides appropriate overhead for inserting a path alarm indication signal (AIS) in the overhead structure. The path AIS notifies downstream nodes of upstream path-level failures so that appropriate actions can be taken. SONET/SDH also improves network management by providing OAM&P (operations, administration, maintenance, and provisioning) data channels in the overhead. These OAM&P data channels enables centralized maintenance. Such remote provisioning ability provides reduced travel for maintenance personnel–which translates to expense savings.

Today, SONET/SDH is the premier backbone transmission technology for leading carriers. It is deployed at all levels of the telephone infrastructure. Because of the large amount of traffic carried in the optical fibers, any fiber cut or nodal failure will be catastrophic unless rapid restoration of service is an integral part of the network design and operations strategies. Thus it is essential for the SONET/SDH networks to provide fast restoration in case of link or nodal failures. Much of the survivable SONET/SDH networks are built around SONET/SDH self-healing rings based on SONET/SDH add-drop multiplexers (ADMs) and SONET/SDH mesh networks based on SONET/SDH digital cross-connects (DXCs). The next two subsections will briefly introduce these two network architectures respectively.

1.2 SONET/SDH SELF-HEALING RINGS

SONET/SDH self-healing rings are the most common network architecture because of its relatively simple implementation and easy management. The key equipment in building SONET/SDH self-healing rings is the SONET/SDH ADM. A SONET/SDH ADM can insert and drop individual STS-1 signals and/or VTs while passing the remaining traffic without any expensive back-to-back multiplexing and inefficient back-hauling. In addition, it has various functions to maintain network survivability. These SONET self-healing rings based on SONET/SDH ADMs are further interconnected as a set of interlocking rings or trees of rings through SONET/SDH DXCs to provide a nation-wide coverage.

Three types of SONET/SDH self-healing rings have been standardized [54, 71]: *unidirectional path-switched ring* (UPSR), *two-fiber bidirectional line-switched ring* (BLSR/2), and *four-fiber bidirectional line-switched ring* (BLSR/4). A UPSR consists of two unidirectional counter-propagating fiber rings as shown in Figure 1.1. The protection is achieved by sending two identical copies of the signal simultaneously on the two rings. The destination end relies on the performance monitoring parameters at the path level to make an autonomous decision as to which of the two alternate rings is providing the higher quality signal. This protection switching, known as *1+1 protection*, does not require any communications between nodes. Thus it can be easily implemented and provides the fastest restoration. However, UPSR cannot use the bandwidth efficiently. The bandwidth requirement of the UPSR is the maximum bandwidth requirement over any span between two nodes in the ring. No reuse of bandwidth is possible on a UPSR network. Moreover, it is impossible to transmit extra traffic because the protection capacity is always used for protection switching operation in a UPSR network. The UPSR architecture achieves its maximum efficiency when all traffic on the ring is between a hub node and other remote nodes, i.e., a centralized traffic pattern. The UPSR architecture is thus suitable for access networks.

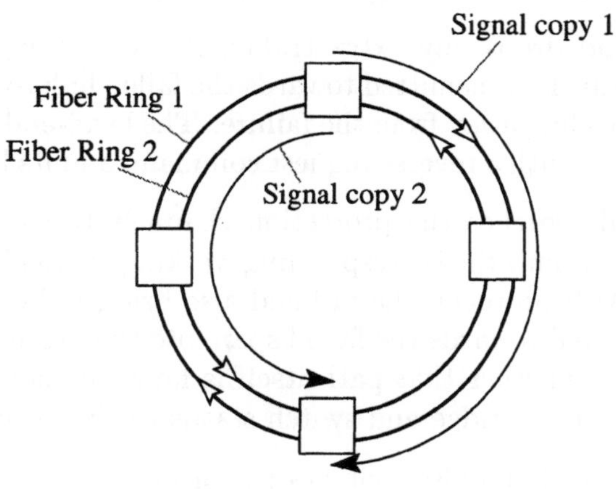

Figure 1.1. Unidirectional path-switched ring.

A BLSR/2 also comprises of two unidirectional counter-propagating fiber rings as well as illustrated in Figure 1.2. Half of capacity in each ring is allocated as a working traffic and the other half is reserved for protection. The working channels in one ring are protected by the protection channels in the other ring. These channels are identified on a time slot basis. For example, in an OC-n ring, working traffic is placed in the first $\frac{n}{2}$ STS-1 time-slots, and the remaining $\frac{n}{2}$ STS-1 time-slots serve as the protection facility, with time slot i of one ring protected with time slot $\frac{n}{2}+i$ of the other. In contrast to 1+1 protection, at any time only one copy of the signals is transmitted at any time. In a normal state, the signals are transmitted using only the work capacity, and the protection capacity can be used to carry low priority extra traffic. The latter feature is also referred to as *protection access*. In case of any link failure, the failure is restored by the *ring switching*. The ring switching relies on the performance monitoring parameters at the optical line level to determine the quality or loss of the received working channel signal. When the working channel signal is find to be lost or have unacceptable quality, protection switching must be coordinated between the two nodes adjacent to the failed link. Such coordination is accomplished by the exchange of the *automatic protection switching* (APS) messages. The APS messages are transported over the *first* STS-1 (within an STS-n) line overhead K1/K2 bytes, and are active only on the channels carrying *protection* traffic. The following is the four steps involved in a ring switching operation.

1. The tail-end of the failed link detects the failure and informs the head-end of the failure along the long path.

2. The head-end drops any extra traffic, if there is any, and bridges working channels transmitted towards the failed link to the protection channels traveling away from the failure. The head-end then responds to the tail-end with a reverse request command and its bridged status.

3. The tail-end switches the protection channels transmitted towards the failed span into the corresponding working channels. In addition, under the APS protocol, the tail end also bridges the working channels transmitted towards the failed span into the protection channels, *regardless of* whether that path itself is faulty or not. The tail-end then conveys this bridge and switch status to the head-end.

4. Finally, the head-end also switches the protection channels transmitted towards the failed link into the working channels. It then completes the entire ring switching operation by informing the tail-end of the current bridge and switch status.

The maximum number of nodes on a BLSR/2 must not exceed 16 due to the 4 bit node identification field in the APS messages. This limitation on ring size also accommodates the protection switching time within 50 ms. The location of each node must be known by every other node on the ring by way of a network map. The BLSR/2 architecture provides some flexibility to share the load away from saturated spans and therefore can potentially support a higher maximum load than a UPSR. Various routing algorithms to minimize the ring load have been developed in [45, 94, 136, 137, 152]. The bandwidth utilization of BLSR/2

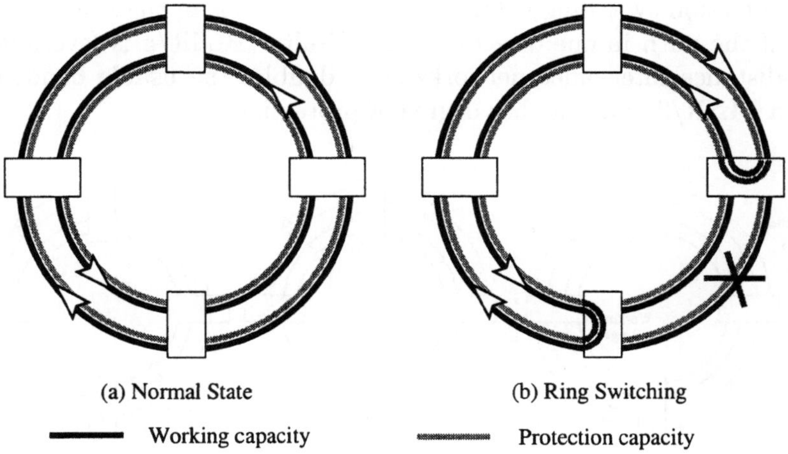

Figure 1.2. Two-fiber bidirectional line-switched ring.

depends on the traffic pattern in the ring. If all traffic homes on a single node, as would be in the case in an access ring, there is no utilization increase over the UPSR; but as the traffic pattern becomes more distributed, as in the case of inter-office ring, the bandwidth utilization feature increases the bandwidth efficient of BLSR/2. Thus, due to its bandwidth efficiency, the best application of the BLSR/2 architecture is for inter-office networks.

A BLSR/4 utilizes four fiber rings as shown in Figure 1.3. Two counter-propagating fiber rings are dedicated to working traffic and two counter-propagating fiber rings are dedicated to protection traffic. The traffic is normally sent along on the working rings, up to the full utilization of these fibers (as opposed to the BLSR/2 case). It also allows protection access as BLSR/2. Because of the separate working and protection rings, BLSR/4 can support both *span switching* (directly between two nodes) as well as ring switching (along the long path between two nodes). The operation of both span switching and ring switching are very similar to four-step process of the ring switching in BLSR/2. The APS messages for the coordination between the two nodes adjacent to a failed link are exchanged over the protection fibers. However, every APS message are always transmitted both along the short span and the long path simultaneously. If both messages are received, then the one arriving later will be ignored. As long path communication is always beaten by the short span communication, the span switching, if it is possible, will always be taken. In addition, under the APS protocol, the protection switching is taken in both directions in the same manner even if the path in one direction is not faulty. BLSR/4 is favorable in long-distance inter-office networks as it doubles the usable bandwidth over a BLSR/2 and a higher degree of protection.

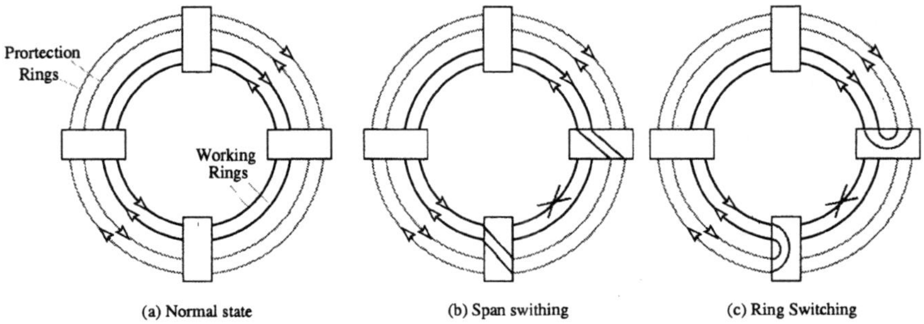

Figure 1.3. Four-fiber bidirectional line-switched ring.

Table 1.5 briefly compares the three types the SONET rings.

Criterion	UPSR	BLSR/2	BLSR/4
Fiber Rings	2	2	4
Working Capacity	full	half	full
Protection Access	none	allowed	allowed
Protection Entity	path	line	line
Protection Mechanism	1+1 switching	ring switching	span and ring switching
Channel Reuse	none	allowed	allowed
Routing	trivial	complicated	same as BLSR/2
Application	access ring	inter-office ring	long-haul inter-office ring
Cost	low	slightly more	twice more
Management	easy	harder	similar to BLSR/2

Table 1.5. Comparison of SONET/SDH self-healing rings.

1.3 SONET MESH NETWORKS

Historically, the core PDH digital transmission networks used general mesh topologies for interconnecting the digital cross-connects by point-to-point fibers. Thus it would be more natural and economical to upgrade the PDH mesh networks to SONET/SDH mesh networks by replacing the PDH DXCs with SONET/SDH DXCs. In addition, SONET/SDH mesh networks promise more cost effective protection and restoration than the interconnected self-healing rings networks. When a failure occurs, connections are switched by the SONET/SDH DXC from the failed links and rerouted over others having spare capacity. The mesh topologies provides flexibility in the use of fiber capacity, in contrast to the requirement of 100% protection capacity in the self-healing rings. The richer connectivity of mesh topologies implies higher survivability than self-healing rings which in general can restore only single nodal or link failure. However, the challenge is to design a restoration strategy that allows the subsecond restoration for the carrier-scale mesh-based networks.

Many restoration approaches have been proposed for generic mesh networks. The restoration approaches can be further classified by their route computation and execution mechanisms as centralized versus distributed, by their type of rerouting as link-based versus path-based, and by their computation timing as precomputed versus real time (after failure has occurred). These various restoration scheme have different trade-offs in restoration speed, restorability, capacity efficiency, and algorithm scalability.

2. OPTICAL TRANSPORT NETWORKS

2.1 DWDM FOR CAPACITY EXPANSION

SONET networks delivered major gains in capacity, reliability, and efficiency. The bad news, however, is that the once seemingly inexhaustible capacity promised by ever increasing SONET/SDH rates is reaching its limit during the mid-1990s. Back in early 1980s when fibers were installed, the traffic was mainly voice and the telecommunications carriers were sufficiently cash-constrained. Consequently, a typical fiber route might have just 16 fibers; fiber counts of 32 or more were relatively rare. However, during the 1990s, the traffic grows at an unprecedented pace, driven mainly by the explosion of Internet, the multimedia applications and the enhancement of network survivability. The bandwidth demand is already approaching the maximum capacity available in some networks [133]. Confronted by this fiber exhaust, carriers have two possible solutions: install new fiber or deploy wavelength division multiplexing (WDM) in combination with TDM.

Laying new fibers is a perfectly adequate solution for new service providers or for those who have plans to grow their embedded base substantially. However, it is too costly and immensely time-consuming for many service providers. The cost of installing new fiber is prohibitively high in many areas, especially in densely populated metropolitan areas, where the anticipated demand is often highest. It can also be a daunting prospect, particularly for carriers with tens of thousands of route miles. In many cases, the right-of-way of the cable route or the premises needed to house transmission equipment is owned by a third party, such as a railroad or even a competitor. Moreover, single-mode fiber is currently in short supply owing to manufacturing limitations, potentially adding to costs and delays. As an consequence, the comprehensive deployment of additional fiber is an impractical, if not impossible, solution for many carriers.

On the other hand, WDM has proven to be an immediate and cost-effective means of increasing the bandwidth of installed fiber plant without significant investment of new fiber routes. As the initial step toward optical networking, WDM can be deployed on individual spans of a SONET network that are congested because of high capacity demands. This is especially helpful where multiple rings intersect between two nodes, resulting in fiber exhaust. The existing SONET network continues to operate as if it were still the only application on the fiber. This feature preserves the investment in existing TDM equipment. With WDM, new applications can be added to new wavelengths without affecting the existing SONET traffic. For example, a number of SONET rings can be stacked over a physical fiber optical ring, each of which is carried in a distinct wavelength. The architecture of the stacked SONET rings can minimize capital expenditure by matching capacity of each individual SONET ring to actual demand. In addition, these stacked SONET rings can be upgraded individually in capacity as capacity requirement grows. Thus the point-to-point WDM provides an economic, scalable and flexible solution for the capacity expansion and network upgrade.

The mature of the optical amplifying technology makes WDM especially attractive over long fiber span. In addition to the savings on the optical fiber, substantial cost savings is obtained by replacing the conventional electronic regenerators with optical amplifiers (OAs). Optical amplifiers are capable of amplifying multiple channels simultaneously. Thus for a W-channel system a single optical amplifier can replace W conventional electronic repeaters at repeater site. Associated with this drastic decreasing of the equipment count is the simplified network operations and increased networking reliability. Such economic and operational advantage has fueled the exceptionally rapid and relatively uniform adoption of the WDM technology in long-haul terrestrial and submarine networks.

The deployment of WDM over individual spans is just the first step toward optical networking. In these initial point-to-point WDM systems, most of the networking functionality including restoration and protection is still accomplished by the underlying TDM systems that use the WDM span. Since TDM terminating equipments such as SONET ADMs and DXCs can each only handle one wavelength of traffic, these TDM terminating equipments would proliferate geometrically in a multiwavelength transport system. For example, in a 80-channel stacked SONET rings, each fiber span must terminate into 80 individual SONET ADMs. The space and power requirements of this equipment mix often create upwardly spiraling operational costs. Furthermore, as the number of WDM channels and links increases, network management grows more complex, making service provisioning even more cumbersome. Therefore, as more wavelengths and/or fibers are deployed, there is an increasing need to manage the capacity and hand off signals at the wavelength channel level.

In much the same way that SONET ADMs and DXCs emerged to manage capacity at the electrical layer, *optical add-drop multiplexers* (OADMs) and *optical cross connects* (OXCs) will emerge to manage capacity at the optical layer. The OADMs will allow individual wavelengths to be added, dropped or bypassed at any site, and each site processes only the wavelengths that it needs to access. Therefore, the cost of each node is somewhat independent of the aggregate speed of the network. OADMs are typically used in WDM ring networks, and similar to their electrical counterparts, will perform performance monitoring and fault restoration. The OXCs offer more bandwidth management capability than OADMs. They performing functions such as wavelength-level switching, wavelength multiplexing and demultiplexing, wavelength signal monitoring, and optionally wavelength conversion. In addition, they will reconfigure networks for new service setup, changes in traffic patterns and fault restoration.

The deployment of OADMs and OXCs for bandwidth management creates truly optical networks. Because of their wavelength-switching feature, these optical networks are referred to as *wavelength-switched optical networks* [29] [118]. Built upon the legacy SONET self-healing ring networks and mesh networks, the wavelength-switched optical networks will also take the OADM-based WDM self-healing ring architecture or

OXC-based mesh architecture. However, it is essential to avoid the trap of assuming that whatever has been done for SONET/SDH is directly applicable to the optical transport networks. Fundamental challenges still remain on the wavelength conversion, optical performance monitoring and measurements and the transport of OAM&P information. These technical issues have great impact on the design of the practical optical transport network architectures.

2.2 WDM SELF-HEALING RINGS

WDM self-healing rings inherit both network architecture and the associated protection switching mechanism from the SONET self-healing rings. However, there do exist some subtle differences, as we will point out.

The WDM UPSR, as the counterpart of SONET UPSR, consists of two counter-propagating fiber rings. The signal is split into two copies by using an optical splitter (and therefore at the same wavelength) at the transmit end. These two copies of the original signal are fed into both fiber rings in opposite direction in the same wavelength, and dropped at the receiving end. However, unlike the SONET 1+1 protection, there are no duplicated electronics. In the normal state, the receiving end connects to only one fiber ring and thus receives one copy of the duplicated signals from this fiber. If a failure occurs, the receiving end switches to the other fiber ring without knowing anything about its state. Since the receiving end does not need to notify the transmitting end of the failure, a internodal signalling channel is not needed.

In WDM BLSR/2, as the analogy of SONET BLSR/2, consists of two counter-propagating fiber rings. Only half of the wavelengths are used in each fiber and the ring switching is used for protection. To avoid wavelength conversion, the working wavelengths on the two fibers are chosen to be disjoint. For example, in a W-wavelength system, the working traffic in one ring uses wavelengths 1 through $\frac{W}{2}$, while the working traffic in the other ring can uses wavelengths $\frac{W}{2}+1$ through W. When a failure occurs, the rerouted traffic will keeps its wavelength and therefore simplify the ring switching operation. Such channel partition is different from that in SONET BLSR/2. As only one copy of signal is transmitted, no duplicate electronics are needed for accessing any optical channel. APS signalling is required by the ring switching. Since the APS signalling is not associated with any particular optical channel, it can be carried over a dedicated wavelength (the optical supervisory channel). This is another difference between WDM BLSR/2 and SONET BLSR/2.

The WDM BLSR/4 is the analogy of SONET BLSR/4. It consists of four fiber rings and provides both span switching and ring switching. All the wavelengths in the two working rings are used to carry the working traffic and the rerouted traffic by both span switching and ring switching keeps the old wavelength. Once again no duplicate electronics are used for each optical channel, and APS signalling can be carried over a dedicated wavelength (the optical supervisory channel.

It should be pointed out that the individual optical channels usually have high-speed transmission capacity in order to make the make the system cost effective. Thus OADMs provide very coarse granularity bandwidth management. To accomplish finer granularity bandwidth management, SONET ADMs have to be used for traffic grooming. The amount of SONET ADMs required by a set of low-speed traffic streams may vary significantly depending on how they are routed and groomed. As these SONET ADMs are the dominant cost factors in the investment capital, it is essential to find efficient algorithms for routing and grooming the low-speed traffic. These algorithms will be developed in Chapter 10, Chapter 11, and Chapter 12.

2.3 WDM MESH NETWORKS

OXC-based optical mesh networks have more flexibility for fault protection and restoration than OADM-based optical self-healing rings. However, the management of wavelengths is much more challenging. The complexity of managing wavelengths depends on whether wavelength conversion is provided by the OXCs. If the OXCs have no wavelength conversion capability, any signal enters and exits the optical network at the same wavelength, without ever changing to a different wavelength throughout the network. Essentially a wavelength is dedicated to connect the two endpoints together, and this requirement is known as the *wavelength-continuity constraint.* The path established for a connection together with the associated wavelength is referred to as a *lightpath.* Figure 1.4 shows an example of such network. The problem of choosing a set of lightpaths using minimal number of wavelengths for a given set of pairs of nodes have been well studied in the literature [4, 13, 19, 20, 59, 60, 96, 128, 129]. Such problem is known as the *optimal route and wavelength assignment problem,* and its special instance in which the path between any pair of nodes is given in advance is known as the *minimal path-coloring problem.* In Chapter 7, some new and efficient algorithms for path-coloring in rings and trees of rings will be presented. Tighter relations among the wavelenth number, load and clique number of a set of paths in these two netwok topologies will also be obtained. If the network can not accommodate all traffic demands simultaneously, it is desirable to support as many traffic demands as possible. This *maximal throughput problem* will be studied in Chapter 8.

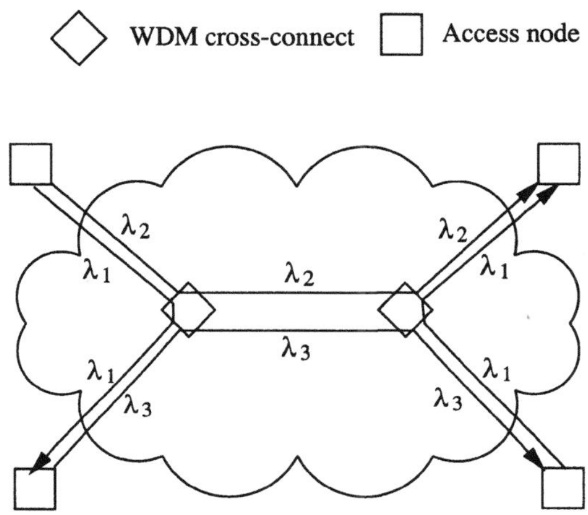

Figure 1.4. An example of wavelength-switched optical network without wavelength conversion.

The introducing of wavelength conversion at certain OXCs can avoid the need for a dedicated wavelength for an end-to-end connection and thus enable more wavelength reuse. If an OXC of the network contains a full-wavelength converter, any path that passes through it may change its wavelength. In the extreme case, each OXCs has full wavelength conversion capability, the number of wavelengths required by any routing is reduced down to the natural *congestion* or *load* bound, defined to be the maximum number of paths passing through any one link in the network. The optimal routing and wavelength assignment problem is thus simply reduced to the well-studied load-balanced routing. However, as observed in [166], placing full-wavelength converters at a carefully selected subset of OXCs, called *sufficient set,* instead of all OXCs is sufficient to ensure that the number of wavelengths required by any set of paths is equal to the load of these paths. In order to minimize the cost associated with the wavelength conversion, it is desirable to find a *minimum sufficient set* for any given network topology. The minimum sufficient set problem has been addressed in [166] and [97]. In Chapter 9, we will extend their work and provide efficient algorithms for this problem in various practical optical networks.

3. OPTICAL ACCESS NETWORKS

Today, most residential connections to the Internet use 56-Kbps dial-up modems on twisted-pair lines that were designed only for narrowband telephony. Such bandwidth bottleneck in the local loop continues to slow Internet access to a crawl, frustrating the ever-increasing number of end users who demand for broadband services. In response to competition and growing demand for broadband services, telephone companies are attempting to leverage their existing copper infrastructure through deployment of various digital subscriber loop (xDSL) technologies. While asymmetric digital subscriber line (ADSL) technology can provide up to 8 Mbps of downstream capacity directly to a customer premises, this is currently the limit for most existing copper loops. Very-high-speed DSL (VDSL) technology supports up to 50-Mbps transmission over twisted copper pairs but only over short distance of less than 1000 feet. Thus, only a very small percentage of subscribers can benefit from this technology.

Extending optical fiber deeper into the local access network can help to overcome the distance/bandwidth problem. Many attempts have been made to extend *fiber to the home* (FTTH), *fiber to the curb* (FTTC) serving approximately 16 customers, (typically within 500 feet from the curb) or *fiber to the cabinet* (FTTCab) serving approximately 100 customers (typically within 3,000 feet from the neighborhood cabinet). These fiber access systems are terminated by one *optical line terminal* (OLT) in the central office and the *optical network units* (ONUs) at or near the subscriber's home. An optical access network is then used to provide the connectivity between the OLT and the ONUs. Such optical access network is also referred to as the *outside plant.* Comparing to transport networks, the access networks serves less customers, and supports lower bandwidth demand. Thus the deployment of optical access networks is extremely cost-sensitive. Typically the optical access network relies on passive branching devices to interconnect the OLT and ONUs. Because of the passiveness of the branching devices, these networks fall in the category of *passive optical networks* (PONs). The fiber path and the passive branching devices between the OLT and ONUs are referred to as the *outside plant.* As no intervening active elements are between the OLT and the ONUs, the number of sites that require costly maintenance, powering, and provisioning is reduced. In addition, PONs also reduce cost by sharing expensive central office optoelectronics and feeder fiber over many ONUs. In this section, we will describe various PON architectures suitable for the outside plant.

3.1 POWER-SPLITTING PON

The power splitter/combiner is the simplest passive branching device that can be deployed in the optical access network. The PONs that uses a single power splitter/combiner or cascaded power splitter/combiners as the branching devices is referred to as the *power-splitting PON* [58]. An early effort to establish a standard for the power-splitting PON was made by the Full Services Access Network (FSAN) Consortium, which was formed by major telephone companies from around the world in the early 1990s. Based on their recommendations, the ITU-T Recommendation G.983.1 was formally approved in October of 1998. The proposed power-splitting PON, known as ATM-PON, transmits all traffic in ATM cells. The power splitter/combiner can support up to 64 ONUs with a reach of up to 20 km. Typically, the upstream signal and the downstream signal are carried on a single fiber on two different wavelengths, namely 1310 nm and 1550 nm, and they are separated at the OLT and the ONUs by CWDM (see Figure 1.5). The upstream transmission uses a time division multiple access (TDMA) protocol. The upstream cells from individual ONUs are combined by the combiner to form an OC-3 stream. This TDMA upstream transmission supports dynamic bandwidth allocation, which is a quite useful feature if the traffic is bursty, or if the need for bandwidth is distributed unevenly among the subscribers. The downstream transmission, at the rate of either OC-3 or OC-12, consists of a continuous time-division multiplexed stream of fixed-length time slots. The downstream cells from the OLT is split into multiple copies, one for each ONU. The ONUs determine which part of the downstream transmission to read based on time slots or ATM cell headers. Such broadcast nature of the downstream transmission is very appealing for telephone companies that plan to introduce cable TV.

The deployment of the power splitters/combiners in the outside plan is very flexible. In each carrier service area, the splitting ratio can be

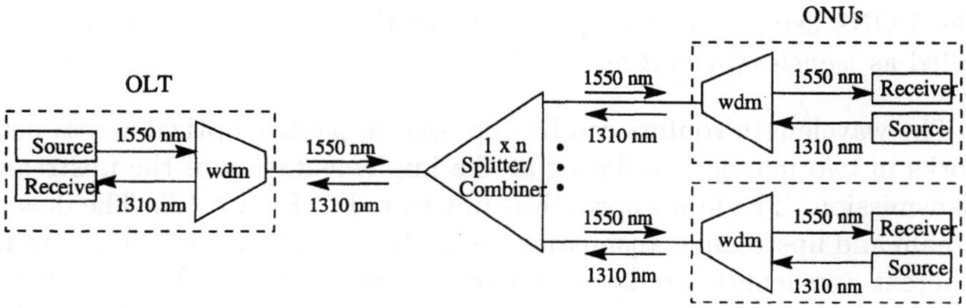

Figure 1.5. Passive optical access network based on power splitter/combiner.

selected to match the actual bandwidth demand and improve the power loss budget. As long as the maximum loss budget of the PON is not exceeded, the transmitters and receivers would not need to be redesigned for the different splitting ratios, and the aggregate bit rates would be unchanged. The power splitters/combiners can also be cascaded quite easily. This could lead to substantial savings in fiber costs for certain layout of the subscribers. However, passive optical access network based on power splitters/combiners have limited network size, mainly due to the high splitting loss of the power splitters, unless expensive amplifiers are used. Since the downstream signal is broadcast to all ONUs, privacy is sacrificed unless encryption is used. The upstream bandwidth sharing in TDM manner implies that electronic and optoelectronic components in the OLT and ONU must run at the aggregate bit rate. For example, a OC-192 line rate is required to give 16 subscribers each an average of OC-12 rate. Thus in addition to the requirement of high-cost ONUs, this bandwidth sharing imposes an upper limit on the available bandwidth to each subscribers and/or the network size.

3.2 WAVELENGTH-ROUTING PON

Wavelength router is another type of passive branching device that can be used in the optical access network. Instead of simple power splitting, the wavelength router demultiplexes a multi-wavelength signal into individual single-wavelength signals, each exiting from a unique output port. Conversely, the wavelength router can also act as a multiplexer to combine multiple signals at different wavelengths into a single multi-wavelength signal. Typically, the wavelength router is simply a waveguide grating router (WGR) [53], also known as arrayed waveguide grating (AWG) multiplexer [147], or optical phased array (OPA) [141].

The PONs using a wavelength router as the passive routing device is called as *wavelength-routing PONs.*

The wavelength routers can be deployed in passive optical access networks in two manners, differing at the implementation of the upstream transmission. The first approach is illustrated in Figure 1.6. The downstream and upstream transmission are multiplexed in a single fiber but in different wavelength windows, and are separated at the OLT and ONUs using WDM. The wavelength router acts as an optical multiplexer for the upstream path, and a demultiplexer for the downstream path. The OLT has a multiwavelength transmitter and a multiwavelength receiver. Each ONU has a transmitter and a receiver, each operating on a single, dedicated wavelength. For the outside plant to operate properly, the wavelength produced by the transmitter at each ONU must match the corresponding port of the router. While fixed-wavelength transmitters specially tuned to meet this wavelength-port matching requirement are practically available, they introduce much inventory, maintenance, and installation difficulties. These difficulties can be avoided by using widely tunable transmitters. However, they are currently far from practical. Such dilemma can be resolved by using the loopback technique, which leads to the second approach to the deployment of wavelength routers, as described in the next.

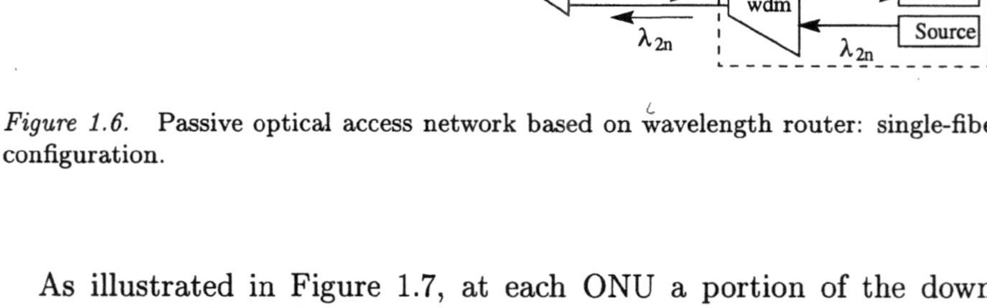

Figure 1.6. Passive optical access network based on wavelength router: single-fiber configuration.

As illustrated in Figure 1.7, at each ONU a portion of the downstream signal is modulated with the upstream traffic by a modulator and reflected back on a second fiber to the OLT [58]. If the wavelengths emitted by the multiwavelength source in the OLT match the router passbands for the downstream direction, they will be the correct wavelengths for the upstream direction as well. Loopback, however, introduce its own set of problems. As separate downstream and upstream fibers are used, both the fiber count and the number of ports on the wavelength router are doubled, increasing the cost of the outside plant and complicating the installation and maintenance. Higher line rate is also needed as the signal is divided between an upstream and a downstream portion. An equal division of the signal, for example, doubles the line rate in both directions. In addition, the power budget on the OLT transmission is increased to accommodate the round-trip propagation and faster line rate.

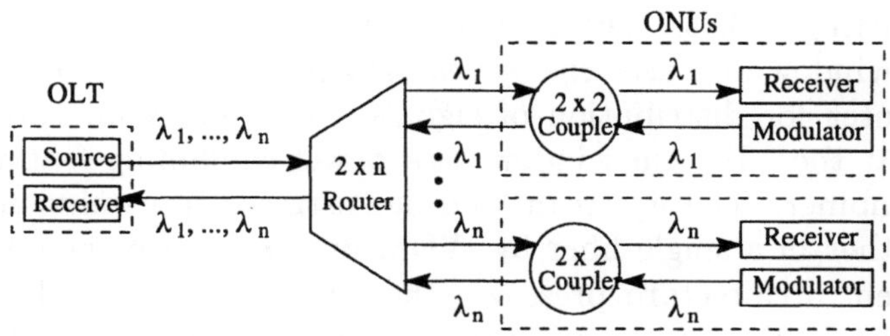

Figure 1.7. Passive optical access network based on wavelength router: two-fiber configuration.

The power splitter/combiner and wavelength router both have pros and cons [58]. The insertion loss of the wavelength router is much smaller than the splitting loss of the power splitter/combiner. The wavelength routing capability of the wavelength routers allows the receivers at ONUs to operate at the per ONU bit rate rather than at the aggregate bit rate. If bandwidth needs become high enough, the lower line rates will result in a lower-cost ONU. The wavelength routing capability of the wavelength routers also enables the privacy of the downstream transmission. However, the deployment of wavelength routes in the outside plant deployment is less flexible than that of power splitter/combiners. While adding new ONUs is trivial in an optical access network, it is rather complicated as the OLT receiver has to be able to receive signals and the wavelength routers must match the channels to the corresponding ONUs. Cascading routers requires much more delicate matching between channels and ports. In addition, if the ONUs' ranges vary significantly, severe crosstalk among the different wavelength channels would be experienced by the OLT's multiwavelength receiver. While the crosstalk problem can be managed, but there will be high costs involved.

3.3 HYBRID PONS FOR OPTICAL ACCESS NETWORKS

Power splitter/combiner and the wavelength router are not exclusive to each other in the optical access networks. In fact they can be used together to achieve various desirable features. Figure 1.8 shows a hybrid system which uses an integrated router/combiner as the passive branching device [58]. The OLT's transmitter and the ONUs' receivers are the same as those in the wavelength-routing PONs. On the other hand, the OLT's receiver and the ONUs' transmitters are the same as those in the power-splitting PONs. The downstream transmission operates at the 1550 nm window and uses the wavelength routing feature of the wavelength router for distribution of signals. The upstream transmission operates at the 1310 nm window uses power-combining feature of the power combiner. The upstream and downstream transmissions can be carried either in a single fiber by WDM or simply two separate fibers. This system architecture preserves the properties of the downstream transmission in the wavelength-routing PONs, and the properties of the upstream transmission in the power-splitting PONs. An optical access network can further cascade the integrated router/combiner with the power splitter/combiner to increase the number of subscribers that can be server in an outside plant as shown in Figure 1.9. The main difficulty of this architecture, however, is the more demanding TDMA upstream transmission, resulting from the increase of the sharing ratio and the added optical losses. Optical amplifiers and high speed electronics may be needed.

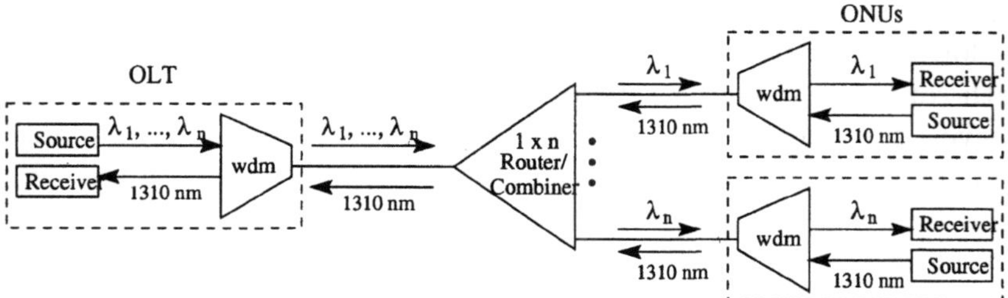

Figure 1.8. A hyprid optical access network based on integrated router/combiner.

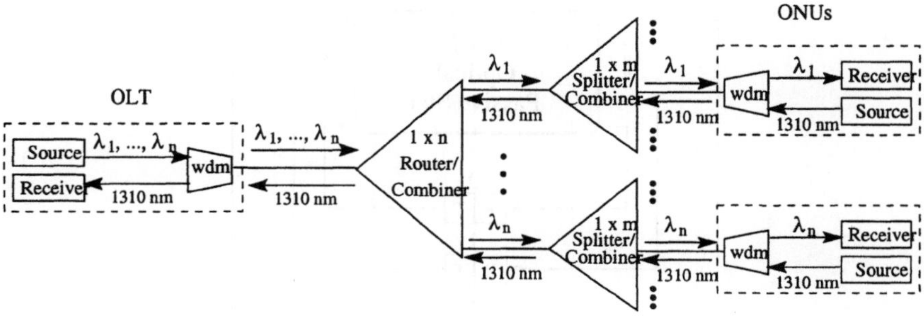

Figure 1.9. The cascading of the integrated router/combiner and power splitter/combiner.

4. OPTICAL PREMISE NETWORKS

As with the access network, PONs are well suited for broadband premise networks, such as LANs and parallel interconnects, as they have very short span and are less sensitive to optical transmission impairments such as power attenuation and dispersion. However, in contrast with the access networks in which the traffic is between one central office and multiple ONUs, LANs and optical interconnect have to support distributed traffic pattern. Thus passive star couplers [52, 104], are the desirable passive branching device to be used in PONs for premise networks. A passive star coupler, or coupler in short, splits the signal from any input port evenly to any output port. A network node attached to the star coupler can then select the desired signal by tuning its receiver to a specific wavelength. Thus the PONs based on passive star couplers are referred to as the *broadcast-and-select PONs*. Figure 1.10 shows a typical WDM network in which N nodes connect to a common optical passive star coupler.

The broadcast-and-select PONs can be classified into two broad categories according whether or not the signals have to be forwarded by any intermediate nodes before reaching the destination: single-hop PONs and multi-hop PONs. In a *single-hop network*, the signal transmitted by any node can arrive to the destination directly. Thus no optical-electronic conversion is involved in the propagation of the signal. Ideally, each node can transmit the data without waiting for a free wavelength channel. Such single-hop network would require $N(N-1)$ wavelengths, one for each ordered pair of nodes, where N is the network size. Accordingly, each node has to use either a tunable transceiver with range

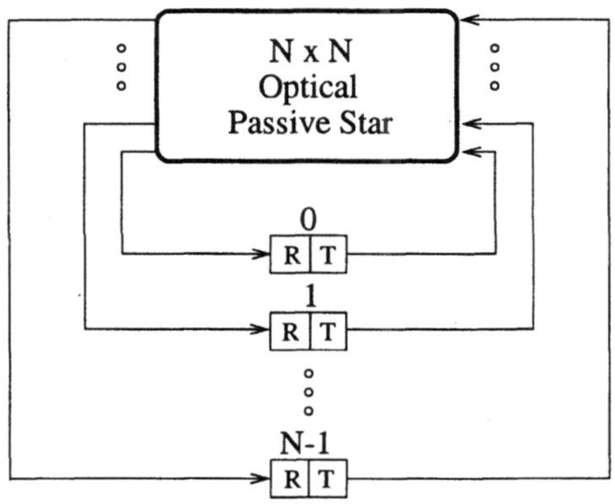

Figure 1.10. A N-node broadcast-and-select PON.

$N(N-1)$ or $N-1$ fixed-wavelength (or slowly-tunable) transceivers. Either configuration, however, is technically impractical and economically prohibitive even with a medium network size. Thus in general it is necessary for the nodes to share the access to the wavelength channels in the time domain. The channel access latency depends on a number of factors, including the number of wavelengths used by the network, the number and type of transmitters and receivers at each node, and the channel allocation scheme. As an extreme example, the pure TDM network uses only one wavelength channel, one fixed-wavelength transmitter and one fixed-wavelength receiver at each node, but has poor performance. In general, multiple wavelength channels are used, and all of them are accessible to each node via transmission and/or reception. For a node to access all wavelength channels by transmission (reception), it either uses a tunable transmitter (receiver) whose tunable range covers all these channels, or an array of fixed-wavelength transmitters (receivers) each of which can access a unique channel. Thus various transceiver configurations are possible.

In a *multi-hop network*, two nodes may communicate via one or more intermediate nodes. Although the end-to-end latency may be increase, each node does not have to be capable of accessing all wavelength channels. Typically each node uses a small number of fixed-wavelength transmitter/receivers, while it is possible for each node to use a single tunable transmitter/receiver with small tunable range. Any wavelength assignment to the fixed-wavelength transceivers determines a *logical topology* over the access nodes as follows: there is a link from a node a to node b in the logical topology if and on if a has a transmitter and b has a receiver which are assigned the same wavelength channel. Figure 1.11 illustrates the concept of logical topology. The logical topology is a directed graph, or a digraph in short, by nature. It is often represented by a bipartite digraph in which one column of nodes corresponds to the transmission part of the stations and the other column of nodes correspond to the reception part of the stations. As the least requirement, the logical topology must be strongly connected, i.e., there is a path from any node to any other node. However, this is not enough for many applications. Additional requirements include short diameters, good fault-tolerance and load balancing, among others. A common approach is to make the logical topology as a supergraph of some regular interconnection topology, which itself meets these requirements. Some of those regular interconnection topologies are described in Appendix A. A wavelength assignment is said to realize a given interconnection topology if the given interconnection topology is a subgraph of the resulting logical topology. Chapter 4 and Chapter 5 will present the realization of various interconnection topologies and the structural and routing properties of the resulting logical topologies.

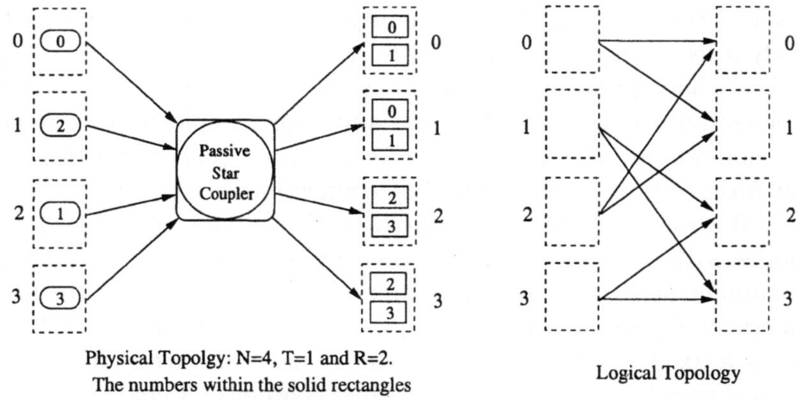

Figure 1.11. The logical topology concept.

In broadcast-and-select PONs, channel access protocol is needed to either prevent or deal with collisions. A *channel collision* happens when two nodes attempt to send message on the same channel at the same time. For networks using single tunable receiver at each node, a *receiver collision* happens when two nodes attempt to send a message to the same destination at the same time. The channel allocation can be done by pre-assigning access at network configuration time, dynamically coordinating access on a demand basis, random access, or by a combination of these. If tunable receivers are used and the transmission schedule is not pre-determined, then control information on which channel to tune to have to be sent to a receiving node. This control information could be transmitted in-band (within the same channel as the payload data), or out-of-band on a separate control channel.

The broadcast-and-select PONs based on a single star coupler can connect around 100 access nodes. Their transmission concurrence is also upper-bounded by the number of available wavelength channels. To expand the network size and/or the transmission concurrence, multiple star couplers can be used to interconnect the access nodes. The *cluster interconnection networks (CINs)* are a class of such multi-star PONs proposed in [11]. In a cluster interconnection network as shown in Figure 1.12, the access nodes are partitioned into clusters of the same size. Each cluster is provided with two passive star couplers, an *output star coupler* which connects the transmitters of all nodes in this cluster, and an *input star coupler* which connects the receivers of all nodes in this cluster. A number of fiber links are then used to provide connections from the output star couplers to the input star couplers according to a properly selected topology, referred to as *cluster interconnection topology*. The least requirement on the cluster interconnection topology is the connectivity from any coupler to any other coupler in one or more hops. Sometimes, the cluster interconnection topology provides *self-cluster links* to provide the single-hop connection among nodes within a cluster. However without these self-cluster links, the connections among the nodes within a cluster can still be achieved, although with possible more number of hops. The transmitters and the receivers at the access nodes are configured in the following way. The transmitter at each node is simply a fixed-wavelength transmitter. The transmitters of all nodes within a cluster have to operate at distinct channels. A channel here can be a reserved time slot, a dedicated wavelength, or a reserved time slot over a given wavelength. The set of channels used by the transmitters in a cluster is called a *channel set*. Two clusters can share a channel set as long as their output star couplers are not adjacent to any common input star coupler. The receiver at each node has to be capable of simultaneously monitor all channels coming into the input star couplers to which it is attached.

One of the most dominant features of the cluster interconnection networks is their ability to share the channels sets in the spatial domain. Figure 1.13 shows a channel set assignment in an CIN based on 3-cube

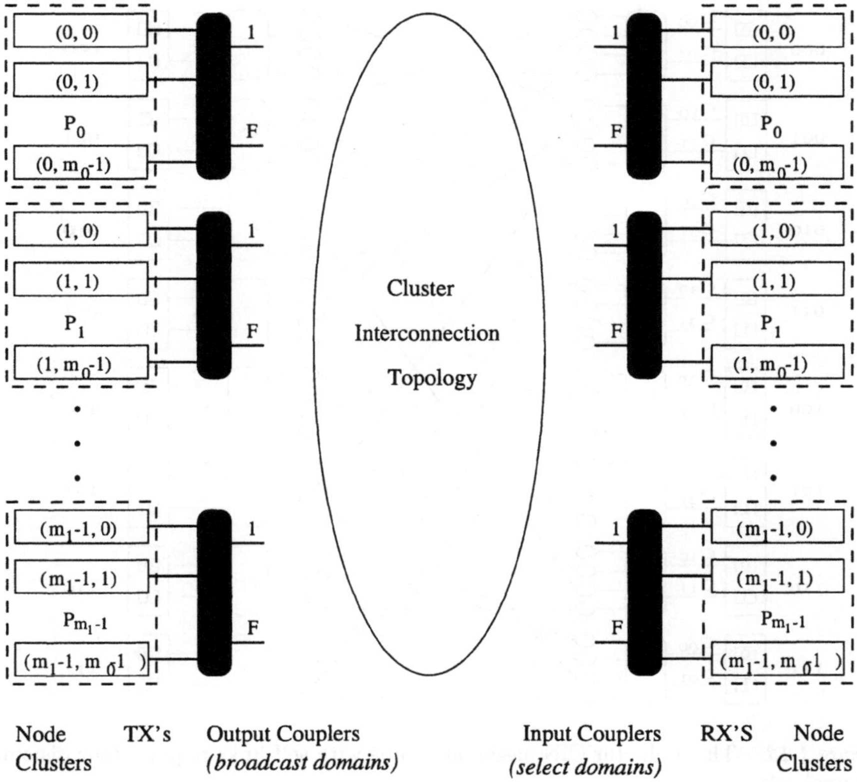

Figure 1.12. Cluster interconnection network with m_1 clusters and m_0 nodes per cluster.

with self links (2 nodes per cluster). Clusters 000 and 111 transmit over channel set $C_0 = \{C_{00}, C_{01}\}$, clusters 001 and 110 transmit over channel set $C_1 = \{C_{10}, C_{11}\}$, clusters 010 and 101 transmit over channel set $C_2 = \{C_{20}, C_{21}\}$, and clusters 011 and 100 transmit over channel set $C_3 = \{C_{30}, C_{31}\}$. Input couplers receive the set of all channels $C = C_0 \cup C_1 \cup C_2 \cup C_3$. Because of this channel sharing, the entire network requires much less channel sets and therefore achieve higher scalability. The minimal number of channel sets required by some cluster interconnection topologies were addressed in [10]. In Chapter 6, we will extend their study to other cluster interconnection topologies including hypercubes, star graphs and rotator digraphs.

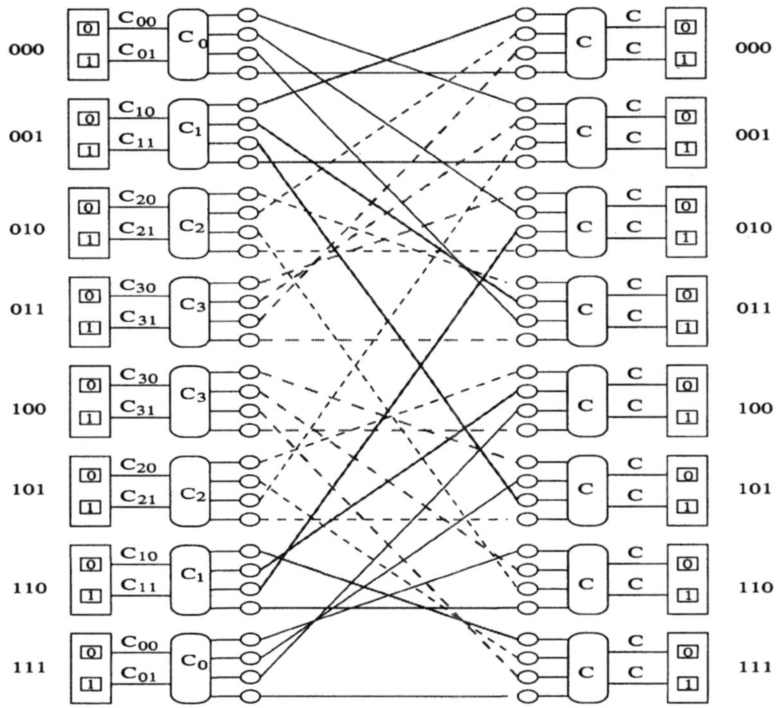

Figure 1.13. The muli-star CIN based on 3-cube with self-links requires four disjoint channel sets.

5. SUMMARY

WDM can not only expand the network capacity cost-effectively, but also simplify the network management. It has been successfully deployed in the transport networks and is currently moving steadily toward the access networks and the premise networks. This chapter provides a brief survey on the their networking architectures. It also introduces various networking design issues that will be studied in this book, together with their background and motivations.

Chapter 2

CONSTRUCTION OF OPTICAL ROUTING SYSTEMS WITH AWGS

1. INTRODUCTION

An *arrayed waveguide grating* (AWG) consists of a number of input/output waveguides, two focal slabs, and an array of grating waveguides with progressively longer optical path length (see Figure 2.1). The two slabs are mirror images of each other, so are the input/output waveguides attached to them. The two boundaries of each slab are two circular arcs of the same radius and their centers are separated by a distance equal to the radius. At the slab interface, the waveguides are perpendicular to the boundaries and have a constant angular separation. A waveguide focal slab constructed in this way becomes a star coupler, and even power split is realized as a result of the properties of a coupled periodic waveguide array [103].

The principle of the AWG can be described as follows [103]. The lightwave from an input waveguide excites the arrayed waveguides as it radiates in the input slab. If there were no differential phase shift in the grating region, the lightwave propagation in the output slab would appear as if it were the reciprocal propagation in the input slab. The input waveguide would thus be imaged at the interface between the output slab and the output waveguides. The imaged input waveguide would be butt-coupled to one of the output waveguides. The linear length difference in the waveguide grating results in a wavelength-dependent tilt of the wave front of the lightwave in the grating waveguides, and thus shifts the input waveguide image to a wavelength-dependent position. As the

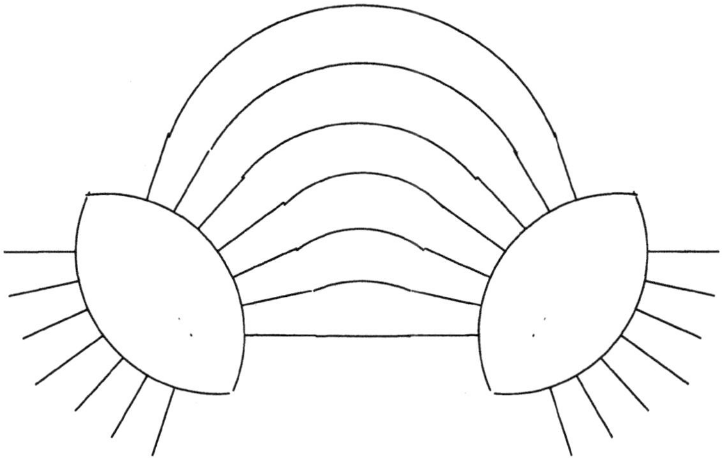

Figure 2.1. Arrayed waveguide grating.

wavelength changes, the input waveguide image sweeps across and couples light into different waveguides. In addition, two lights of the same wavelength from any two different input waveguides will not interfere with each other due to the additional phase difference created by the distance between the two input waveguides. They will be combined in the grating waveguides but will be separated again at the output slab and directed to different output waveguides.

One fundamental property of the AWG is the *reciprocity*: if a signal on a wavelength propagates from one input port to an output port, then any signal on the same wavelength injected at that output port will propagate backward to the input port in exactly the same way. Another fundamental property of the AWG is *periodicity* with the frequency, i.e., wavelengths in each frequency period follow the same routing pattern. The frequency span of one period is called the *free spectral range* (FSR). In general, an $N \times N$ AWG supports N wavelength channels, denoted by $\{0, 1, \cdots, N-1\}$, within a FSR. These wavelength channels have constant frequency spacing, rather than constant wavelength spacing. The wavelength $0 \leq w \leq N-1$ from input port $0 \leq i < N$ is routed to the output port $(w - i + c) \bmod N$ where c is a constant integer, referred to as the FSR constant, and its value depends on the selection of the FSR. Regardless of the value of c, the AWG has the following three wavelength-routing properties:

- *cyclic wavelength-shifting*: If an input signal on wavelength w leaves the AWG from a port j, then any input signal on wavelength $w+1$ entering the AWG from the same port will leave the AWG from the port $(j+1) \bmod N$.
- *cyclic port-shifting*: If an input signal enters form a port i and leaves from a port j, then any input signal of the same wavelength entering from port $(i+1) \bmod N$ will leave the AWG from the port $(j-1) \bmod N$.
- *symmetry*: If an input signal enters form a port i and leaves from a port j, then any input signal of the same wavelength entering from port j will leave the AWG from the port i.

AWGs are typically fabricated by depositing thin layers of silica glass onto silicon wafers in much the same way that electronics are integrated on a computer chip [103]. The fine silica glass layers are processed at high temperatures and patterned into waveguide circuits by using photolithography before the wafers are diced into individual circuit chips. The great precision of photolithography enables compact size and accurate channel spacing by precise control of the relative delay and coupling between different optical paths. In addition, the cost of AWG is not proportional to channel count. These advantages are particularly prominent for DWDM applications with a large number of wavelength channels.

AWGs have a broad range of applications [142]. In addition to the basic functions of wavelength multipexting/demultplexing, filtering and static routing/sorting, they are also highly versatile in integration with active optical components to create more complex and powerful functions. AWGs have been integrated with detectors in multiwavelength receivers [173], with optical amplifiers in multiwavelength lasers [149], and with optical switches in wavelength add-drop multiplexers (WADMs) [85, 148] and in optical cross-connects (OXCs) [146]. In this chapter, we will focus on various configurations of WADMs, OXCs and loopless star couplers based on a *single* AWG. In particular, we will describe in detail how to explore the rich wavelength routing properties of AWG to avoid crossovers of waveguides.

2. COMPATIBLE PORTS

In many practical applications, only a subset of ports of an $N \times N$ AWG are actually used for the input or output of signals. On the other hand, the number of ports N is often required to be larger than the number of wavelengths, denoted by W, that are actually used in all input signals. For example, in order to manipulate individual wavelength components in all input signals, each wavelength component should be routed in an exclusive output port. Thus an L-line W-wavelength system would then requires the number of ports N to be at least LW. Throughout of this chapter, we will use $\mathcal{W}$ to denote the set of W wavelengths used by the input signals. A set of ports is said to be *compatible* with respect to $\mathcal{W}$ if any pair of wavelengths in $\mathcal{W}$ from any two of these ports are routed to two different ports. The concept of compatible ports play an important role in the design of WADMs and OXCs. Of practical interest, we will identify compatible ports with respect to a set of wavelengths which have equal channel spacing. A set $S \subseteq \{0, 1, \cdots, N-1\}$ is said to have an equal spacing of k if it can be represented as

$$S = \{s + ik : 0 \leq i < |S|\}$$

We will begin with the characterization of compatible ports with respect to a set of wavelengths whose channel spacing is one.

LEMMA 2.1 *Let $0 \leq i < j \leq N-1$ be any two ports. Suppose that the channel spacing of $\mathcal{W}$ is one and $N \geq 2W$. Then they are compatible with respect to $\mathcal{W}$ if and only if $W \leq j - i \leq N - W$. In particular, when $N = 2W$, they are compatible with respect to $\mathcal{W}$ if and only if $j - i = W$.*

Proof. Assume that

$$\mathcal{W} = \{(w + x) \bmod N : 0 \leq x < W\}$$

for some integer w. We only need to prove that there do not exist two integers $0 \leq x, y \leq W - 1$ such that

$$(w + x - i + c) \bmod N = (w + y - j + c) \bmod N$$

if and only if $W \leq j - i \leq N - W$. We consider the following three cases.

Case 1: $W \leq j - i \leq N - W$. Consider any two integers $0 \leq x, y \leq W - 1$. Since

$$1 - W \leq x - y \leq W - 1,$$

we have

$$1 \leq (x - y) + (j - i) \leq N - 1.$$

Thus

$$\begin{aligned} &[(w + x - i + c) \bmod N - (w + y - j + c) \bmod N] \bmod N \\ &= [(w + x - i + c) - (w + y - j + c)] \bmod N \\ &= [(x - y) + (j - i)] \bmod N \\ &= [(x - y) + W] \bmod 2W \\ &> 0. \end{aligned}$$

So,

$$(w + x - i + c) \bmod 2W \neq (w + y - j + c) \bmod 2W.$$

Case 2: $0 < j - i < W$. For any

$$0 \leq x \leq W - 1 - (j - i),$$

let

$$y = x + (j - i).$$

Then

$$0 < j - i \leq y \leq W - 1$$

and

$$(w + x - i + c) \bmod N = (w + y - j + c) \bmod N.$$

Case 3: $j - i > N - W$. Then $0 < N - (j - i) < W$. For any

$$0 \leq y \leq (j - i) - (N - W + 1),$$

let

$$x = N - (j - i) + y.$$

Then

$$0 < N - (j - i) \leq x \leq W - 1$$

and

$$(w + x - i + c) \bmod N = (w + y - j + c) \bmod N.$$

Therefore, there do not exist two integers $0 \leq x, y \leq W-1$ such that

$$(w + x - i + c) \bmod N = (w + y - j + c) \bmod N$$

if and only if $w \leq j - i \leq N - W$. ∎

As a corollary of Lemma 2.1, we have the following characterization of compatible sets when $\mathcal{W}$ consists of W resolvable wavelengths.

COROLLARY 2.2 *Suppose that the channel spacing of* $\mathcal{W}$ *is one and* $N \geq LW$. *Let* $\mathcal{P} = \{p_i : 0 \leq i \leq L-1\}$ *be a set of* L *ports with*

$$0 \leq p_0 < p_1 < \cdots < p_{L-1} < N.$$

Then $\mathcal{P}$ *is compatible with respect to* $\mathcal{W}$ *if and only if*

$$p_i - p_{i-1} \geq W, \forall 1 \leq i \leq L-1$$

and

$$p_{L-1} - p_0 \leq N - W.$$

In particular, if $N = LW$, $\mathcal{P}$ *is compatible with respect to* $\mathcal{W}$ *if and only if* $\mathcal{P}$ *has an equal spacing of* W.

From the above lemma is that when the channel spacing of $\mathcal{W}$ is one and $N = LW$, there are exactly W sets of L compatible ports.

In the next, we will characterize compatible ports with respect to a set of wavelengths whose channel spacing is two.

LEMMA 2.3 *Suppose that the channel spacing of* $\mathcal{W}$ *is two,* N *is even and* $N \geq 2W$. *Then any two ports with different parities are compatible with respect to* $\mathcal{W}$. *Furthermore, if* $N \leq 4W-2$, *then any two compatible ports with respect to* $\mathcal{W}$ *must also have different parities.*

Proof. Assume that

$$\mathcal{W} = \{(w + 2x) \bmod N : 0 \leq x < W\}$$

for some integer w. Let i and j be any two ports with $0 \leq i < j < N$, and x be any integer with $0 \leq x < W$. Because N is even, the number $(w + 2x - i + c) \bmod N$ has the same parity as the number $w - i + c$, and the number $(w + 2x - j + c) \bmod N$ has the same parity as the number $w - j + c$. If i and j have different parities, the two numbers $w-i+c$ and $w-j+c$ also have different parities, so are the two numbers $(w + 2x - i + c) \bmod N$ has $(w + 2x - j + c) \bmod N$. Therefore,

$$\{(w + 2x - i + c) \bmod N : 0 \leq x < W\} \cap \\ \{(w + 2x - j + c) \bmod N : 0 \leq x < W\} = \emptyset,$$

which implies i and j are compatible.

Now assume that $N \leq 4W - 2$ and i have the same parity as j. Then all numbers in the set

$$\{(w + 2x - i + c) \bmod N : 0 \leq x < W\} \cup \\ \{(w + 2x - j + c) \bmod N : 0 \leq x < W\}$$

have the same parity. Therefore,

$$|\{(w + 2x - i + c) \bmod N : 0 \leq x < W\} \cup \\ \{(w + 2x - j + c) \bmod N : 0 \leq x < W\}| \\ \leq \frac{N}{2} \leq 2W - 1.$$

This implies that

$$|\{(w + 2x - i + c) \bmod N : 0 \leq x < W\} \cap \\ \{(w + 2x - j + c) \bmod N : 0 \leq x < W\}| \\ = |\{(w + 2x - i + c) \bmod N : 0 \leq x < W\}| + \\ |\{(w + 2x - j + c) \bmod N : 0 \leq x < W\}| - \\ |\{(w + 2x - i + c) \bmod N : 0 \leq x < W\} \cup \\ \{(w + 2x - j + c) \bmod N : 0 \leq x < W\}| \\ \geq 2W - (2W - 1) = 1.$$

which means i and j are not compatible. ∎

From the above lemma, when the channel spacing of $\mathcal{W}$ is two and N is equal to $2W$, the number of compatible pairs with respect to $\mathcal{W}$ is W^2. This is much more than that when the channel spacing is one.

The next lemma given a sufficient condition for a set of ports to be compatible with respect to $\mathcal{W}$ whose channel spacing is L. It partially generalizes Lemma 2.3.

LEMMA 2.4 *Suppose that the channel spacing of* $\mathcal{W}$ *is* L, $N \geq LW$ *and* $N \bmod L = 0$. *Let* $\mathcal{P} = \{p_i : 0 \leq i \leq L-1\}$ *be any set of* L *ports. If*

$$p_i \bmod L \neq p_j \bmod L, \forall 0 \leq i < j < L$$

then $\mathcal{P}$ *is compatible with respect to* $\mathcal{W}$. *In particular, any* L *consecutive ports are compatible with respect to* $\mathcal{W}$.

Proof. Let

$$\mathcal{W} = \{(w + Lx) \bmod N : 0 \leq x \leq W - 1\}$$

for some integer w. Suppose that there exist $0 \leq x, y \leq W-1$ such that

$$\begin{aligned}&(w + Lx - p_i + c) \bmod N\\&= (w + Ly - p_j + c) \bmod N.\end{aligned}$$

Then

$$\begin{aligned}&(w + Lx - p_i + c) - (w + Ly - p_j + c)\\&= L(x - y) + (p_j - p_i)\end{aligned}$$

would be a multiple of N and therefore also a multiple of L, as N is a multiple of L. This implies that $p_j - p_i$ is a multiple of L, which contradicts to the assumption that

$$p_i \bmod L \neq p_j \bmod L$$

for any $0 \leq i < j \leq L-1$. ∎

3. SELF-BLOCKING PORTS

When only a single AWG is used in WADMs or OXCs, the same set of ports opposite to the input ports should be reserved as the output ports, and hence can not be used for intermediate processing. A set $\mathcal{P}$ of ports is said to be *self-blocking* with respect to $\mathcal{W}$ if

$$\mathcal{P} \cap \{(w - i + c) \bmod N : i \in \mathcal{P}, w \in \mathcal{W}\} = \emptyset.$$

Thus the input ports should be not only compatible but also self-blocking. A necessary condition for the existence of a set of L self-blocking and compatible ports with respect to $\mathcal{W}$ is $N \geq L(W+1)$.

The necessary and sufficient for the existence of a single self-blocking port when $N = W + 1$ is given in the next lemma.

LEMMA 2.5 *Suppose that*

$$\mathcal{W} = \{w : 0 \leq w \leq W\} \setminus \{w'\}$$

for some $0 \leq w' \leq W$ *and* $N = W + 1$. *Then for any port* $0 \leq q \leq W$, *it is self-blocking with respect to* $\mathcal{W}$ *if and only if*

$$(w' + c - 2q) \bmod (W + 1) = 0.$$

Proof. It's obvious that port q is self-blocking with respect to $\mathcal{W}$ if and only if the AWG routes the wavelength w' from port q to port q, i.e.,

$$q = (w' - q + c) \bmod (W + 1)$$

or

$$(w' + c - 2q) \bmod (W + 1) = 0.$$

■

Despite of its simplicity, the previous proof contains a general idea for the selection of an arbitrary number of self-blocking ports. The generalization relies on the following concept. For any wavelength $0 \leq w \leq N - 1$,a port set $\mathcal{P}$ is said to be *invariable* with respect to w if the wavelength w from any port $\mathcal{P}$ in is always routed to some port also in $\mathcal{P}$, i.e.,

$$\mathcal{P} = \{(w - i + c) \bmod N : i \in \mathcal{P}\}.$$

It's obvious that if a port set $\mathcal{P}$ is invariable with respect to some wavelength $w \notin \mathcal{W}$ and is compatible with respect to $\mathcal{W} \cup \{w\}$, then $\mathcal{P}$ is compatible and self-blocking with respect to $\mathcal{W}$. The next lemma gives necessary and sufficient conditions for two types of port sets to be invariable with respect to a wavelength.

LEMMA 2.6 *Let* $0 \leq w < N$ *be any wavelength.*

1. *For any* $0 \leq q < N$, *the port set*

$$\mathcal{P} = \{(q + i) \bmod N : 0 \leq i < L\}$$

is invariable with respect to w if and only if

$$[w - (2q + L - 1) + c] \bmod N = 0.$$

2. Suppose that $N \bmod W = 0$. Then for any $0 \le q < W$, the port set

$$\mathcal{P} = \{0 \le i < N : i \bmod W = q\}$$

is invariable with respect to w if and only if

$$(w - 2q + c) \bmod W = 0.$$

Proof. (1). $\mathcal{P}$ is invariable with respect to w if and only if the wavelength w from port q is routed to port $(q + L - 1) \bmod N$, i.e.,

$$(w - q + c) \bmod N = (q + L - 1) \bmod N,$$

or equivalently

$$[w - (2q + L - 1) + c] \bmod N = 0.$$

(2). The lemma is obviously true when $N = W$. So we assume that $N > W$. For any port $i \in \mathcal{P}$,

$$\begin{aligned} &[(w - i + c) \bmod N] \bmod W \\ &= (w - i + c) \bmod W \quad (as\ N \bmod W = 0) \\ &= (w - q + c) \bmod W \quad (as\ i \bmod W = q). \end{aligned}$$

Therefore, $(w - i + c) \bmod N \in \mathcal{P}$ if and only if

$$(w - q + c) \bmod W = q,$$

or equivalently

$$(w - 2q + c) \bmod W = 0.$$

■

Now are ready to present sufficient conditions for a port set to be compatible and self-blocking with respect to a given set of wavelengths. We will first consider the conditions when the set of wavelengths whose channel spacing is one.

LEMMA 2.7 *Suppose that*

$$\mathcal{W} = \{(w + x) \bmod N : 0 \le x < W\}$$

for some integer w*, and* $N \bmod (W+1) = 0$*. Let*

$$\mathcal{P} = \{0 \leq i < N : i \,\mathrm{mod}(W+1) = q\}$$

for some $0 \leq q \leq W$*. If*

$$[w - (2q+1) + c] \,\mathrm{mod}(W+1) = 0,$$

then $\mathcal{P}$ *is compatible and self-blocking with respect to* $\mathcal{W}$ *.*

Proof. From Corollary 2.2, for any $0 \leq q \leq W$, $\mathcal{P}$ is compatible with respect to $\mathcal{W} \cup \{(w+W) \bmod N\}$. As

$$[w - (2q+1) + c] \,\mathrm{mod}(W+1) = 0,$$

and $N \bmod (W+1) = 0$,

$$\begin{aligned}
&[(w+W) \bmod N - 2q + c] \,\mathrm{mod}(W+1) \\
&= [(w+W) - 2q + c] \,\mathrm{mod}(W+1) \\
&= [w + (W+1) - (2q+1) + c] \,\mathrm{mod}(W+1) \\
&= 0.
\end{aligned}$$

Thus from Lemma 2.6, $\mathcal{P}$ is invariable with respect to the wavelength $(w+W) \bmod N$. Therefore $\mathcal{P}$ is compatible and self-blocking with respect to $\mathcal{W}$. ∎

Finally, we consider compatible and self-blocking ports with respect to a set of wavelengths with equal channel spacing of arbitrary value.

LEMMA 2.8 *Suppose that* $N \bmod L = 0$ *and* $N \geq L(W+1)$*,*

$$\mathcal{W} = \{(w + Lx) \bmod N : 0 \leq x < W\}$$

for some integer $0 \leq w \leq N-1$*. Let*

$$\mathcal{P} = \{(q+i) \bmod N : 0 \leq i \leq L-1\}$$

for some $0 \leq q < N$*. If*

$$[w - (2q + 2L - 1) + c] \bmod N = 0$$

or

$$[w + L(W-1) - (2q-1) + c] \bmod N = 0,$$

then $\mathcal{P}$ is compatible and self-blocking with respect to $\mathcal{W}$.

Proof. We first assume that

$$[w - (2q + 2L - 1) + c] \bmod N = 0.$$

From Lemma 2.4, $\mathcal{P}$ is compatible with respect to $\mathcal{W} \cup \{(w - L) \bmod N\}$. As

$$\begin{aligned}
&[(w - L) \bmod N - (2q + L - 1) + c] \bmod N \\
&= [(w - L) - (2q + L - 1) + c] \bmod N \\
&= [w - (2q + 2L - 1) + c] \bmod N \\
&= 0,
\end{aligned}$$

$\mathcal{P}$ is also invariable with respect to $(w - L) \bmod N$ by Lemma 2.6. Therefore $\mathcal{P}$ is compatible and self-blocking with respect to $\mathcal{W}$.

Now we assume that

$$[w + L(W - 1) - (2q - 1) + c] \bmod N = 0.$$

From Lemma 2.4, $\mathcal{P}$ is compatible with respect to

$$\mathcal{W} \cup \{(w + LW) \bmod N\}.$$

As

$$\begin{aligned}
&[(w + LW) \bmod N - (2q + L - 1) + c] \bmod N \\
&= [(w + LW) - (2q + L - 1) + c] \bmod N \\
&= [w + L(W - 1) - (2q - 1) + c] \bmod N \\
&= 0.
\end{aligned}$$

$\mathcal{P}$ is also invariable with respect to $(w + LW) \bmod N$ by Lemma 2.6. Therefore $\mathcal{P}$ is compatible and self-blocking with respect to $\mathcal{W}$. ■

Note that the two conditions in the above lemma become identical when $N = L(W + 1)$.

4. WAVELENGTH ADD/DROP MULTIPLEXERS

WADMs have one input line and one output line. They provide the functionality to drop/add or bypass individual wavelengths independently. The conventional WADM is based on the *cascaded demux/mux* configuration, as illustrated in Figure 2.2 [113]. It consists of back-to-back wavelength-division multiplexer and demultiplexer in conjunction with an optical space switch on each wavelength. An incoming multi-wavelength signal is demultiplexed by the optical demultiplexer. Switch elements can be dynamically configured to drop and add the selected wavelength channels. This configuration requires separate multiplexer and demultiplexer, and thus requires precise matching of the device characteristics of the demultiplexer and the multiplexer. In this section, we will study various improvements by using a *single* AWG as both multiplexer and demultiplexer. In all these configurations, the signal always passes *twice* through the AWG before entering the output fiber. In the first pass, the AWG acts as a demultiplexer; in the second pass the AWG acts as a multiplexer. As single AWG is used, potential wavelength misalignment is then avoided.

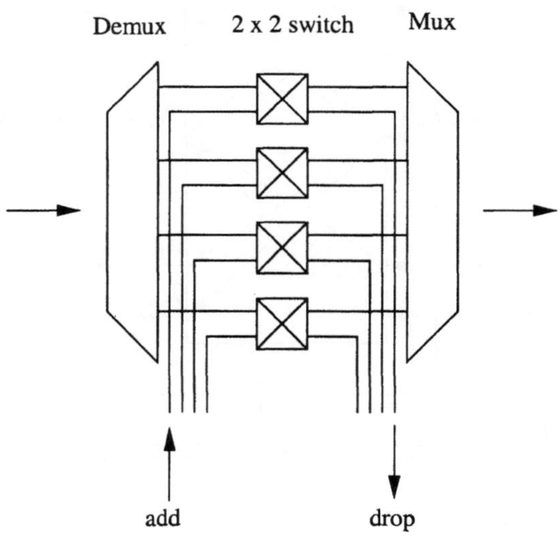

Figure 2.2. A reconfigurable WADM uses space-division switches between a demultiplexer and a multiplexer.

4.1 LOOP-BACK CONFIGURATION

In the *loop-back configuration* as illustrated in Figure 2.3, a pair of opposite ports of the AWG, which are *self-blocking* with respect to the set of wavelengths contained in the input signal, are reserved for the input/output lines. All other pairs of opposite ports are each connected by a loop-back optical path, which contains an 2×2 optical space switch. Because the two ports for the transmission lines are self-blocking, the demultiplexed signals are all looped back after passing through the respective optical space switches, with some signals possibly dropped/added. Due to the symmetric and reciprocal nature of the AWG, the looped-back signals are automatically multiplexed again and fed into the output line. Therefore, each wavelength passed through the AWG twice in the *same* direction before entering the output fiber.

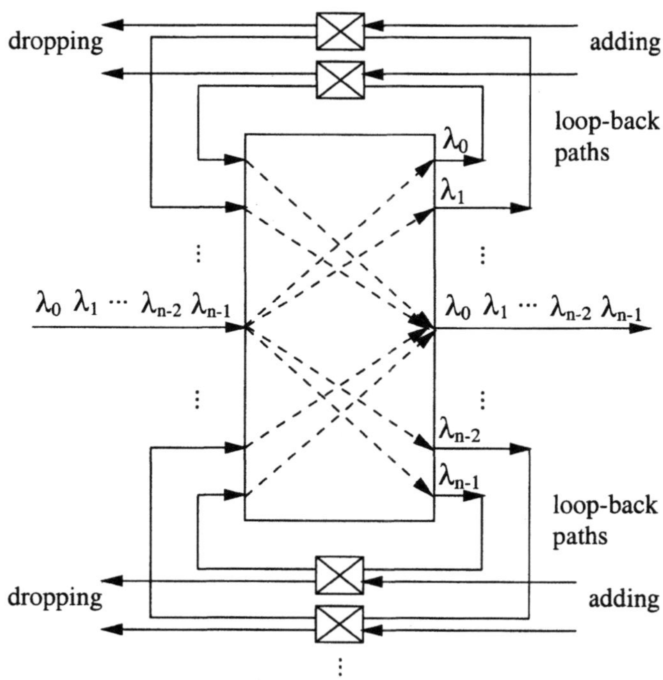

Figure 2.3. A schematic illustration of a loop-back configuration of WADM.

The operation principle of the loop-back configuration implies that the AWG must have at least $W + 1$ ports each side. On the other hand, a $(W + 1) \times (W + 1)$ AWG is sufficient to build a W-wavelength WADM from Lemma 2.5, and the input/output ports can be uniquely determined by the wavelength set and the FSR constant c. In general, the AWG is most effective when the signal passes through the middle waveguides, and provides the highest refractive efficiency and the least signal attenuation. Therefore, the central regions are the most preferable for connecting to input or output line. There are many ways to make the input/output lines located in the central region. For example, by setting $\mathcal{W} = \{w : 0 \leq w < W\}$ and $c = 0$ (or 1 respectively) when W is even (or odd respectively), the input line and the output line are then both attached to the port $\lceil \frac{W}{2} \rceil$ in the opposite sides.

The loop-back optical paths can be implemented by separate fibers or waveguides. However, if they are implemented as waveguides, some crossovers can not be avoided. The loop-back configuration is also prone to coherent crosstalk; scattered light from the input port directly enters the port for the output and beats with the optical signal. These problems can be reduced by applying the AWG in a fold-back configuration.

4.2 FOLD-BACK CONFIGURATION

In the *fold-back configuration*, two *compatible* ports with respect to $\mathcal{W}$, say i and j, at the same side, say left side, of an $N \times N$ AWG are reserved for the input line and output line respectively. For any $w \in \mathcal{W}$, there is an optical path called fold-back path, which is similar to that in the loop-back configuration, from the right-hand port $(w - i + c) \bmod N$ to the right-hand port $(w - j + c) \bmod N$. The demultiplexed input signals are thus all folded back along the fold-back paths to the right-hand ports after passing through the respective optical space switches, with some signals possibly dropped/added. Due to the symmetric and reciprocal nature of the AWG, the signals folded back are automatically multiplexed again and fed out to the output line. Therefore, each wavelength passes through the AWG twice in *opposite* directions before entering the output fiber.

This operation principle of the fold-back configuration implies that N must be at least $2W$. On the other hand, $N = 2W$ is sufficient to build a W-wavelength WADM from Lemma 2.1 and Lemma 2.3, and the input/output ports can be selected in many ways. To minimize the cost and size of the WADMs, we will assume that $N = 2W$. The fold-back configuration proposed in the literature [61, 85] can be regarded as the straightforward applications of Lemma 2.1. Here $N = 2W$ and $\mathcal{W}$ consists of W *consecutive* wavelengths, i.e.,

$$\mathcal{W} = \{(w + x) \bmod N : 0 \leq x < W\}$$

for some integer w. As long as i and j are W-apart, they are compatible. Figure 2.4 shows the configuration with $i = \lfloor \frac{W}{4} \rfloor$ and $j = i + W$.

However, the drawback of such configuration is that the W fold-back paths have too many crossovers.

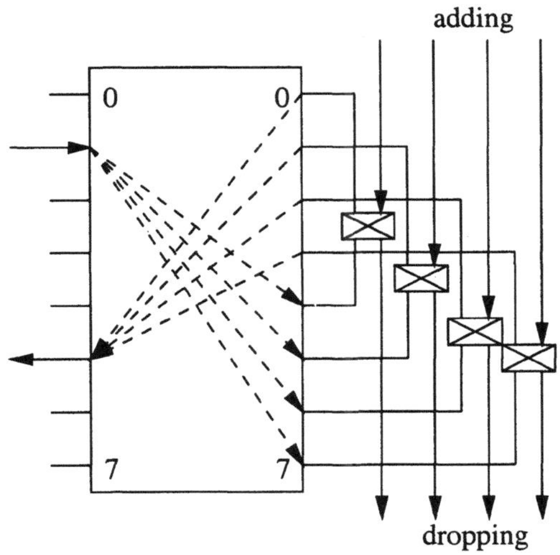

Figure 2.4. A non-planar fold-back configuration for a 4-wavelength WADM with crossovers.

In order to eliminate these crossovers, each the fold-back optical path has to connect two neighboring ports (port 0 and port $N-1$ not considered to be *neighboring*). Thus we for any $0 \leq k < W$, a fold-back path connects two ports $2k$ and $2k+1$. Correspondingly, $\mathcal{W}$ has to consist of W *alternating* wavelengths, i.e.,

$$\mathcal{W} = \{(w + 2x) \bmod N : 0 \leq x \leq W - 1\}$$

for some integer w, and the output port j has to be either $(i+1) \bmod N$ or $(i-1) \bmod N$. Without of loss generality, we set $j = (i+1) \bmod N$. Then to maintain proper operation, $(w - i + c) \bmod N$ has to be an odd number. Thus the input port i can be any number that has different parity from the number $w+c$. In to maximize the performance, i should be further refined to be either W if W and $w + c$ have different parities or $W - 1$ otherwise. Figure 2.5 shows an example of such planar configuration.

Comparing to the loop-back configuration, a major advantage of the fold-back configuration is that scattered light from the input fibers cannot directly enter the output fiber channels; thus, coherent crosstalk

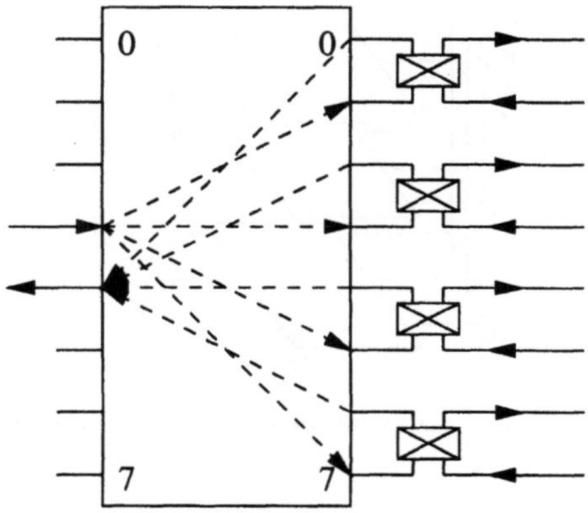

Figure 2.5. A planar fold-back configuration for a 4-wavelength WADM.

is minimized. A disadvantage of the fold-back configuration is that it requires larger AWG.

5. OPTICAL CROSS-CONNECT

The conventional configuration for an L-line W-wavelength OXC is shown in Figure 2.6. In this configuration, each input line is attached to a dedicated demultiplexer, each output line is attached to a dedicated multiplexer, and each wavelength channel requires a dedicated $L \times L$ optical space switch. The switch settings determine which signals are transmitted to each output fiber. Thus this configuration requires L separate demultiplexers, L separate multiplexers, and W optical space switches. As with WADMs, the OXC can be made more robust, compact and economical by utilizing AWGs.

5.1 CASCADED CONFIGURATION

The cascaded configuration uses two $N \times N$ AWGs, one as the demultiplexer for all input lines and the other as the multiplexer for all output lines. The two AWGs, the input ports and the output ports, and all internal connections are symmetric. We use $\mathcal{P}$ to denote the set of L input/output ports. The port set $\mathcal{P}$ must be compatible with respect to $\mathcal{W}$, and thus N should be at least LW. The first AWG simultaneously

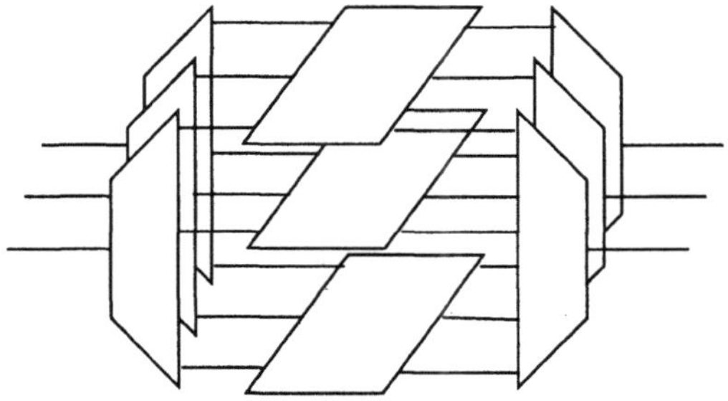

Figure 2.6. Conventional OXC configuration.

demultiplexes the L input signals into LW component wavelengths. For any wavelength $w \in \mathcal{W}$, all the L components with wavelength w are routed to an $L \times L$ optical space switch. Each $L \times L$ optical space switch provides a cross-connect from a set of L ports of the first AWG to the *same* set of ports of the second AWG. From the AWG's symmetric property, the wavelengths collected by the second AWG from the W space-division switches are multiplexed onto the L output lines.

To minimize the cost and complexity, we set $N = LW$. One may choose $\mathcal{W}$ to consist of W consecutive wavelengths, and $\mathcal{P}$ to be any L ports with equal spacing of W. Obviously, $\mathcal{P}$ is compatible with respect to $\mathcal{W}$. However the resulting configuration, as illustrated in Figure 2.7, contains two many crossovers among the waveguides between the AWGs and the optical space switches. So we will seek some other configurations.

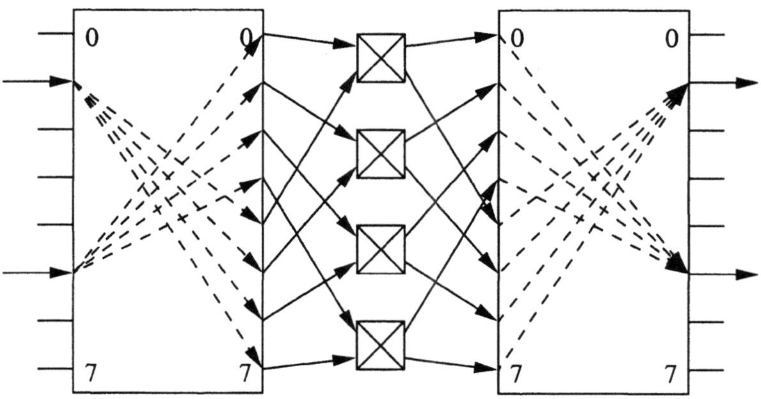

Figure 2.7. A cascaded configuration for a 2-line 4-wavelength OXC with crossovers.

If the wavelength set $\mathcal{W}$ has a channel spacing of L, say

$$\mathcal{W} = \{(w + Lx) \bmod N : 0 \leq x < W\}$$

for some integer $0 \leq w < L$, then any port set $\mathcal{P}$ consisting of L consecutive ports is compatible with respect to $\mathcal{W}$. Let

$$\mathcal{P} = \{(q + i) \bmod N : 0 \leq i < L\}$$

for some $0 \leq q < N$. It's obvious that no waveguides interconnecting the AWGs and the space switches cross with each other if and only if the wavelength w from port $(q + L - 1) \bmod N$, is routed to a port which is a multiple of L, i.e.,

$$\{[w - (q + L - 1) \bmod N + c] \bmod N\} \bmod L = 0.$$

As N is a multiple of L,

$$\begin{aligned}
&\{[w - (q + L - 1) \bmod N + c] \bmod N\} \bmod L \\
&= \{[w - (q + L - 1) + c] \bmod N\} \bmod L \\
&= [w - (q + L - 1) + c] \bmod L \\
&= (w + c + 1 - q) \bmod L.
\end{aligned}$$

Therefore, q should be selected such that

$$(w + c + 1 - q) \bmod L = 0.$$

The solutions to this equation are quite many. For an example, when L is equal to two, any q which has different parity from $w + c$ is a solution. In order to maximize the performance, q should be selected such that $\mathcal{P}$ is around the central region of the AWGs. Thus when W is odd, we choose

$$q = \frac{(W - 1)\, L}{2},$$
$$(w + c + 1) \bmod L = 0.$$

When both W and L are even, we choose

$$q = \frac{(W - 1)\, L}{2},$$
$$(w + c) \bmod L = \frac{L}{2} - 1.$$

When W is even but L is odd, we either choose

$$q = \frac{(W-1)\,L-1}{2},$$
$$(w+c) \bmod L = \frac{L-1}{2} - 1.$$

or choose

$$q = \frac{(W-1)\,L+1}{2},$$
$$(w+c) \bmod L = \frac{L-1}{2}.$$

Figure 2.8 shows an example of such configuration for 2-line 4-wavelength OXC with $w = c = 0$ and $q = 3$.

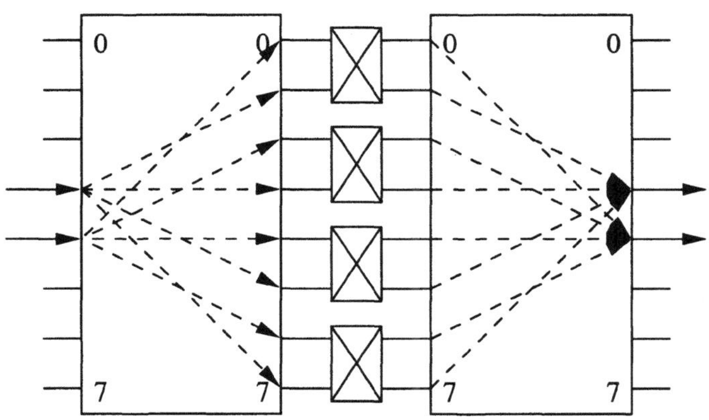

Figure 2.8. A cascaded configuration for a 2-line 4-wavelength OXC without crossover.

5.2 LOOP-BACK CONFIGURATION

The loop-back configuration uses a single $N \times N$ AWG as both the demultiplexer for the input lines and the multiplexer for the output lines. The signals enter the AWG from one side and leave the AWG from the other side. The input ports and the output ports are symmetric, and we denote by $\mathcal{P}$ the set of L input/output ports. The port set $\mathcal{P}$ must be compatible and self-blocking with respect to $\mathcal{W}$, and this implies that N should be at least $L(W+1)$. The AWG first simultaneously demultiplexes the L input signals into LW component wavelengths. For each wavelength $w \in \mathcal{W}$, the L components with wavelength w are routed to an $L \times L$ space-division switch. Each $L \times L$ space-division switch provides a cross-connect from a set of L ports of the AWG to the *same* set of ports of the AWG at the opposite side. From the AWG's symmetric property, the wavelengths looped back from the W space-division switches are multiplexed onto the L output lines.

To minimize the cost, we set $N = L(W+1)$. We first choose

$$\mathcal{W} = \{(w+x) \bmod N : 0 \leq x < W\},$$
$$\mathcal{P} = \{0 \leq i < N : i \bmod (W+1) = q\}$$

for some $0 \leq w < N$ and $0 \leq q \leq W$ with

$$[w - (2q+1) + c] \bmod (W+1) = 0.$$

From Lemma 2.7, the port set $\mathcal{P}$ is compatible and self-blocking with respect to the wavelength set $\mathcal{W}$. The loop-back configuration, however, contains many crossovers. One such configuration for a 2-line 3-wavelength OXC with $w = 0$, $c = 1$ and $q = 0$ is illustrated in Figure 2.9.

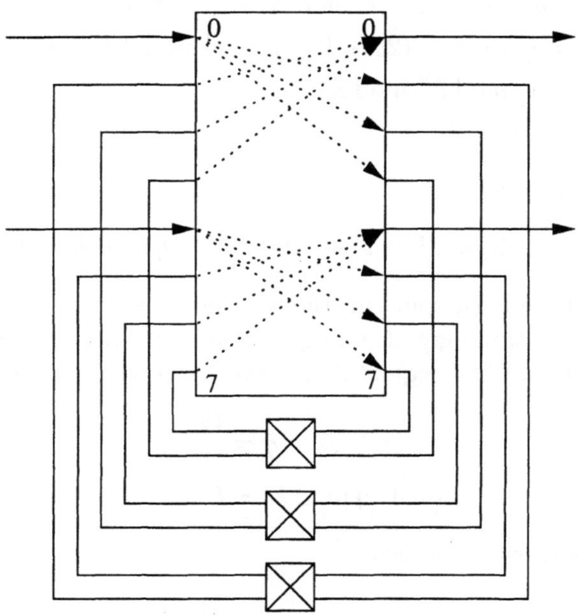

Figure 2.9. A loop-back configuration for a 2-line 3-wavelength OXC with crossovers.

Alternatively, we select

$$\mathcal{W} = \{(w + Lx) \bmod N : 0 \le x < W\},$$
$$\mathcal{P} = \{(q + i) \bmod N : 0 \le i < L\}$$

for some $0 \le w < L$ and $0 \le q < N$ with

$$[w - (2q + 2L - 1) + c] \bmod N = 0.$$

From Lemma 2.8, the port set $\mathcal{P}$ is also compatible and self-blocking with respect to the wavelength set $\mathcal{W}$. In order to avoid any crossover, it is necessary and sufficient to further require the wavelength w from port $(q + L - 1) \bmod N$ to be routed to a port which is a multiple of L, i.e.,

$$\{[w - (q + L - 1) \bmod N + c] \bmod N\} \bmod L = 0.$$

As N is a multiple of L and

$$[w - (2q + 2L - 1) + c] \bmod N = 0,$$

we have

$$\begin{aligned}
&\{[w - (q + L - 1) \bmod N + c] \bmod N\} \bmod L \\
&= \{[q + w - (2q + L - 1) + c] \bmod N\} \bmod L \\
&= (q \bmod N) \bmod L \\
&= q \bmod L.
\end{aligned}$$

Therefore, as long as

$$[w - (2q + 2L - 1) + c] \bmod N = q \bmod L = 0,$$

the resulting loop-back configuration has no crossover. Among all solutions to this equation may also be quite many, we prefer those such that $\mathcal{P}$ is around the central region of the AWG. So when W is even, we set

$$q = \frac{W}{2}L,$$
$$(w + c) \bmod N = L - 1.$$

When W is odd, we either set

$$q = \frac{W-1}{2}L,$$
$$(w + c + 1) \bmod N = 0;$$

or set

$$q = \frac{W+1}{2}L,$$

$$(w + c) \bmod N = 2L - 1.$$

Figure 2.10 shows a 2-line 4-wavelength OXC with $w = 0$, $c = 1$ and $q = 4$.

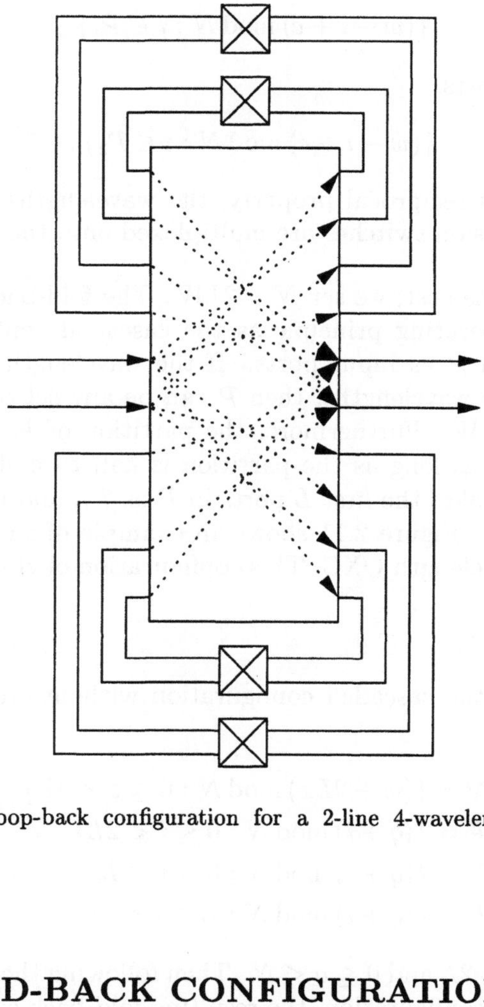

Figure 2.10. A loop-back configuration for a 2-line 4-wavelength OXC without crossover.

5.3 FOLD-BACK CONFIGURATION

The fold-back configuration also uses a single $N \times N$ AWG as both the demultiplexer for the L input lines and the multiplexer for the L output lines. But the L input lines and the L output lines are attached to the *same* side of the AWG. We use $\mathcal{P}_1$ and $\mathcal{P}_2$ to denote the set of L input ports and the set of L output ports respectively, and let $\mathcal{P} = \mathcal{P}_1 \cup \mathcal{P}_2$. The port set $\mathcal{P}$ must be compatible with respect to $\mathcal{W}$, which implies that $N \geq 2LW$. The AWG first simultaneously demultiplexes the L input signals into LW component wavelengths. For each wavelength $w \in \mathcal{W}$, all the L components of wavelength w are routed to an $L \times L$ optical space switch. The $L \times L$ space switch corresponding to wavelength w provides a cross-connect from the set of L ports

$$\{(w - i + c) \bmod N : i \in \mathcal{P}_1\}$$

to the set of L ports

$$\{(w - i + c) \bmod N : i \in \mathcal{P}_2\}.$$

From the AWG's reciprocal property, the wavelengths fold back from the W space-division switches are multiplexed onto the L output lines.

To minimize the cost, we set $N = 2LW$. The fold-back configuration has the same operating principle as the cascaded configuration if we treat all ports in $\mathcal{P}$ as input ports. If the wavelength set $\mathcal{W}$ consists of W consecutive wavelengths, then $\mathcal{P}$ can be any set of $2L$ ports with equal spacing of W. Furthermore, the partition of $\mathcal{P}$ into $\mathcal{P}_1$ and $\mathcal{P}_2$ can be arbitrary as long as the partition is half to half. The common implementation takes the first L ports in $\mathcal{P}$ as $\mathcal{P}_1$, and the remaining L ports in $\mathcal{P}$ as $\mathcal{P}_2$. Figure 2.11 shows an example of such configuration for a 2-line 3-wavelength OXC. This configuration obviously has a lot of crossovers.

Motivated by the cascaded configuration without crossovers, we choose

$$\begin{aligned}
\mathcal{W} &= \{(w + 2Lx) \bmod N : 0 \leq x < W\}, \\
\mathcal{P} &= \{(q + i) \bmod N : 0 \leq i < 2L\}, \\
\mathcal{P}_1 &= \{(q + i) \bmod N : 0 \leq i < L\}, \\
\mathcal{P}_2 &= \{(q + i) \bmod N : L \leq i < 2L\}.
\end{aligned}$$

for some $0 \leq w < 2L$ and $0 \leq q < N$. Then following the same argument for the cascaded configuration, the resulting fold-back configuration has no crossover. In addition, q should be selected such that $\mathcal{P}$ is within the central region of the AWG. Thus we set

$$q = (W - 1)\, L,$$

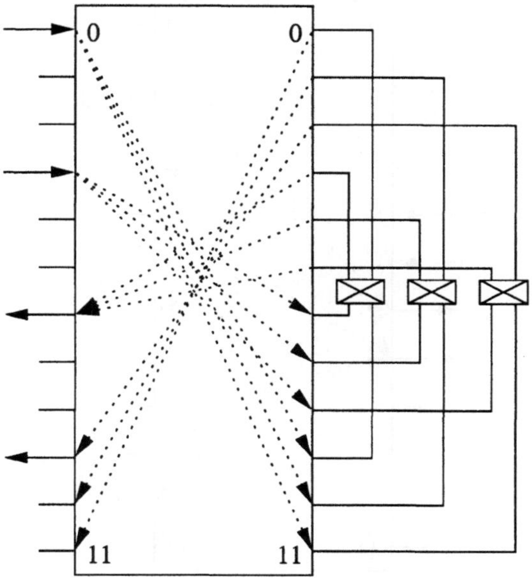

Figure 2.11. A fold-back configuration for a 2-line 3-wavelength OXC with crossovers.

and

$$(w + c) \bmod 2L = L - 1$$

if W is even, or

$$(w + c + 1) \bmod 2L = 0$$

if W is odd. Figure 2.12 shows an example of such fold-back configuration for a 2-line 3-wavelength OXC with $w = 0$, $c = 1$ and $q = 4$.

5.4 HYBRID CONFIGURATION

The hybrid configuration combines both the loop-back configuration and fold-back configuration. But it only applies to even-line OXCs, i.e. L is even. Similar to the loop-back configuration and the fold-back configuration, the hybrid configuration uses a single $N \times N$ AWG as both the demultiplexer for the L input lines and the multiplexer for the L output lines. However, half of the input lines enter the AWG from the one side, and another half enter the AWG from the *opposite*

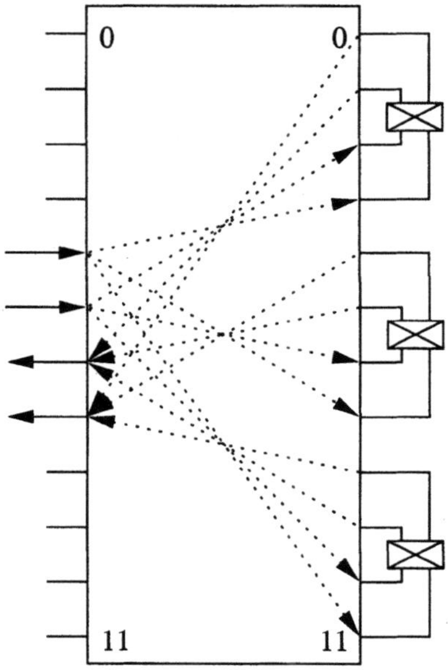

Figure 2.12. A fold-back configuration for a 2-line 3-wavelength OXC without crossover.

ports. Similarly, half of the output lines leave the AWG from the one side, and another half leave the AWG from the *opposite* ports. We use $\mathcal{P}_1$ and $\mathcal{P}_2$ to denote the set of $\frac{L}{2}$ input ports and the set of $\frac{L}{2}$ output ports respectively, and let $\mathcal{P} = \mathcal{P}_1 \cup \mathcal{P}_2$. The port set $\mathcal{P}$ must be compatible and self-blocking with respect to $\mathcal{W}$, and thus N must be at least $L(W+1)$. The AWG first simultaneously demultiplexes the L input signals into LW component wavelengths. For each wavelength $w \in \mathcal{W}$, the L components with wavelength w, from $\frac{L}{2}$ pairs of opposite ports in $\mathcal{P}_1$, are routed to an $L \times L$ space-division switch. The $L \times L$ optical space switch corresponding to wavelength w provides a cross-connect from $\frac{L}{2}$ pairs of opposite ports

$$\{(w - i + c) \bmod N : i \in \mathcal{P}_1\}$$

to $\frac{L}{2}$ pairs of opposite ports

$$\{(w - i + c) \bmod N : i \in \mathcal{P}_2\}.$$

From the AWG's symmetric and reciprocal properties, the wavelengths folded or looped back from the W optical space switches are multiplexed onto the L output lines.

To minimize the cost, we set $N = L(W+1)$. The hybrid configuration has the very similar operating principle as the loop-back configuration. We can choose

$$\mathcal{W} = \{(w+x) \bmod N : 0 \le x < W\},$$
$$\mathcal{P} = \{0 \le i < N : i \bmod (W+1) = q\},$$

for some $0 \le w < N$ and $0 \le q \le W$ with

$$[w - (2q+1) + c] \bmod (W+1) = 0.$$

The port set $\mathcal{P}$ must be compatible and self-blocking with respect to $\mathcal{W}$. The partition of $\mathcal{P}$ into $\mathcal{P}_1$ and $\mathcal{P}_2$ can be arbitrary as long as the partition is half to half. The common implementation takes the first $\frac{L}{2}$ ports in $\mathcal{P}$ as $\mathcal{P}_1$, and the remaining $\frac{L}{2}$ ports in $\mathcal{P}$ as $\mathcal{P}_2$, i.e.,

$$\mathcal{P}_1 = \left\{q + i(W+1) : 0 \le i < \frac{L}{2}\right\},$$
$$\mathcal{P}_2 = \left\{q + i(W+1) : \frac{L}{2} \le i < L\right\}.$$

Figure 2.13 shows an example of such configuration for a 2-line 3-wavelength OXC with $w = 0$, $c = 1$ and $q = 0$. This configuration has a lot of crossovers.

Similarly, we can choose

$$\mathcal{W} = \{(w+Lx) \bmod N : 0 \le x < W\},$$
$$\mathcal{P} = \{(q+i) \bmod N : 0 \le i < L\},$$

for some $0 \le w < L$ and $0 \le q < N$ with

$$[w - (2q+2L-1) + c] \bmod N = 0.$$

The partition of $\mathcal{P}$ into $\mathcal{P}_1$ and $\mathcal{P}_2$ is also arbitrary as long as the partition is half to half. The resulting loop-back configuration has no crossover if and only if

$$q \bmod L = 0.$$

To avoid crossover and make $\mathcal{P}$ be within the central region of the AWG, the parameters w, c, and w are set as follows. When W is even,

$$q = \frac{W}{2}L,$$
$$(w+c) \bmod N = L-1.$$

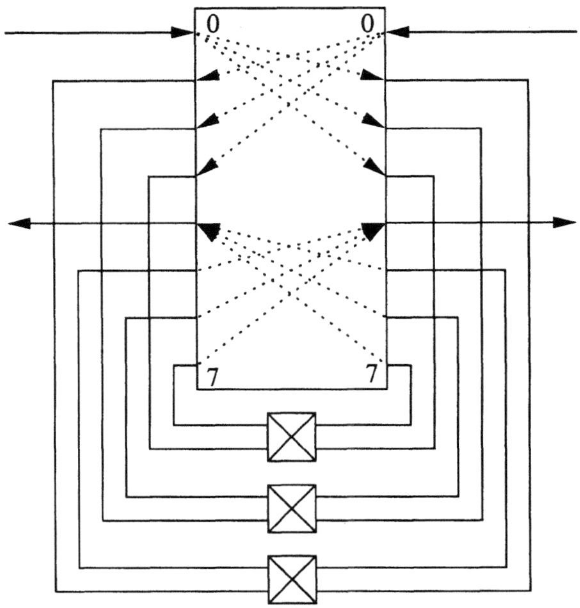

Figure 2.13. A hybrid configuration for a 2-line 3-wavelength OXC with crossovers.

When W is odd, either

$$q = \frac{W-1}{2}L,$$
$$(w + c + 1) \bmod N = 0;$$

or

$$q = \frac{W+1}{2}L,$$
$$(w + c) \bmod N = 2L - 1.$$

Figure 2.14 shows an example of such hybrid configuration for a 2-line 4-wavelength OXC with $w = 0$, $c = 1$ and $q = 4$.

6. LOOP-FREE STAR COUPLERS

In a broadcast-and-select PON based on star coupler, a signal sent from one node is always received by the node itself. Such loop is usually unnecessary, and may even cause some undesirable consequences. For

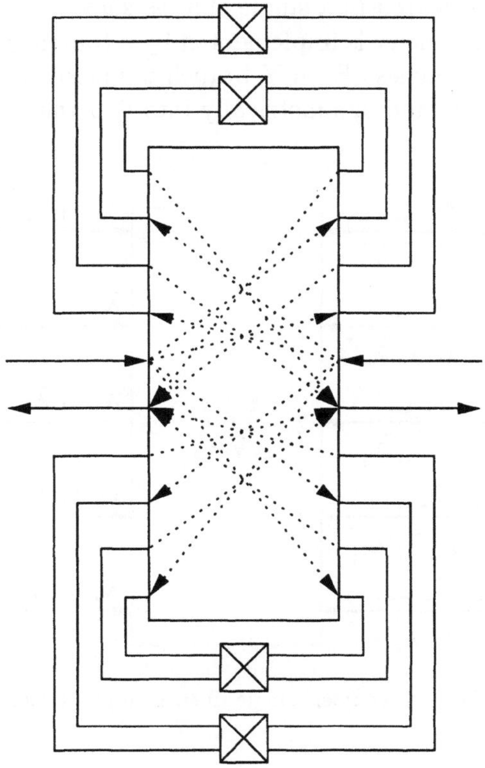

Figure 2.14. A hybrid configuration for a 2-line 4-wavelength OXC without crossover.

example, due to the wavelength overlapping between the transmitted signal and the received signal caused by such loop, each node has to connect to the star coupler via two separate fibers. In this section, we will describe how to revise the star coupler to avoid this situation. We will use an $N \times N$ AWG and a $1 \times (N-1)$ power splitter. The operating principle is as follows. The N ports at one side, say left side, of the AWG are used for both signal input and output. Each port at the left side is assigned with a dedicated unique wavelength for input signal such that all the N input signals would be multiplexed into a single port at the right side. The multiplexed signal is then forwarded to the $1 \times (N-1)$ power splitter. The $N-1$ copies of the multiplexed signal are then folded back to the remaining $N-1$ ports at the right side. Then each left port will receive all wavelengths but its input wavelength. Thus the input signal and the output signal will propagate in opposite direction in one fiber/waveguide without interference. The entire loop-free star coupler can be fabricated in a single chip as shown in Figure 2.15. The $1 \times (N-1)$ power splitter is implemented by a (partial) balanced binary tree of $N-2$ Y-branches. Each Y-branch is made by splitting a single waveguide gradually and symmetrically into two waveguides.

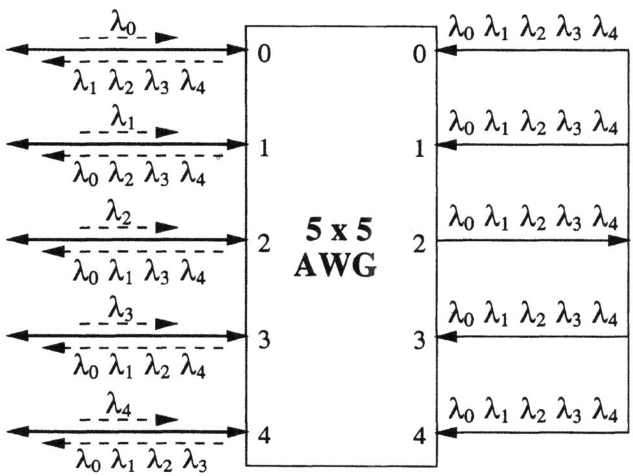

Figure 2.15. Loop-free star coupler: single fiber for input signal and output signal.

If the input signals and output signals have to be carried in separate waveguides/fibers, similar configuration can be obtained, but with $N = 2W$ (see Figure 2.16).At the left side of the AWG, the even ports are used for input ports, and the odd ports are used for output ports. Each input port is assigned with a unique wavelength such that all input signals are multiplexed at port, say k, at the right side. The multiplexed signal is then forwarded to the $1 \times (W-1)$ power splitter. The $W-1$ copies of the multiplexed signal are then folded back to the $W-1$ ports at the right side which have different parity from k but are not equal to $(k-1) \bmod N$. Then for any $0 \leq i < W$, the port $2i+1$ at the left receives all wavelengths except the input wavelength of port $2i$.

7. SUMMARY

Silica-based optical integrated waveguide is a very promising technology for manufacturing large-scale and compact optical devices. As

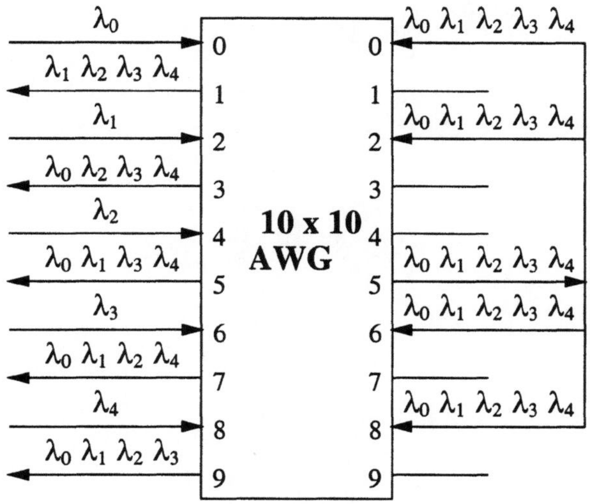

Figure 2.16. Loop-free star coupler: separate fibers for input signal and output signal.

its basic application, AWG has many elegant wavelength-routing properties. In addition, it can be integrated with other optical devices to achieve various optical functions such as multiwavelength transmission and reception, wavelength adding and dropping, wavelength switching, and wavelength broadcasting. This chapter has described various configurations of these integrations. In particular, the proposed layouts of these configurations avoid as much crossovers of waveguides as possible.

Chapter 3

RELAXATION ON TUNABILITY

1. INTRODUCTION

In WDM optical networks, the multi-wavelength transmission and reception at each access node is realized through either an array of fixed-wavelength transceivers or a tunable transceiver. A transceiver array has the advantage of being able to select the wavelength for each transceiver independently. How it becomes bulky as the number of wavelengths increases. On the other hand, with the current technology the tunable transceiver has very limited tunable range. Its tunability is typically realized via the tunable optical filters. Many tunable optical filters have a very limited set of resolvable wavelength channels. These wavelength channels usually are contiguous, for an example, within a free spectral range [70]. In the previous studies, each access node is required to access all wavelength channels used in the network. Under such requirement, the number of working channels in network is equal to the number of resolvable wavelength channels of individual tunable transceivers. To support a large number of working channels, the transceivers at each access node must be upgraded correspondingly. Unfortunately, this is not practical most of the time.

At the first sight, one might feel pessimistic unless the transmission capacity of the tunable transceivers can be increased. However, careful re-examining of the network operations reveals some optimistic discoveries. In most situations, it is an overkill to require access nodes to access all working wavelength channels in the network. For example, in many real applications, an access node only communicates with a small subset of access nodes among a potentially large population in the entire network. Thus its tunable transceiver only has to access the wavelength channels used by these subset of access nodes. The tunability requirement can be further relaxed if multi-hop communications are allowed. In this chapter, we assume that each access node has a tunable transmitter and a fixed-wavelength receiver (the configuration with transmitter is fixed-wavelength and tunable receiver can be dealed with in the same way). The tunable range of each tunable transmitter is represented by a *waveband* which consists of a number of contiguous wavelength channels. The wavebands of all transmitters have the same size but may start from different wavelengths. To support a communication link in a communication topology (either physical or logical), the wavelength of the receiver at the receiving node must be within the waveband of the transmitter at the transmitting node. With this relaxation, it's possible that a large number of wavelength channels can be utilized by the networks even though each transmitter has a waveband of small size. Therefore, two natural design problems arise on the waveband/wavelength assignment to the transmitters/receivers. First, if each receiver needs a unique wavelength, what is the smallest size of the wavebands that the transmitters can have? Secondly, if the size of the wavebands of all transmitters is given, what is the maximal number of wavelength channels that the receivers in the network can utilize? The former is referred to as the *minimum tunability problem*, and the latter is referred to as the *maximum concurrence problem.*

In this chapter, we will address the above two problems in various communication topologies. Section 2. formulate them into graph-theoretic optimization problems. Section 3. studies their computational complexities in general communication topologies and provides approximation algorithms with provable performance. Section 4. presents the optimal solutions in complete graphs, de Bruijin/Kautz/shuffle digraphs, and rings respectfully, and tight lower bounds on minimum tunability in hypercubes. Finally Section 5. summarizes this chapter.

2. GRAPH-THEORETIC FORMULATION

The given communication topology in a WDM network is represented by a graph $G = (V, E)$ where $V = \{0, 1, \cdots, N-1\}$. Depending on whether the communications are bidirectional or unidirectional, G is expressed as an undirected graph or directed graph correspondingly. The wavelength channels are indexed by nonnegative integers, and a waveband is then represented by a closed interval between two integers.

First of all, we observe that we can reduce the waveband/wavelength assignment to transmitters/receivers to the wavelength assignment to receivers alone. A wavelength assignment to the receivers can be represented as a function λ from V to nonnegative integers. The minimum waveband size required by a wavelength assignment λ can be calculated by

$$t(\lambda, G) = 1 + \max_{v \in V} \left\{ \max_{(v,u) \in E} \lambda(u) - \min_{(v,u) \in E} \lambda(u) \right\}.$$

From this wavelength assignment λ, we can assign a waveband of size $t(\lambda, G)$ to each transmitter that covers all the wavelength channels of the adjacent receivers.

Secondly, if a wavelength assignment to the receivers uses w different wavelengths, we can assume that these wavelengths are $0, 1, \cdots, w-1$. In fact, such wavelength assignment can be obtained from any one as follows. Consider a wavelength assignment which uses wavelengths $i_0 < i_1 < \cdots < i_{w-1}$. Let j be the minimum integer with $i_j > j$. For any $j \leq k < w$, we subtract each wavelength i_k by $i_j - j$. The minimum waveband size required by this new wavelength assignment does not exceed that of the original wavelength assignment. Such procedure can be repeated until $i_k = k$ for all $0 \leq k < w$.

Now we consider the formulation of the minimum tunability problem. From the above argument, we assume each receiver is assigned with a unique wavelength in $\{0, 1, \cdots, N-1\}$. Thus each wavelength assignment corresponds to a permutation over the set $\{0, 1, \cdots, N-1\}$. Let P_n denote the set of all permutations on $\{0, 1, \cdots, N-1\}$. Then the minimum tunability is given by

$$\Phi(G) = \min_{\lambda \in P_n} t(\lambda, G).$$

The minimum tunability problem is then to find $\Phi(G)$ for any given graph G.

Finally, we give the formulation of the maximum concurrence problem. Let w be the tunability of all transmitters. The maximum concurrence is given by

$$\Lambda_w(G) = \max_{\lambda : t(\lambda, G) \leq w} |\{\lambda(v) \mid v \in V\}|.$$

The maximum concurrence problem is then to find $\Lambda_w(G)$ for any given graph G and any given integer $1 \leq w \leq |V|$.

Obviously

$$\Phi(G) = \min\{1 \le w \le |V| : \Lambda_w(G) = |V|\}.$$

So in order to calculate $\Phi(G)$, one may first calculate $\Lambda_w(G)$ for each $1 \le w \le |V|$. Thus from the computational complexity point of view, the maximum concurrence problem is at least as hard as the minimum tunability problem. In the next section, we will study the computational complexity of the minimum tunability problem in arbitrary topologies.

3. ARBITRARY TOPOLOGIES

In this section, we first prove that the minimum tunability problem is NP-hard in general. A reduction will be made from the well-known *minimum bandwidth problem [92]*: Given a graph $G = (V, E)$, find

$$BW(G) = 1 + \min_{\pi \in P_{|V|}} \max_{(u,v) \in E} |\pi(u) - \pi(v)|.$$

THEOREM 3.1 *The minimum tunability problem is NP-Hard. It is even NP-Hard to approximate it within absolute error $|V|^{1-\epsilon}$ for any $\epsilon > 0$.*

Proof. We reduce the problem of minimum bandwidth of cobipartite graph to the minimum tunability problem [92]. Let $G = (U, V, E)$ be any cobipartite graph. Then for any $\epsilon \ge 0$, it is NP-hard to approximate $BW(G)$ within absolute error of $(|U| + |V|)^{1-\epsilon}$ [92]. Without loss of generality, we assume that $|U| \le |V|$. We construct graph $H = (V(H), E(H))$ as follows:

$$\begin{aligned} U' &= \{u' : u \in U\}, \\ V(H) &= \{a, b\} \cup U \cup U' \cup V, \\ E(H) &= \{(a,u) : u \in U\} \cup \{(v,b) : v \in V\} \cup \\ &\quad \{(u,u') : u \in U\} \cup \{(u',v) : (u,v) \in E(G)\}. \end{aligned}$$

Let $K = (V(K), E(K))$ be the graph with

$$\begin{aligned} V(K) &= \{a, b\} \cup U', \\ E(K) &= \{(a,u') : u \in U\} \cup \{(u',b) : u \in U\} \cup \{(u_1', u_2') \\ &\quad : u_1, u_2 \in U; \exists v \in V, (u_1, v), (u_2, v) \in E(G)\}. \end{aligned}$$

Then it's easy to see that

$$\Phi(H) = \max\{BW(G), BW(K)\}.$$

Obviously,

$$BW(K) \leq 1 + |U'| = 1 + |U| \leq 1 + |V|$$

and

$$BW(G) \geq |V|.$$

Thus

$$BW(G) \leq \Phi(H) \leq 1 + BW(G).$$

So the minimum tunability problem is at least as hard as the minimum bandwidth problem over cobipartite graphs. Therefore, the lemma follows from the NP-hardness and approximality of the minimum bandwidth problem over cobipartite graphs. ∎

In the next, we seek approximation algorithms for the minimum tunability problem in general graphs. The approximation algorithms can be obtained by reducing the minimum tunability problem to minimum bandwidth problem. For any graph $G = (V, E)$, let $\overline{G} = (V, \overline{E})$ be the graph in which

$$\overline{E} = \{(v_1, v_2) : \exists v \in V, (v_1, v) \in E \wedge (v_2, v) \in E\}.$$

It's obvious that

$$\Phi(G) = BW(\overline{G}).$$

Thus any approximation algorithms for the minimum bandwidth problem can be applied to $\overline{G}$ to obtain an approximation algorithm for the minimum tunability in G. There are many approximation results on the minimum bandwidth problem that can be readily adopted. For examples, if the nodal degree is $\Theta(|V|)$, the minimum bandwidth problem is approximable within 3 [93]. If G is a caterpillar, the minimum bandwidth problem is approximable within $O(\log|V|)$ [72]. If G is asteroidal triple-free, the minimum bandwidth problem is approximable within 2 [98]. However, there are also some inapproximality results. For any $\varepsilon > 0$, the minimum bandwidth problem is not approximable within $1.5 - \varepsilon$ [27], and not approximable with an absolute error guarantee of $|V|^{1-\varepsilon}$ [92]. Even if G is a tree, it is still not approximable within $1.332 - \varepsilon$ for any $\varepsilon > 0$ [27].

4. REGULAR TOPOLOGIES

4.1 COMPLETE GRAPHS

The complete graph corresponds to a single-hop network or a all-to-all personalized communication request. We use K_N to denote the complete graph of N vertices. We will prove the following Theorem.

THEOREM 3.2 *For any* $1 \leq w \leq N$,

$$\Lambda_w (K_N) = w.$$

Consequently,

$$\Phi (K_N) = N.$$

Proof. Obviously, $\Lambda_w (K_N) \geq w$. On the other hand, consider any feasible wavelength assignment $\lambda(\cdot)$. Let node a be the one assigned with the smallest wavelength $\lambda(a)$, and node b the one assigned with the largest wavelength $\lambda(b)$. As a and b are both neighbors of any other node other than a and b, $\lambda(b) - \lambda(a) < w$. Thus $\Lambda_w (K_N) \leq w$. Therefore $\Lambda_w (K_N) = w$. This implies that $\Phi (K_N) = N$. ∎

Since $\Lambda_w (K_N) = w$, any wavelength assignment which uses w wavelengths is optimal. To make the wavelengths sharing the same workload, we should equally partition the nodes into w subsets, and assign a distinct wavelength to all nodes in the same subset. Theorem 3.2 implies that the maximal number of working channels is at most the number of resolvable wavelength channels of each tunable transmitter.

4.2 DE BRUIJIN/KAUTZ/SHUFFLE DIGRAPHS

A de Bruijin digraph [80, 131] (or Kautz digraph [81], shuffle respectively) of size N and degree p is denoted by $D(N,p)$ (or $K(N,p)$, $S(N,p)$ respectively). We assume that in these graphs, the degree p is always selected to be a factor of the size N.

THEOREM 3.3 *For any* $1 \leq w \leq N$,

$$\Lambda_w\left(D\left(n,p\right)\right) = \Lambda_w\left(K\left(n,p\right)\right) = \Lambda_w\left(S\left(n,p\right)\right) = \min\left\{N, w\frac{N}{p}\right\}.$$

Consequently,

$$\Phi\left(D\left(n,p\right)\right) = \Phi\left(K\left(n,p\right)\right) = \Phi\left(S\left(n,p\right)\right) = p.$$

Proof. For each $0 \leq i < \frac{N}{p}$, let

$$\mathcal{R}_i = \{ip + k \mid 0 \leq k < p\}.$$

Then $\mathcal{R}_i$ consists of exactly the immediate successors of some node in $D(N,p)$, $K(N,p)$ or $S(N,p)$, and thus can be assigned with at most $\min\{p,w\}$ wavelengths. Furthermore, in any optimal wavelength assignment the two sets of wavelengths assigned to $\mathcal{R}_i$ and $\mathcal{R}_j$ should be disjoint for any $0 \leq i < j < \frac{N}{p}$. So the total number of wavelength used is at most

$$\frac{N}{p}\min\{p,w\} = \min\left\{N, w\frac{N}{p}\right\}.$$

On the other hand, there are many wavelength assignments which use $\min\left\{N, w\frac{N}{p}\right\}$ wavelengths. For an example, for each $0 \leq i < \frac{N}{p}$, we partition $\mathcal{R}_i$ into $\min\{p,w\}$ groups as equally as possible and then any $\min\{p,w\}$ contiguous wavelengths are assigned to these $\min\{p,w\}$ groups. Therefore,

$$\Lambda_w\left(D\left(n,p\right)\right) = \Lambda_w\left(K\left(n,p\right)\right) = \Lambda_w\left(S\left(n,p\right)\right) = \min\left\{N, w\frac{N}{p}\right\}.$$

This implies that

$$\Phi\left(D\left(n,p\right)\right) = \Phi\left(K\left(n,p\right)\right) = \Phi\left(S\left(n,p\right)\right) = p.$$

■

Thus even with fixed-wavelength transmitters, certain degree of concurrence $(\frac{N}{p})$ is achievable. In particular, if the nodal degree is two, we can still achieve half of the full concurrence. As long as the tunability of each transmitter is no less than the degree of these graphs, full concurrence can be achieved. In particular, if the degree is equal to two, then any tunable transmitters are sufficient to achieve full concurrence.

4.3 RINGS

We begin with unidirectional rings. Let UR_N denote the unidirectional ring of N vertices $\{0, 1, \cdots, N-1\}$.

THEOREM 3.4 *For any* $1 \leq w \leq N$,

$$\Lambda_w(UR_N) = N.$$

Therefore

$$\Phi(UR_N) = 1.$$

Proof. As each vertex has only one immediate successor, we can always assign each vertex a distinct wavelength. Thus $\Lambda_w(UR_N) = N$, and $\Phi(UR_N) = 1$. ∎

The above theorem implies that in unidirectional rings, there is no need to use tunable transmitters at all. The fixed-wavelength transmitters is sufficient to achieve full concurrence.

Now we consider bidirectional rings. Let BR_N denote the bidirectional ring of N vertices $\{0, 1, \cdots, N-1\}$. Without loss of generality, we assume that the vertices are numbered in the clockwise order. The wavelength assignment in bidirectional rings will be reduced to the following optimal ring labeling problem in the bidirectional rings.

Optimal ring labeling problem: Given a ring of size N and an integer $0 < w < N$, assign a label $\ell(i)$ to each $0 \leq i < N$ such that

$$|\ell(i) - \ell((i+1) \bmod N)| < w$$

and $|\{\ell(i) : 0 \leq i < N\}|$ is maximized. The maxima is denoted by $\Gamma_w(N)$.

The next lemma describes the relations between the optimal wavelength assignment in bidirectional rings and optimal ring labeling problem.

LEMMA 3.5 *For any* $1 \leq w \leq N$,

$$\Lambda_w(BR_N) = \begin{cases} 2\Gamma_w\left(\frac{N}{2}\right) & \textit{if } N \textit{ is even,} \\ \Gamma_w(N) & \textit{if } N \textit{ is odd.} \end{cases}$$

Proof. We first assume that N is even. Consider the two rings consisting of all even nodes and all odd nodes respectively. We will treat the wavelengths and labels to be the same. It can be shown that a wavelength assignment in the original ring is feasible if and only if the corresponding wavelength assignment to each ring is a ring labeling with respect to the same w. Thus, to maximize the number of wavelength used, the wavelengths/labels assigned to these two small rings should be disjoint. Hence $\Lambda_w(BR_N)$ is twice of $\Gamma_w\left(\frac{N}{2}\right)$.

Now we assume that N is odd. Consider the ring G in which nodes are arranged clockwise in the following order

$$0, 2, 4, \cdots, N-3, N-1, 1, 3, 5, \cdots, N-4, N-2$$

i.e., the first $\left\lceil \frac{N}{2} \right\rceil$ nodes are all even numbers in increasing order and the remaining $\left\lfloor \frac{N}{2} \right\rfloor$ nodes are all odd numbers in increasing order. It can be shown that a wavelength assignment in the original ring BR_N if and only the corresponding wavelength assignment in the ring G is a ring labeling with respect to the same w. Thus $\Lambda_w(BR_N) = \Gamma_w(N)$. ■

In the next we construct the optimal ring labeling.

LEMMA 3.6 $\Gamma_w(N) = \begin{cases} 1 & \textit{if } w = 1, \\ \left\lceil \frac{N+1}{2} \right\rceil & \textit{if } w = 2, \\ N & \textit{if } w > 2. \end{cases}$

Proof. If $w = 1$, then all nodes in the ring must have the same labeling and thus $\Gamma_1(N) = 1$.

Now we assume that $w > 2$. If N is even, we consider the following ring labeling $\ell(\cdot)$:

$$\ell(0) = 0,$$

$$\ell\left(\frac{N}{2}\right) = N - 1,$$

$$\ell(i) = 2i - 1, \text{for any } 0 < i < \frac{N}{2},$$

$$\ell(i) = 2(N - i), \text{for any } \frac{N}{2} < i < N.$$

If N is odd, we consider the following ring labeling $\ell(\cdot)$:

$$\begin{aligned}
&\ell(0) = 0, \\
&\ell(i) = 2i - 1, \text{for any } 0 < i \leq \left\lfloor \frac{N}{2} \right\rfloor, \\
&\ell(i) = 2(N - i), \text{for any } \left\lceil \frac{N}{2} \right\rceil \leq i < N.
\end{aligned}$$

In both cases,

$$|\ell(i) - \ell((i+1) \bmod N)| \leq 2 \leq w - 1,$$

and

$$|\{\ell(i) : 0 \leq i < N\}| = N.$$

Thus $\Gamma_w(N) = 2$ if $w > 2$.

Finally we assume that $w = 2$. Then in any ring labeling $\ell(\cdot)$, the labels of any two consecutive nodes in the ring must also have the consecutive labels. This implies that for any $0 < i < N$,

$$|\ell(i) - \ell(0)| \leq \min\{i, N - i\} \leq \left\lfloor \frac{N}{2} \right\rfloor.$$

Thus

$$\Gamma_2(N) \leq 1 + \left\lfloor \frac{N}{2} \right\rfloor = \left\lceil \frac{N+1}{2} \right\rceil.$$

So to prove the lemma, we only have to give a ring labeling that uses $\lceil \frac{N+1}{2} \rceil$ different labels. If N is odd, we consider the following ring labeling $\ell(\cdot)$:

$$\begin{aligned}
&\ell(0) = 0, \\
&\ell(i) = \ell(N - i) = i, \text{for any } 0 < i \leq \left\lfloor \frac{N}{2} \right\rfloor.
\end{aligned}$$

If N is even, we consider the following ring labeling $\ell(\cdot)$:

$$\begin{aligned}
&\ell(0) = 0, \\
&\ell\left(\frac{N}{2}\right) = \frac{N}{2} \\
&\ell(i) = \ell(N - i) = i, \text{for any } 0 < i < \frac{N}{2}.
\end{aligned}$$

In both cases,

$$|\ell(i) - \ell((i+1) \bmod N)| \leq 1,$$

and

$$|\{\ell(i) : 0 \leq i < N\}| = \left\lceil \frac{N+1}{2} \right\rceil.$$

Thus $\Gamma_2(N) = \lceil \frac{N+1}{2} \rceil$ for any N. ■

Now we are ready to present the optimal wavelength assignment in bidirectional rings.

THEOREM 3.7 *For any* $N > 0$,

$$\Lambda_w(BR_N) = \begin{cases} 2 - N \bmod 2 & \text{if } w = 1, \\ 2\lfloor \frac{N}{4} \rfloor + \lceil \frac{(N-1) \bmod 4}{4} \rceil + 1 & \text{if } w = 2, \\ N & \text{if } w > 2. \end{cases}$$

$$\Phi(BR_N) = \begin{cases} 2 & \text{if } N = 2 \text{ or } 4, \\ 3 & \text{if } N \geq 5 \text{ or } N = 3. \end{cases}$$

Thus to achieve the full concurrence the tunability can be as low as three. If each transmitter can tune to only two contiguous wavelengths, we can still achieve the full concurrence when $N = 2$ or 4, and around half of the full concurrence if $N \geq 5$ or $N = 3$. If the transmitters are fixed, the concurrence is very poor.

4.4 HYPERCUBES

Let C_n to denote the n-dimensional hypercube. Without loss of generality, we assume that the wavelength is 0 which is assigned to node 0. We first show that

$$\Phi(C_n) \geq 1 + \binom{n}{2}.$$

Consider any optimal wavelength assignment in C_n which achieves full concurrence while requires minimum tunability. Note that there are $\binom{n}{2}$ nodes which have distance of two from node 0, and all of them have a unique larger wavelength. One of them must have wavelength which is at least $\binom{n}{2}$. Hence

$$\Phi\left(C_n\right) \geq 1 + \binom{n}{2}.$$

This lower bound is also sufficient when $n \leq 4$. If $n = 2$, $1 + \binom{n}{2} = 2$. The following feasible wavelength assignment achieves full concurrence, and the difference between the wavelengths assigned to two neighbors of any node is at most 2:

$$\lambda(00) = 0, \lambda(11) = 1, \lambda(01) = 2, \lambda(10) = 3.$$

If $n = 3$, $1 + \binom{n}{2} = 4$. The following feasible wavelength assignment achieves full concurrence, and the difference between the wavelengths assigned to two neighbors of any node is at most 4:

$$\lambda(000) = 0, \lambda(011) = 1, \lambda(101) = 2, \lambda(110) = 3,$$
$$\lambda(001) = 4, \lambda(010) = 5, \lambda(100) = 6, \lambda(111) = 7.$$

If $n = 4$, $1 + \binom{n}{2} = 7$. The following feasible wavelength assignment achieves full concurrence, and the difference between the wavelengths assigned to two neighbors of any node is at most 7:

$$\begin{array}{llll}
\lambda(0000) = 0 & \lambda(0011) = 1 & \lambda(0101) = 2 & \lambda(0110) = 3 \\
\lambda(1001) = 4 & \lambda(1010) = 5 & \lambda(1100) = 6 & \lambda(1111) = 7 \\
\lambda(0001) = 8 & \lambda(0010) = 9 & \lambda(0100) = 10 & \lambda(0111) = 11 \\
\lambda(1000) = 12 & \lambda(1011) = 13 & \lambda(1101) = 14 & \lambda(1110) = 15
\end{array}$$

When $n \geq 5$, we have

$$\Phi\left(C_n\right) \geq \binom{n}{2} + \left\lfloor \frac{\binom{n-2}{2}}{2} \right\rfloor + 1.$$

We also prove it by contradiction. Suppose to the contrary. Then all $\binom{n}{2}$ nodes which contain exactly two 1's must have wavelengths of no more than $\binom{n}{2} + \left\lfloor \frac{\binom{n-2}{2}}{2} \right\rfloor - 1$. Let a be the node with the smallest wavelength among these $\binom{n}{2}$ nodes. Then the wavelength assigned to a is at most $\left\lfloor \frac{\binom{n-2}{2}}{2} \right\rfloor$. Among the wavelengths assigned to those $\binom{n-2}{2}$ nodes which contain exactly four 1's and have distance of two from a, the maximum is at least $\binom{n}{2} + \binom{n-2}{2}$. Therefore,

$$\begin{aligned}
\Phi(C_n) &\geq 1 + \binom{n}{2} + \binom{n-2}{2} - \left\lfloor \frac{\binom{n-2}{2}}{2} \right\rfloor \\
&= \binom{n}{2} + \left\lceil \frac{\binom{n-2}{2}}{2} \right\rceil + 1 \\
&\geq \binom{n}{2} + \left\lfloor \frac{\binom{n-2}{2}}{2} \right\rfloor + 1
\end{aligned}$$

which is a contradiction.

In the next, we present another stronger lower bound when n is large. We consider the wavelengths assigned to all 2^{n-1} nodes of even parity in any optimal wavelength assignment. All these nodes have unique wavelengths. Suppose that node a has the lowest wavelength and node b has the largest wavelength. Let m be the Hamming distance between a and b. Let

$$a = v_0, v_1, \cdots, v_{\frac{m}{2}} = b$$

be the nodes of even parity along a shortest path from a to b. Then for any $0 \leq i < \frac{m}{2}$, the difference of the two wavelengths assigned to v_i and v_{i+1} is a lower bound of $\Phi(C_n) - 1$. On the other hand, since the differences between the two wavelengths assigned to $a = v_0$ and $v_{\frac{m}{2}} = b$ is at least $2^{n-1} - 1$, there exist some $0 \leq i < \frac{m}{2}$ such that the difference of the two wavelengths assigned to v_i and v_{i+1} is at least

$$\left\lceil \frac{2^{n-1} - 1}{\frac{m}{2}} \right\rceil = \left\lceil \frac{2^n - 2}{m} \right\rceil.$$

Since $m \leq n$ when n is even and $m \leq n - 1$ when n is odd, we have

$$\Phi(C_n) \geq \left\lceil \frac{2^n - 2}{n} \right\rceil + 1$$

when n is even and

$$\Phi(C_n) \geq \left\lceil \frac{2^n - 2}{n - 1} \right\rceil + 1$$

when n is odd.

The next theorem summarizes our preliminary results in this section.

THEOREM 3.8 *For any* $2 \leq n \leq 4$,

$$\Phi(C_n) = \binom{n}{2} + 1.$$

For any $n \geq 5$, *if* n *is even,*

$$\Phi(C_n) \geq \max\left\{\binom{n}{2} + \left\lfloor \frac{\binom{n-2}{2}}{2} \right\rfloor, \left\lceil \frac{2^n - 2}{n} \right\rceil\right\} + 1$$

and if n *is odd,*

$$\Phi(C_n) \geq \max\left\{\binom{n}{2} + \left\lfloor \frac{\binom{n-2}{2}}{2} \right\rfloor, \left\lceil \frac{2^n - 2}{n-1} \right\rceil\right\} + 1.$$

5. CONCLUSION

This chapter studies the minimum tunability requirement of a WDM network with a tunable transmitter and a fixed-wavelength receiver at each access node to achieve full transmission concurrence. The problem is proved to be NP-hard in general. In addition, it is even NP-Hard to approximate it within absolute error $N^{1-\epsilon}$ for any $\epsilon > 0$. However, polynomial time approximation algorithms can be obtained by reducing it to the well-studied minimum bandwidth problem. When the communication topologies are complete graphs, de Bruijin digraphs, Kautz digraphs, shuffle or rings, the optimal solutions are provided. Finally, we present tight lower bounds when the communication topology is a hypercube.

II

BROADCAST-AND-SELECT PASSIVE OPTICAL NETWORKS

Chapter 4

MULTI-TRANSCEIVER PONS

1. INTRODUCTION

This chapter studies broadcast-and-select passive optical networks which use a small number of fixed-wavelength transmitters and receivers at each access node. As described in Chapter 1, any wavelength assignment to the fixed-wavelength transceivers determines a logical topology over the access nodes. Typically the logical topology is a supergraph of some regular interconnection topology. A wavelength assignment is said to realize a given interconnection topology if the given interconnection topology is a subgraph of the resulting logical topology. A general approach to realize a given interconnection topology is to partition the outgoing (incoming respectively) links at each node into a number of groups with one for each transmitter (receiver respectively). The partition determines a *transmission graph*, a bipartite digraph whose vertex sets are all transmitters and all receivers, and there is a link from a transmitter to a receiver if and only if they are responsible for one common link in the given interconnection topology. It's obvious that the transmission graph and the interconnection topology have the same number of links. Since the transmitters and receivers have fixed wavelengths, any transmitter (receiver) and its adjacent receivers (transmitters) in the transmission graph are forced to have the same wavelength channels of the transmitter (receiver). Therefore any pair of transceivers must have the same wavelength channel if there is path between them assuming the links in the transmission graph are bidirectional. This key observation leads to the concept of *subnetworks*. In the transmission graph, a set of transmitters and receivers form a subnetwork if there is a path between any two of them if we ignore the unidirectional nature of the links[1]. Thus all transmitters and receivers in the same subnetwork must have the same wavelength channel.

The concept of subnetwork is a cornerstone for the design of multi-hop broadcast-and-select PONs with fixed-wavelength transceivers. Let's first examine some of its applications in the next. The first application is the wavelength assignment to the transmitters and receivers. As all transceivers in the same subnetwork must have the same wavelength channel, the wavelength assignment can be conducted at the much coarse subnetwork level, instead of individual transmitters/receivers. If the number of subnetworks is no more than the number of available wavelength channels, each subnetwork can be assigned a unique wavelength channel. If the number of subnetworks are more than the number of available wavelength channels, the available wavelength channels are then *evenly* shared among all subnetworks.

The second application of the subnetworks is the transmission scheduling. The scheduling can be confined within all transceivers in one or more subnetworks that share a wavelength, instead of all transceivers in the entire network. If the static round robin policy is adopted, the scheduling becomes trivial as we only have to make sure that all transmitters in the subnetworks that share a common wavelength transmit at different time slots.

The third application of the subnetworks is the selection of right number of transmitters and receivers at each access node. As the network cost is mainly determined by the number of transmitters and receivers, we would like to minimize the usage of transceivers while maintaining the desired performance. For example, if an access node has more than one transmitters (or receivers) in the same subnetwork, we can simply replace them by a single transmitter (or receiver) to take care of all links assigned to them. Such adjustment reduces the cost of the network, decreases the number transmitters in the same subnetwork, and maintain the same transmission concurrence in terms of the wavelength channels utilized.

Forth, the subnetwork structure provides a way to determine the logical topology of a wavelength assignment. The full characterization of

[1]The subnetwork concept is similar to the concept of connected component in the graph theory. The only difference here is that we ignore the unidirectional nature of the links.

the logical topology of a wavelength assignment is very useful in the design of optimal routing in multi-hop PONs. The characterization of the subnetworks can provide insight to the structure of the logical topology. For example, if each subnetwork has a unique wavelength, then there is a link from node a to node b in the logical topology if and only if a has a transmitter and b has a receiver which are in the same subnetwork. The logical topology and the optimal routing in the logical topology will be addressed in Chapter 5.

Finally, the decomposition of the PON into a number of subnetworks also has the application in constructing multi-star PONs. Because the dimension of a passive star coupler is limited due to the power split and other factors, the single-star passive PONs have poor scalability. One way to overcome this limitation is to use multiple passive star couplers to interconnect the access nodes as the network size becomes very large. The design on which star coupler to interconnect which set of access nodes can borrow the results from the decomposition of a PON into subnetworks. By properly selecting the number of transmitters and receivers such that the size of the subnetworks is smaller than the dimension of each star coupler, each star coupler can be used to interconnection all access nodes in one subnetwork.

Because of the important roles of subnetworks, this chapter will determine the subnetwork structures when various regular interconnection topologies are to be realized. Section 2. describes a uniform consecutive link partition scheme assuming that the number of transmitters and the number of receivers at each access node are both factors of the nodal degree of the regular interconnection topology to be realized. Such partition depends on the ordering of the links at each access node. After the discussion of general properties of the subnetworks in Section 3., we study the impact of the link ordering, or the partition equivalently, on the subnetwork structure and cost-performance relation in Section 4.. Section 5. through Section 9. then characterize the subnetwork structures of the uniform partitions to realize the generalized de Bruijin digraph, generalized Kautz digraph, hypercube, star graph and rotator digraph [154, 157, 159, 164, 125]. In each of these sections, we also discuss how to select the number of transmitters and receivers so as to maximize the cost-effectiveness. Finally Section 10. considers the general consecutive partition allowing arbitrary number of transmitters and receivers at each access node, and generalizes the results on the subnetwork structures in previous sections.

2. UNIFORM CONSECUTIVE PARTITION

Let N be the number of access nodes in the network. The access nodes are indexed by numbers from 0 to $N-1$. Each access node is equipped with T transmitters and R receivers. The set of transmitters at node a is denoted by

$$\{(a,t) \mid 0 \le t < T\}.$$

The set of receivers at node a is denoted by

$$\{(a,r) \mid 0 \le r < R\}.$$

For each transmitter (a,t), a and t are called its *node index* and *local index* respectively. Similarly, for each receiver (a,r), a and r are also called its *node index* and *local index* respectively. The distinguishing between the transmitter and the receiver is specified either explicitly or by the context.

Let G be the regular interconnection topology over N nodes. If G is an undirected graph, we treat each edge in it as two unidirectional links. Let d be the nodal degree of G. As a regular interconnection topology, G typically has an intrinsic ordering of the outgoing links and the incoming links at each vertex. Such ordering is givens as a numbering of the outgoing (incoming) links at each vertex by the numbers 0 through $d-1$. The link ordering is usually *symmetric* in the sense that if a link is the i-th outgoing link of its source node, it is also the i-th incoming link of its destination node.

We first assume that both T and R are factors of d. This assumption will be relaxed later in Section 10.. To realize G, the d outgoing links at each node a are divided consecutively into T groups of equal size, with the t-th group consisting of outgoing links

$$\left\{(a, t\frac{d}{T}+i) \mid 0 \le i < \frac{d}{T}\right\}.$$

The t-th group of outgoing links at node a is then associated with the transmitter (a,t). Figure 4.1 illustrates this partition. Similarly, the d incoming links at each node a are divided consecutively into R groups of equal size, with the r-th group consisting of incoming links

$$\left\{(a, r\frac{d}{R}+i) \mid 0 \le i < \frac{d}{R}\right\};$$

The r-th group of incoming links at node a is then associated with the receiver (a, r). This partition is referred to as *uniform consecutive partition.*

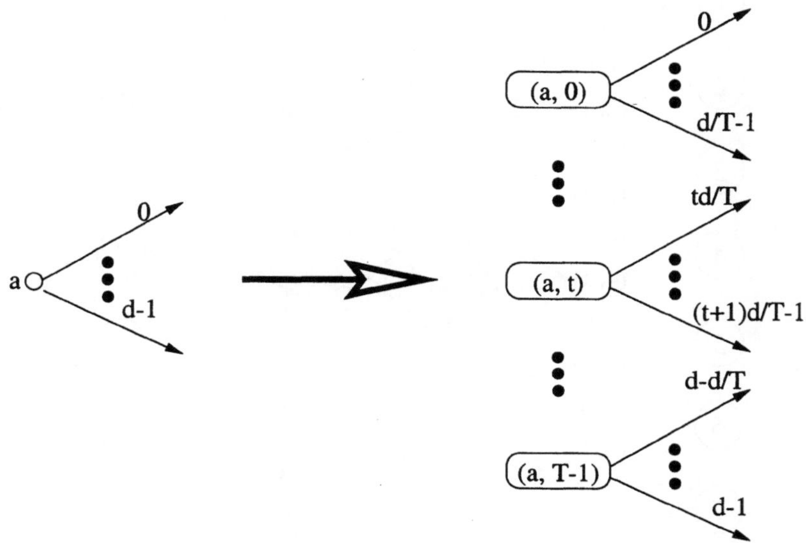

Figure 4.1. Partition the outgoing links at each station a into T groups, and associate each group with a transmitter.

The uniform consecutive partition scheme can be intuitively interpreted as *virtually* breaking each source (destination) node into T (R) small nodes, with each small node implementing d/T (d/R) outgoing (incoming) links. Such interpretation is illustrated in Figure 4.2, which shows the uniform consecutive partition to realize the generalized de Bruijin digraph $D(8, 4)$ with $T = R = 2$.

Now let's examine the transmission graph induced by the uniform consecutive partition. The vertex set of the transmission graph consists of the transmitter set

$$\{(a, t) \mid 0 \leq a \leq N - 1, 0 \leq t \leq T - 1\}$$

and the receiver set

$$\{(b, r) \mid 0 \leq b \leq N - 1, 0 \leq r \leq R - 1\}.$$

Each link of the transmission graph is from a vertex(or transmitter) in the transmitter set to a vertex (or receiver) in the receiver set. Each

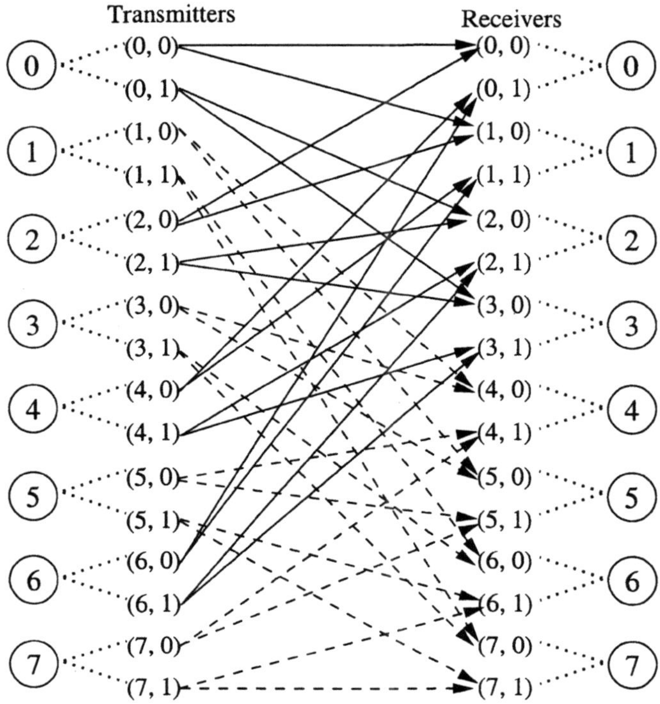

Figure 4.2. Virtual breaking of nodes and the transmission graph.

transmitter has $\frac{d}{T}$ outgoing links, and each receiver has $\frac{d}{R}$ incoming links. Figure 4.2 also shows the transmission graph when the interconnection topology is the generalized de Bruijin digraph $D(8,4)$ with $T = R = 2$.

There is an one-to-one correspondence between the links in the regular interconnection topology G and the links in the transmission graph. For any link

$$a \longrightarrow b$$

in G, if this link is the i-th outgoing link of a and j-th incoming link of b, then the corresponding link in the transmission graph is

$$\left(a, \left\lfloor \frac{i}{\frac{d}{T}} \right\rfloor\right) \longrightarrow \left(b, \left\lfloor \frac{j}{\frac{d}{R}} \right\rfloor\right).$$

Recall that a set of transmitters and receivers in the transmission graph form a subnetwork if there is a path between any two of them if we ignore the unidirectional nature of the links. Figure 4.3 displays all subnetworks in the transmission graph in Figure 4.2. We denote by $L(T,R)$ the number of transmitters in each subnetwork, and denote by $W(T,R)$ the number of subnetworks in the transmission graph. As the total number of transmitters is NT, $L(T,R)$ and $W(T,R)$ have the following relation

$$L(T,R) \cdot W(T,R) = NT.$$

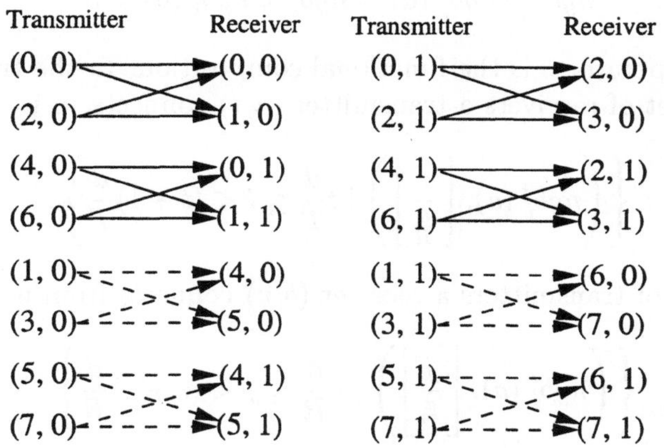

Figure 4.3. The structure of subnetworks in the transmission graph in Figure 4.2.

When $\max(T,R) = d$, each subnetwork consists of $\frac{d}{R}$ transmitters and $\frac{d}{T}$ receivers no matter which topology is to be realized. Therefore

$$L(T,R) = \frac{d}{R},$$
$$W(T,R) = N \min(T,R).$$

In the rest of this chapter, we assume that $\max(T,R) < d$. For simplicity of our description, throughout of this chapter, the long straight arrows $\longrightarrow$, $\searrow$ and $\nearrow$ are used to connect two adjacent transmitter and receiver in the transmission graph; the symbol $\leadsto$ (read as "leads to") is used between two transmitters or two receivers to indicate that the two transmitters or two receivers are in the same subnetwork.

3. SYMMETRIC LINK ORDERING

In this section, we will explore the structural properties of the local indices of the transceivers in a subnetwork when the link ordering of the regular interconnection topology is symmetric. For each node a and each $0 \leq i < d$, we use $ngb_i^+(a)$ to denote the destination node of the i-th outgoing link of node a, and $ngb_i^-(a)$ to denote the source node of the i-th incoming link of node a. If the link ordering is symmetric, then for any node a and any $0 \leq i < d$,

$$ngb_i^+ \circ ngb_i^-(a) = ngb_i^- \circ ngb_i^+(a) = a$$

where the operator $\circ$ is the functional composition. In the transmission graph the set of receivers a transmitter (a, t) connects to is

$$\left\{ \left(ngb_i^+(a), \left\lfloor \frac{i}{\frac{d}{R}} \right\rfloor \right) \mid t\frac{d}{T} \leq i < (t+1)\frac{d}{T} \right\},$$

and the set of transmitters a receiver (b, r) connects from is

$$\left\{ \left(ngb_i^-(a), \left\lfloor \frac{j}{\frac{d}{T}} \right\rfloor \right) \mid r\frac{d}{R} \leq j < (r+1)\frac{d}{R} \right\}.$$

We denote by m the least common multiple of $\frac{d}{T}$ and $\frac{d}{R}$. Let

$$T' = \frac{m}{\frac{d}{T}} = \frac{T}{\frac{d}{m}},$$
$$R' = \frac{m}{\frac{d}{R}} = \frac{R}{\frac{d}{m}}.$$

Then

$$\frac{T}{T'} = \frac{R}{R'} = \frac{d}{m}.$$

For any $0 \leq t < T$ and $0 \leq i < \frac{d}{T}$,

$$\left\lfloor \frac{\left\lfloor \frac{t\frac{d}{T}+i}{\frac{d}{R}} \right\rfloor}{R'} \right\rfloor = \left\lfloor \frac{t\frac{d}{T}+i}{m} \right\rfloor = \left\lfloor \frac{\frac{t\frac{d}{T}+i}{\frac{d}{T}}}{\frac{m}{\frac{d}{T}}} \right\rfloor = \left\lfloor \frac{t}{T'} \right\rfloor.$$

Similarly, for any $0 \le r < R$ and $0 \le i < \frac{d}{R}$,

$$\left\lfloor \frac{\left\lfloor \frac{r\frac{d}{R}+i}{\frac{d}{T}} \right\rfloor}{T'} \right\rfloor = \left\lfloor \frac{r\frac{d}{R}+i}{m} \right\rfloor = \left\lfloor \frac{\frac{r\frac{d}{R}+i}{\frac{d}{R}}}{\frac{m}{\frac{d}{R}}} \right\rfloor = \left\lfloor \frac{r}{R'} \right\rfloor .$$

Therefore, for any subnetwork there exists an integer $0 \le k < \frac{d}{m}$ such that for any transmitter (a,t) and receiver (b,r) in this subnetwork,

$$\left\lfloor \frac{t}{T'} \right\rfloor = \left\lfloor \frac{r}{R'} \right\rfloor = k.$$

The following lemma summarizes this local property.

LEMMA 4.1 *If the link ordering is symmetric, then for any subnetwork there exists an integer $0 \le k < \frac{d}{m}$ such that for any transmitter (a,t) and receiver (b,r) in this subnetwork,*

$$\left\lfloor \frac{t}{T'} \right\rfloor = \left\lfloor \frac{r}{R'} \right\rfloor = k.$$

Now we show that in such a subnetwork, the set of local indices of all transmitters is actually

$$\{t \mid kT' \le t < (k+1)T'\}$$

and the set of local indices of all receivers is actually

$$\{r \mid kR' \le r < (k+1)R'\}.$$

LEMMA 4.2 *Let a be the index of any node.*

1. *If $t \bmod T' > 0$, then for any $\lfloor \frac{t}{T'} \rfloor \le t' < t$, the two transmitters (a,t) and $\left(\left(\prod_{i=t'+1}^{t}\left(ngb^{-}_{i\frac{d}{T}-1} \circ ngb^{+}_{i\frac{d}{T}}\right)\right)(a), t'\right)$ are in the same subnetwork, where the product $\prod$ and the operator $\circ$ are all functional compositions.*
2. *If $r \bmod R' > 0$, then for any $\lfloor \frac{r}{R'} \rfloor \le r' < r$, the two receivers (a,r) and $\left(\left(\prod_{i=r'+1}^{r}\left(ngb^{+}_{i\frac{d}{R}-1} \circ ngb^{-}_{i\frac{d}{R}}\right)\right)(a), r'\right)$ are in the same*

subnetwork, where the product $\prod$ and the operator $\circ$ are all functional compositions.

Proof. (1). The lemma is trivial when $T' = 1$. So we assume that $T' > 1$. The condition $t \bmod T' > 0$ implies that

$$\left(t\frac{d}{T}\right) \bmod \frac{d}{R} > 0.$$

First assume that $t' = t - 1$. Consider the following two links:

$$\left(\left(ngb^{-}_{t\frac{d}{T}-1} \circ ngb^{+}_{t\frac{d}{T}}\right)(a), t-1\right) \searrow$$

$$\left(ngb^{+}_{t\frac{n}{T}}(a), \left\lfloor \frac{t\frac{n}{T}-1}{\frac{n}{R}} \right\rfloor\right)$$

$$= \left(ngb^{+}_{t\frac{n}{T}}(a), \left\lfloor \frac{t\frac{n}{T}}{\frac{n}{R}} \right\rfloor\right)$$

$$(a,t) \nearrow$$

Thus the two transmitters (a,t) and $\left(\left(ngb^{-}_{t\frac{d}{T}-1} \circ ngb^{+}_{t\frac{d}{T}}\right)(a), t-1\right)$ connects to one common receiver, and therefore are in the same subnetwork. If $\lfloor \frac{t}{T'} \rfloor \leq t' < t-1$, then we can apply the previous argument for $t - t'$ times. Figure 4.4 illustrates the idea, and we omit the details here.

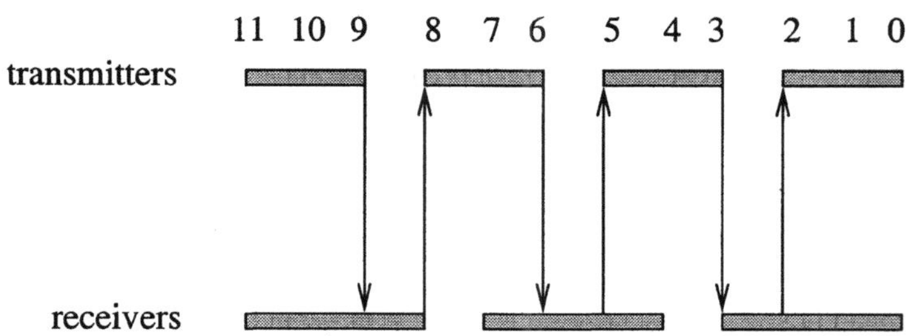

Figure 4.4. If $t \bmod T' > 0$ and $\lfloor \frac{t}{T'} \rfloor \leq t' < t$, the two transmitters (a,t) and $\left(\left(\prod_{i=t'+1}^{t}\left(ngb^{-}_{i\frac{d}{T}-1} \circ ngb^{+}_{i\frac{d}{T}}\right)\right)(a), t'\right)$ are in the same subnetwork. In this example, $d = 12, T = 4, R = 3, t = 3$ and $t' = 0$.

(2). The proof is similar to (1). ■

From Lemma 4.1 and Lemma 4.2, we can completely determine the structure of the set of local indices of all transmitters/receivers in the same subnetwork:

COROLLARY 4.3 *For any subnetwork, there exists an integer* $0 \leq k < \frac{d}{m}$ *such that in this subnetwork*

- *the set of local indices of all transmitters is*

$$\{t \mid kT' \leq t < (k+1)T'\},$$

  and for any $kT' \leq t, t' < (k+1)T'$ *the number of transmitters with local index* t *and the number of transmitters with local index* t' *are equal;*

- *the set of local indices of all receivers is*

$$\{r \mid kR' \leq r < (k+1)R'\},$$

  and for any $kR' \leq r, r' < (k+1)R'$ *the number of receivers with local index* r *and the number of receivers with local index* r' *are equal.*

4. IMPACT OF LINK ORDERING

The link ordering of the regular interconnection topology affects the uniform consecutive partition. In this section, we will show that different link orderings and therefore different partitions can result in dramatically different subnetwork structures. The regular interconnection topology to be investigated is the complete graph with self-loop, i.e., there is a link from each node to itself. The resulting network is therefore single-hop.

4.1 SINGLE-HOP PONS WITH ASYMMETRIC LINK ORDERING

This section consider the following asymmetric link ordering. For each node $0 \leq a \leq N-1$, its i-th outgoing link is

$$a \rightarrow i,$$

and its i-th incoming link is

$$i \rightarrow a,$$

where $0 \leq i \leq N-1$. In the transmission graph of the uniform consecutive partition based on this asymmetric link ordering, the set of receivers a transmitter (a, t) connects to is

$$\left\{\left(t\frac{N}{T}+i, \left\lfloor\frac{a}{\frac{N}{R}}\right\rfloor\right) \mid 0 \le i < \frac{N}{T}\right\},$$

and the set of transmitters a receiver (b, r) connects from is

$$\left\{\left(r\frac{N}{R}+i, \left\lfloor\frac{b}{\frac{N}{T}}\right\rfloor\right) \mid 0 \le i < \frac{N}{R}\right\}.$$

For any $0 \le t < T$ and $0 \le r < R$, let

$$A_{t,r} = \{(r\frac{N}{R}+i, t) \mid 0 \le i < \frac{N}{R}\},$$
$$B_{t,r} = \{(t\frac{N}{T}+j, r) \mid 0 \le j < \frac{N}{T}\}.$$

Then it's easy to verify that

- the set of receivers that each transmitter in $A_{t,r}$ connects to in the transmission graph is exactly $B_{t,r}$;
- the set of transmitters that each receiver in $B_{t,r}$ connects from is exactly $A_{t,r}$.the structure of the subnetworks can be stated as follows.

Thus we have the following theorem.

THEOREM 4.4 *For any $0 \le t < T$ and $0 \le r < R$, the transmitters in the set $A_{t,r}$ and the receivers in the set $B_{t,r}$ constitute a subnetwork. Therefore, $L(T, R) = \frac{N}{R}$ and $W(T, R) = TR$.*

From Theorem 4.4, $W(T, R)$ does not depend on the network size N. Since T and R are usually small numbers, this partition provides great scalability with respect to the number of available wavelength channels. Another nice property of this partition is that $L(T, R)$ does not depend on T. Thus decreasing the number of transmitters (even to one) can reduces the network cost linearly while still maintaining the same number of transmitters in each subnetwork, but it also decreases the transmission concurrence linearly.

4.2 SINGLE-HOP PONS WITH SYMMETRIC LINK ORDERING

Now we consider another link ordering which is symmetric. For each node $0 \leq a < N$, its i-th outgoing link is

$$a \rightarrow (a+i) \bmod N,$$

and its i-th incoming link is

$$(a-i) \bmod N \rightarrow a,$$

where $0 \leq i < N$. In the transmission graph of the uniform consecutive partition based on this symmetric ordering, the set of receivers a transmitter (a,t) connects to is

$$\left\{ \left((a+i) \bmod N, \left\lfloor \frac{i}{\frac{N}{R}} \right\rfloor \right) \mid t\frac{N}{T} \leq i \leq (t+1)\frac{N}{T} - 1 \right\},$$

and the set of transmitters a receiver (b,r) connects from is

$$\left\{ \left((b-j) \bmod N, \left\lfloor \frac{j}{\frac{N}{T}} \right\rfloor \right) \mid r\frac{N}{R} \leq j \leq (r+1)\frac{N}{R} - 1 \right\}.$$

Let m be the least common multiple of $\frac{N}{T}$ and $\frac{N}{R}$, and let

$$T' = \frac{m}{\frac{N}{T}} = \frac{T}{\frac{N}{m}},$$

$$R' = \frac{m}{\frac{N}{R}} = \frac{R}{\frac{N}{m}}.$$

As the link ordering is symmetric, from Corollary 4.3, for any subnetwork, there exists a unique integer $0 \leq k < \frac{n}{m}$ such that in this subnetwork the set of local indices of all transmitters is

$$\{t \mid kT' \leq t < (k+1)T'\},$$

and the set of local indices of all receivers is

$$\{r \mid kR' \leq r < (k+1)R'\}.$$

Furthermore, from Lemma 4.2,

- for any $kT' \leq t' < t < (k+1)T'$, the two transmitters (a,t) and $((a+t-t') \bmod N, t')$ are in the same subnetwork;
- for any $kR' \leq r' < r < (k+1)R'$, the two receivers (b,r) and $((b-r+r') \bmod N, t')$ are in the same subnetwork.

In addition, the subnetworks have the following properties.

LEMMA 4.5 *For any* $0 \leq a \leq N-1$,

1. *if* $t \bmod T' = 0$, *then the two transmitter* (a,t) *and* $((a+1) \bmod N, t)$ *are in the same subnetwork;*
2. *if* $r \bmod R' = 0$, *then the two receivers* (b,r) *and* $((b-1) \bmod N, r)$ *are in the same subnetwork.*

Proof. (1). Consider the following two links

$$(a,t) \longrightarrow \left(\left(a + t\frac{N}{T} + 1\right) \bmod N, \left\lfloor \frac{t\frac{N}{T}+1}{\frac{N}{R}} \right\rfloor\right),$$

$$((a+1) \bmod N, t) \longrightarrow \left((a + t\frac{N}{T} + 1) \bmod N, \left\lfloor \frac{t\frac{N}{T}}{\frac{N}{R}} \right\rfloor\right).$$

If $t \bmod T' = 0$, then

$$\left(t\frac{N}{T}\right) \bmod \frac{N}{R} = 0.$$

As $\frac{N}{R} > 1$,

$$\left\lfloor \frac{t\frac{N}{T}+1}{\frac{N}{R}} \right\rfloor = \left\lfloor \frac{t\frac{N}{T}}{\frac{N}{R}} \right\rfloor.$$

This means that the two transmitters (a,t) and $((a+1) \bmod N, t)$ connects to the same receiver and therefore are in the same subnetwork;

(2). The proof is similar to (1). ∎

For any $0 \leq k < \frac{N}{m}$, let

$$A_k = \left\{(a,t) \mid 0 \leq a \leq N-1, \left\lfloor \frac{t}{T'} \right\rfloor = k\right\},$$

$$B_k = \left\{(b,r) \mid 0 \leq b \leq N-1, \left\lfloor \frac{r}{R'} \right\rfloor = k\right\}.$$

Then the subnetworks can be characterized in the following theorem.

THEOREM 4.6 *For any* $0 \le k < \frac{N}{m}$, *the transmitters in* A_k *and the receivers in* B_k *constitutes a subnetwork. Therefore,*

$$L(T, R) = NT',$$
$$W(T, R) = \frac{N}{m}.$$

Proof. From Lemma 4.5, we can prove that

- all transmitters with local indices t are in the same subnetwork if $t \bmod T' = 0$;
- all receivers with local indices r are in the same subnetwork if $r \bmod R' = 0$.

In particular, all transmitters with local indices kT' are in the same subnetwork. As for any $kT' \le t' < t < (k+1)T'$ the two transmitters (a, t) and $((a + t - t') \bmod N, t')$ are in the same subnetwork, all transmitters in A_k are in the same subnetwork. By similar argument, all receivers in B_k are also in the same argument. Since the set of receivers any transmitter in A_k connects to are in B_k, and the set of transmitters any receiver in B_k connects from are in A_k, the transmitters in A_k and the receivers in B_k constitute a subnetwork. This implies that

$$L(T, R) = NT',$$
$$W(T, R) = \frac{N}{m}.$$

■

Note that when $T' > 1$ (or $R' > 1$ respectively) an access node has more than one transmitters (or receivers respectively) in the same subnetwork, which is a waste of the resource as they can't work at the same time. Thus these transmitters (or receivers respectively) can be replaced by a single transmitter (or receiver). In this way, the number of subnetworks remains the same, but the cost is reduced and the number of transmitters sharing a wavelength channel is also decreased. After the replacement, we have $T' = R' = 1$, or equivalently $T = R$. There are T subnetworks and each subnetwork consists of exactly one transmitter from each access node and exactly one receiver from each access node. Therefore each subnetwork provide the full connectivity of network. So there are T redundant copies of the network. Unless such redundancy is exploited to provide high reliability or parallel transmission to improve the end-to-end delay, it is usually useless. So to further reduce the cost while maintaining the same performance, each access node can just use one transmitter and one receiver. In this case, this network becomes the pure TDM single-hop network.

5. PONS BASED ON GENERALIZED DE BRUIJIN DIGRAPH

The links at each node in generalized de Bruijin digraph $D(N, p)$ are ordered as follows. At each vertex $0 \leq a \leq N-1$, its i-th outgoing link is

$$a \to (ap + i) \bmod N,$$

and its i-th incoming link is

$$\left\lfloor \frac{iN + a}{p} \right\rfloor \to a,$$

where $0 \leq i \leq p - 1$. This link ordering is asymmetric. One property of such ordering is that if the link $a \to b$ is the i-th outgoing link at a and j-th incoming link at b, then

$$ap + i = jN + b,$$

which implies that

$$i = \frac{jN + b}{a} \bmod p$$

and

$$j = \left\lfloor \frac{ap + i}{N} \right\rfloor.$$

In the transmission graph of the uniform consecutive partition based on the above link ordering, each transmitter has $\frac{p}{T}$ outgoing links, and each receiver has $\frac{p}{R}$ incoming links. The set of receivers a transmitter (a, t) connects to is

$$\left\{ \left(\left(ap + t\frac{p}{T} + i \right) \bmod N, \left\lfloor \frac{ap + t\frac{p}{T} + i}{N\frac{p}{R}} \right\rfloor \right) \mid 0 \leq i < \frac{p}{T} \right\}.$$

The set of transmitters a each receiver (b, r) connects from is

$$\left\{\left(\left\lfloor \frac{(r\frac{p}{R}+j)N+b}{p} \right\rfloor, \left\lfloor \frac{((r\frac{p}{R}+j)N+b) \bmod p}{\frac{p}{T}} \right\rfloor\right) \mid 0 \leq j < \frac{p}{R}\right\}.$$

To characterize the structure of subnetworks , we relabel each transmitter (a, t) by $aT + t$, and relabel each receiver (b, r) by $rN + b$. Then the transmitter set of the transmission graph is

$$\{x \mid 0 \leq x \leq NT - 1\}$$

and the receiver set is

$$\{y \mid 0 \leq b \leq NR - 1\}.$$

The set of receivers a transmitter x connects to is

$$\left\{N \left\lfloor \frac{x\frac{p}{T}+i}{N\frac{p}{R}} \right\rfloor + \left(x\frac{p}{T}+i\right) \bmod N \mid 0 \leq i \leq \frac{p}{T} - 1\right\}.$$

The set of transmitters a receiver y connects from is

$$\left\{\left\lfloor \frac{(\lfloor \frac{y}{N} \rfloor \frac{p}{R}+j)N+y \bmod N}{p} \right\rfloor T + \left\lfloor \frac{((\lfloor \frac{y}{N} \rfloor \frac{p}{R}+j)N+y \bmod N) \bmod p}{\frac{p}{T}} \right\rfloor \right.$$
$$\left. \mid 0 \leq j \leq \frac{p}{R} - 1.\right\}$$

In Subsection 5.1 we study the case that $NT \bmod p = 0$. In Subsection 5.2, we study the case that $R < p$ and $NT \bmod p \neq 0$.

5.1 SUBNETWORK STRUCTURE WHEN NT MOD $P = 0$

As T is a factor of p, $NT \bmod p = 0$ is equivalent to that N is a multiple of $\frac{p}{T}$. Thus

$$\left\lfloor \frac{x\frac{p}{T}+i}{N\frac{p}{R}} \right\rfloor = \left\lfloor \frac{\left\lfloor \frac{x\frac{p}{T}+i}{\frac{p}{T}} \right\rfloor}{\frac{N}{\frac{p}{T}}\frac{p}{R}} \right\rfloor = \left\lfloor \frac{x}{\frac{NT}{R}} \right\rfloor.$$

So the vertex adjacency of the transmission graph can be reduced as follows. For each transmitter $0 \leq x \leq NT - 1$, the set of receivers it connects to is

$$\left\{N \left\lfloor \frac{x}{\frac{NT}{R}} \right\rfloor + \left(x\frac{p}{T}+i\right) \bmod N \mid 0 \leq i \leq \frac{p}{T} - 1\right\}.$$

For any $0 \leq i \leq R-1$ and any $0 \leq j \leq \frac{NT}{p} - 1$, let

$$A_{i,j} = \left\{ i\frac{NT}{R} + k\frac{NT}{p} + j \mid 0 \leq k \leq \frac{p}{R} - 1 \right\},$$
$$B_{i,j} = \left\{ Ni + \left(j\frac{p}{T} + k \right) \bmod N \mid 0 \leq k \leq \frac{p}{T} - 1 \right\}.$$

It's easy to verify that the following:

1. The set of receivers that any transmitter in $A_{i,j}$ connects to is $B_{i,j}$.
2. The set

$$\{A_{i,j} \mid 0 \leq i \leq R-1, 0 \leq j \leq \frac{NT}{p} - 1\}$$

is a partition of the set of all transmitters, and the set

$$\{B_{i,j} \mid 0 \leq i \leq R-1, 0 \leq j \leq \frac{NT}{p} - 1\}$$

is a partition of the set of all receivers.

Therefore, for any $0 \leq i \leq R-1$ and any $0 \leq j \leq \frac{NT}{p} - 1$, the set of transmitters in $A_{i,j}$ and the set of receivers in $B_{i,j}$ constitute a subnetwork. Thus the subnetworks have the following structure.

THEOREM 4.7 *Suppose that $NT \bmod p = 0$. Then for any $0 \leq i \leq R-1$ and any $0 \leq j \leq \frac{NT}{p} - 1$, the set of transmitters in $A_{i,j}$ and the set of receivers in $B_{i,j}$ constitute a subnetwork. So*

$$L(T,R) = \frac{p}{R},$$
$$W(T,R) = \frac{NTR}{p}.$$

5.2 SUBNETWORK STRUCTURE WHEN NT MOD $P \neq 0$

For any integer a and b, let $\gcd(a,b)$ denote the greatest common divisor of a and b. The following lemma states a relation between the node indices of any pair of adjacent transmitter and receiver.

LEMMA 4.8 *Let x be any transmitter and y be any receiver which are adjacent to each other. Then*

$$\left\lfloor \frac{x}{\frac{NT}{\gcd(NT,R)}} \right\rfloor = \left\lfloor \frac{y}{\frac{NR}{\gcd(NT,R)}} \right\rfloor .$$

Proof. Suppose that

$$y = N \left\lfloor \frac{x\frac{p}{T}+i}{N\frac{p}{R}} \right\rfloor + \left(x\frac{p}{T}+i\right) \bmod N,$$

for some $0 \le i \le \frac{p}{T} - 1$. Then

$$\left\lfloor \frac{y}{\frac{NR}{\gcd(NT,R)}} \right\rfloor = \left\lfloor \frac{N\left\lfloor \frac{x\frac{p}{T}+i}{N\frac{p}{R}} \right\rfloor + \left(x\frac{p}{T}+i\right) \bmod N}{\frac{NR}{\gcd(NT,R)}} \right\rfloor$$

$$= \left\lfloor \frac{\left\lfloor \frac{N\left\lfloor \frac{x\frac{p}{T}+i}{N\frac{p}{R}} \right\rfloor + \left(x\frac{p}{T}+i\right) \bmod N}{N} \right\rfloor}{\frac{R}{\gcd(NT,R)}} \right\rfloor = \left\lfloor \frac{\left\lfloor \frac{x\frac{p}{T}+i}{N\frac{p}{R}} \right\rfloor}{\frac{R}{\gcd(NT,R)}} \right\rfloor$$

$$= \left\lfloor \frac{x\frac{p}{T}+i}{\frac{Np}{\gcd(NT,R)}} \right\rfloor = \left\lfloor \frac{x\frac{p}{T}+i}{\frac{p}{T}\frac{NT}{\gcd(NT,R)}} \right\rfloor$$

$$= \left\lfloor \frac{\left\lfloor \frac{x\frac{p}{T}+i}{\frac{p}{T}} \right\rfloor}{\frac{NT}{\gcd(NT,R)}} \right\rfloor = \left\lfloor \frac{x}{\frac{NT}{\gcd(NT,R)}} \right\rfloor .$$

■

For any $0 \le m \le \gcd(NT, R) - 1$, let

$$X_m = \left\{ x \mid 0 \le x \le NT - 1, \left\lfloor \frac{x}{\frac{NT}{\gcd(NT,R)}} \right\rfloor = m \right\},$$

$$Y_m = \left\{ y \mid 0 \le y \le NR - 1, \left\lfloor \frac{y}{\frac{NR}{\gcd(NT,R)}} \right\rfloor = m \right\}.$$

Then the set

$$\{X_m \mid 0 \le m \le \gcd(NT, R) - 1\}$$

is a partition of the set of transmitters, and the set

$$\{Y_m \mid 0 \leq m \leq \gcd(NT, R) - 1\}$$

is a partition of the set of receivers. From Lemma 4.8, we immediately have the following corollary.

COROLLARY 4.9 *For any $0 \leq m \leq \gcd(NT, R) - 1$, a transmitter is in X_m if and only if it connects to some receiver in Y_m, and conversely a receiver is in Y_m if and only if it connects from some transmitter in X_m. Therefore, any two transmitters in the same subnetwork must both be in a set X_m for some $0 \leq m \leq \gcd(NT, R) - 1$.*

In the remaining of this subsection, we will prove along the reverse direction that all transmitters in X_m are in the same subnetwork. If this is true, then we can draw the conclusion that all transmitters in X_m and all receivers in Y_m constitute a subnetwork. Thus

$$L(T, R) = \frac{NT}{\gcd(NT, R)},$$
$$W(T, R) = \gcd(NT, R).$$

LEMMA 4.10 *Let x be any transmitter and i be any integer with $0 \leq i \leq p - 1$. If*

$$\left\lfloor \frac{x\frac{p}{T} + i}{N} \right\rfloor \bmod \frac{p}{R} = 0,$$

then for all $i \leq j \leq p - 1$,

$$\left\lfloor \frac{x\frac{p}{T} + j}{N\frac{p}{R}} \right\rfloor = \left\lfloor \frac{x\frac{p}{T} + i}{N\frac{p}{R}} \right\rfloor.$$

Proof. Suppose that

$$\left\lfloor \frac{x\frac{p}{T} + i}{N} \right\rfloor = k\frac{p}{R}$$

for some k. Then

$$kN\frac{p}{R} \leq x\frac{p}{T} + i \leq kN\frac{p}{R} + N - 1.$$

So for any $i \leq j \leq p-1$,

$$kN\frac{p}{R}+j-i \leq x\frac{p}{T}+j \leq kN\frac{p}{R}+N+j-i-1.$$

Since $\frac{p}{R} \geq 2$,

$$0 \leq j-i \leq p-1 \leq N\frac{p}{R}-1,$$

$$0 \leq N+j-i-1 \leq N+p-2 \leq N\frac{p}{R}-1.$$

Therefore,

$$\left\lfloor \frac{x\frac{p}{T}+j}{N\frac{p}{R}} \right\rfloor = k.$$

■

Let $q = \left\lfloor \frac{N}{\frac{p}{T}} \right\rfloor$ and $r = N \bmod \frac{p}{T}$. Then $N = q\frac{p}{T}+r$ and $0 < r \leq \frac{p}{T}-1$.

LEMMA 4.11 *Let x be any transmitter.*

(1). If $x \geq q+1$, then the two transmitters x and $x-(q+1)$ do not connects to any common receiver only if

$$\left\lceil \frac{kNT}{R} \right\rceil \leq x \leq \left\lfloor \frac{kNT}{R} \right\rfloor + q$$

for some integer k .

(2). If $x \geq q$, then the two transmitters x and $x-q$ do not connects to any common receiver only if

$$\left\lceil \frac{kNT}{R} - \frac{rT}{p} \right\rceil \leq x \leq \left\lfloor \frac{kNT}{R} + \frac{rT}{p} \right\rfloor + q - T$$

for some integer k.

Proof. (1). Suppose that the two transmitters x and $x-(q+1)$ do not connects to any common receiver. We first prove by contradiction that for all $0 \leq i \leq r-1$,

$$\left\lfloor \frac{x\frac{p}{T}+i}{N} \right\rfloor \bmod \frac{p}{R} = 0.$$

Assume that there exists some $0 \leq i \leq r-1$ with

$$\left\lfloor \frac{x\frac{p}{T}+i}{N} \right\rfloor \bmod \frac{p}{R} \neq 0.$$

Let $j = i + \frac{p}{T} - r$. Then

$$\frac{p}{T} - r \leq j \leq \frac{p}{T} - 1.$$

Consider the following path:

$$\begin{aligned}
&x-(q+1)\\
&\quad\searrow N\left\lfloor \frac{(x-(q+1))\frac{p}{T}+j}{N\frac{p}{R}} \right\rfloor + \left((x-(q+1))\frac{p}{T}+j\right) \bmod N\\
&\quad= N\left\lfloor \frac{x\frac{p}{T}-N+i}{N\frac{p}{R}} \right\rfloor + \left(x\frac{p}{T}-N+i\right) \bmod N\\
&\quad= N\left\lfloor \frac{\left\lfloor \frac{x\frac{p}{T}+i}{N} \right\rfloor - 1}{\frac{p}{R}} \right\rfloor + \left(x\frac{p}{T}+i\right) \bmod N\\
&\quad= N\left\lfloor \frac{\left\lfloor \frac{x\frac{p}{T}+i}{N} \right\rfloor}{\frac{p}{R}} \right\rfloor + \left(x\frac{p}{T}+i\right) \bmod N\\
&\quad= N\left\lfloor \frac{x\frac{p}{T}+i}{N\frac{p}{R}} \right\rfloor + \left(x\frac{p}{T}+i\right) \bmod N\\
&x \nearrow
\end{aligned}$$

Therefore the two transmitters x and $x-(q+1)$ both connect to the receiver $N\left\lfloor \frac{x\frac{p}{T}+i}{N\frac{p}{R}} \right\rfloor + \left(x\frac{p}{T}+i\right) \bmod N$, which is a contradiction to the assumption. Hence

$$\left\lfloor \frac{x\frac{p}{T}+i}{N} \right\rfloor \bmod \frac{p}{R} = 0$$

for all $0 \leq i \leq r-1$.

Let

$$k = \frac{\left\lfloor \frac{x\frac{p}{T}}{N} \right\rfloor}{\frac{p}{R}}.$$

Then from Lemma 4.10,

$$\left\lfloor \frac{x\frac{p}{T}+i}{N} \right\rfloor = k\frac{p}{R}$$

for all $0 \le i \le r-1$. So

$$kN\frac{p}{R} \le x\frac{p}{T} + i \le kN\frac{p}{R} + N - 1,$$

for all $0 \le i \le r-1$. This implies that

$$kN\frac{p}{R} \le x\frac{p}{T} \le kN\frac{p}{R} + N - r.$$

Therefore,

$$\left\lceil \frac{kNT}{R} \right\rceil \le x \le \left\lfloor \frac{kNT}{R} \right\rfloor + q.$$

(2). Suppose that the two transmitters x and $x-q$ do not connects to any common receiver. We prove by contradiction that

$$\left\lfloor \frac{x\frac{p}{T}+i}{N} \right\rfloor \bmod \frac{p}{R} = 0$$

for all $r \le i \le p-1$. Assume that there exists some $r \le i \le p-1$ with

$$\left\lfloor \frac{x\frac{p}{T}+i}{N} \right\rfloor \bmod \frac{p}{R} \ne 0.$$

Let $j = i - r$. Then $0 \le j \le \frac{p}{T} - r - 1$. Consider the following path:

$$\begin{aligned}
&x-q\\
&\qquad \searrow N\left\lfloor \frac{(x-q)\frac{p}{T}+j}{N\frac{p}{R}} \right\rfloor + \left((x-q)\frac{p}{T}+j\right) \bmod N\\
&\qquad = N\left\lfloor \frac{x\frac{p}{T}-N+i}{N\frac{p}{R}} \right\rfloor + \left(x\frac{p}{T}-N+i\right) \bmod N\\
&\qquad = N\left\lfloor \frac{\left\lfloor \frac{x\frac{p}{T}+i}{N} \right\rfloor - 1}{\frac{p}{R}} \right\rfloor + \left(x\frac{p}{T}+i\right) \bmod N\\
&\qquad = N\left\lfloor \frac{\left\lfloor \frac{x\frac{p}{T}+i}{N} \right\rfloor}{\frac{p}{R}} \right\rfloor + \left(x\frac{p}{T}+i\right) \bmod N\\
&\qquad = N\left\lfloor \frac{x\frac{p}{T}+i}{N\frac{p}{R}} \right\rfloor + \left(x\frac{p}{T}+i\right) \bmod N\\
&x \nearrow
\end{aligned}$$

Therefore the two transmitters x and $x-q$ both connect to the receiver $N\left\lfloor\frac{x\frac{p}{T}+i}{N\frac{p}{R}}\right\rfloor+\left(x\frac{p}{T}+i\right) \bmod N$, which is a contradiction to the assumption. Thus

$$\left\lfloor\frac{x\frac{p}{T}+i}{N}\right\rfloor \bmod \frac{p}{R}=0$$

for all $r \leq i \leq p-1$.

Let

$$k=\frac{\left\lfloor\frac{x\frac{p}{T}+r}{N}\right\rfloor}{\frac{p}{R}}.$$

Then from Lemma 4.10,

$$\left\lfloor\frac{x\frac{p}{T}+i}{N}\right\rfloor=k\frac{p}{R}$$

for all $r \leq i \leq p-1$. So,

$$kN\frac{p}{R} \leq x\frac{p}{T}+i \leq kN\frac{p}{R}+N-1$$

for all $r \leq i \leq p-1$. This implies that

$$kN\frac{p}{R}-r \leq x\frac{p}{T} \leq kN\frac{p}{R}+N-p.$$

Therefore,

$$\left\lceil\frac{kNT}{R}-\frac{rT}{p}\right\rceil \leq x \leq \left\lfloor\frac{kNT}{R}+\frac{rT}{p}\right\rfloor+q-T.$$

■

Let

$$a_k=\left\lfloor\frac{kNT}{R}\right\rfloor.$$

From the above lemma, we can immediately draw the following two corollaries.

COROLLARY 4.12 *Let k be any integer with $0 \leq k \leq R-1$. Then for all $a_{k-1}+q+1 \leq x \leq a_k-1$ the two transmitters x and $x-(q+1)$*

connect to a common receiver. In addition, if $k \bmod \frac{R}{\gcd(NT,R)} > 0$*, the two transmitters* a_k *and* $a_k - (q+1)$ *also connect to a common receiver.*

Proof. Notice that

$$\left\lceil \frac{kNT}{R} \right\rceil = \begin{cases} a_k + 1 & \text{if } k \bmod \frac{R}{\gcd(NT,R)} > 0, \\ a_k & \text{if } k \bmod \frac{R}{\gcd(NT,R)} = 0. \end{cases}$$

Therefore from the first part of Lemma 4.11, the Corollary is true. ■

COROLLARY 4.13 *Let* k *be any integer with* $0 \le k \le R-1$*. Then for all* $a_k + q - T + 2 \le x \le a_{k+1} - 1$*, the two transmitters* x *and* $x - q$ *connect to a common receiver. In addition, if* $k \bmod \frac{R}{\gcd(NT,R)} = 0$*, the two transmitters* $a_k + q - T + 1$ *and* $a_k - T + 1$ *also connect to a common receiver.*

Proof. First of all, for any $0 < r \le \frac{p}{T} - 1$,

$$\left\lceil \frac{kNT}{R} - \frac{rT}{p} \right\rceil \ge a_k.$$

Secondly, if $k \bmod \frac{R}{\gcd(NT,R)} > 0$

$$\left\lfloor \frac{kNT}{R} + \frac{rT}{p} \right\rfloor \le a_k + 1$$

and if $k \bmod \frac{R}{\gcd(NT,R)} = 0$

$$\left\lfloor \frac{kNT}{R} + \frac{rT}{p} \right\rfloor = a_k.$$

Thus from the second part of Lemma 4.11, the Corollary is true. ■

The following lemma gives a lower bound on the distance between two consecutive a_k's.

LEMMA 4.14 *For any* $1 \le k \le R-1$*,* $a_k - a_{k-1} \ge 2q$*. Furthermore, if* $\frac{p}{R} > 2$ *or* $rT \ge R$*,* $a_k - a_{k-1} \ge 2q+1$*.*

Proof. Since

$$a_k = \left\lfloor \frac{k(q\frac{p}{T} + r)T}{R} \right\rfloor = \left\lfloor \frac{kpq + krT}{R} \right\rfloor = k\frac{p}{R}q + \left\lfloor \frac{krT}{R} \right\rfloor,$$

we have that

$$a_{k+1} - a_k = \frac{p}{R}q + \left\lfloor \frac{(k+1)rT}{R} \right\rfloor - \left\lfloor \frac{krT}{R} \right\rfloor \geq \frac{p}{R}q + \left\lfloor \frac{rT}{R} \right\rfloor.$$

As $\frac{p}{R} \geq 2$, $a_k - a_{k-1} \geq 2q$. If $\frac{p}{R} > 2$ or $rT \geq R$, then $a_k - a_{k-1} \geq 2q+1$. ■

Now we can give the following important lemma.

LEMMA 4.15 *Suppose that* $1 \leq k \leq R-1$. *Then all transmitters*

$$a_{k-1}, a_{k-1}+1, \cdots, a_k - 1$$

are in the same subnetwork. In addition, if $k \bmod \frac{R}{\gcd(NT,R)} > 0$ *then all transmitters*

$$a_{k-1}, a_{k-1}+1, \cdots, a_k - 1, a_k$$

are in the same subnetwork.

Proof. We first prove that for any k, the set of transmitters

$$\{x \mid a_{k-1} \leq x \leq a_k - 1\} \setminus \{a_{k-1} + q\}$$

are in the same subnetwork. In fact, from Corollary 4.13 and Corollary 4.12, if

$$a_{k-1} + q + 1 \leq x \leq a_k - 1,$$

the three transmitters $x, x-q, x-(q+1)$ are in the same subnetwork. So for any

$$a_{k-1} + q + 1 \leq x \leq a_k - 2,$$

the three transmitters $x, x+1, x-q$ are in the same subnetwork. Hence the set of transmitters

$$\{x \mid a_{k-1} + q + 1 \leq x \leq a_k - 1\}$$

are in the same subnetwork. Again because for each transmitter x with

$$a_{k-1} + q + 1 \leq x \leq a_k - 1$$

the three transmitters $x, x-q, x-(q+1)$ are in the same subnetwork, the set of transmitters

$$\{x \mid a_{k-1} \leq x \leq a_k - q - 1\}$$

are in the same subnetwork as the transmitters in the set

$$\{x \mid a_{k-1} + q + 1 \le x \le a_k - 1\}.$$

As $a_k - a_{k-1} \ge 2q$ by Lemma 4.14,

$$a_{k-1} + q - 1 \le a_k - q - 1.$$

Thus, the set of transmitters

$$\{x \mid a_{k-1} \le x \le a_{k-1} + q - 1\} \cup \{x \mid a_{k-1} + q + 1 \le x \le a_k - 1\}$$

are in the same subnetwork.

Next we will prove that the transmitter $a_{k-1} + q$ is also in the same subnetwork with some transmitter in the set

$$\{x \mid a_{k-1} \le x \le a_{k-1} + q - 1\} \cup \{x \mid a_{k-1} + q + 1 \le x \le a_k - 1\}.$$

We prove it by considering the following three possible cases.

Case 1: $T > 1$. In this case,

$$a_{k-1} + q - T + 2 \le a_{k-1} + q \le a_k - 1,$$

and therefore by Corollary 4.13, the two transmitters $a_{k-1} + q$ and a_{k-1} are in the same subnetwork.

Case 2: $a_k - a_{k-1} \ge 2q + 1$. In this case,

$$a_{k-1} + q - T + 2 \le a_{k-1} + q + 1 \le a_{k-1} + 2q \le a_k - 1.$$

Therefore by Corollary 4.13, the transmitter $a_{k-1} + q$ is in the same subnetwork with the transmitter $a_{k-1} + 2q$.

Case 3: $T = 1$ *and* $a_k - a_{k-1} = 2q$. From 4.14, this case happens if and only if $T = 1$, $\frac{p}{R} = 2$ and $r < R$. So in this case

$$\begin{aligned}
&(a_{k-1} + q)p \\
&= p \left\lfloor \frac{(k-1)N}{R} \right\rfloor + qp \\
&= p \left\lfloor \frac{(k-1)qp + (k-1)r}{R} \right\rfloor + N - r \\
&= 2(k-1)qp + p \left\lfloor \frac{(k-1)r}{R} \right\rfloor + N - r \\
&= 2(k-1)N - 2(k-1)r + 2R \left\lfloor \frac{(k-1)r}{R} \right\rfloor + N - r \\
&= (2k-1)N - 2(((k-1)r) \bmod R)) - r.
\end{aligned}$$

Thus the set of receivers that the transmitter $a_{k-1}+q$ connects to is

$$\left\{N\left\lfloor\frac{(a_{k-1}+q)p+i}{2N}\right\rfloor+((a_{k-1}+q)p+i) \bmod N \mid 0\le i\le p-1\right\}$$
$$=\left\{(k-1)N+N\left\lfloor\frac{N-2(((k-1)r) \bmod R)-r+i}{2N}\right\rfloor\right.$$
$$\left.+\left(-2kr+2R\left\lfloor\frac{kr}{R}\right\rfloor-r+i\right) \bmod N \ \mid 0\le i\le p-1\right\}$$
$$=\{(k-1)N+(i-2(((k-1)r) \bmod R)-r) \bmod N \mid 0\le i\le p-1\}.$$

The last equation holds because

$$\begin{aligned}&N-2(((k-1)r) \bmod R)-r+i\\ &\ge N-2(R-1)-r\\ &\ge N-p-r+2\\ &=(q-1)p+2\\ &\ge 2\end{aligned}$$

and

$$\begin{aligned}&N-2(((k-1)r) \bmod R)-r+i\\ &\le N-r+p-1\\ &\le 2N-1.\end{aligned}$$

Let

$$b=\min_{0\le i\le p-1}(i-2(((k-1)r) \bmod R)-r) \bmod N.$$

Then $0\le b\le N-p$. The receiver $(k-1)N+b$ connects from two transmitters $x=\left\lfloor\frac{2(k-1)N+b}{p}\right\rfloor$ and $x'=\left\lfloor\frac{(2k-1)N+b}{p}\right\rfloor$. Obviously, one of x and x' must be equal to $a_{k-1}+q$ as $\frac{p}{R}=2$. Since

$$\begin{aligned}&a_{k-1}+q\\ &=\left\lfloor\frac{(k-1)N}{R}\right\rfloor+\left\lfloor\frac{N}{p}\right\rfloor\\ &=\left\lfloor\frac{2(k-1)N}{p}\right\rfloor+\left\lfloor\frac{N}{p}\right\rfloor\\ &\le\left\lfloor\frac{(2k-1)N}{p}\right\rfloor\end{aligned}$$

and

$$a_{k-1} \leq x < x' = \left\lfloor \frac{(2k-1)N + b}{p} \right\rfloor \leq \left\lfloor \frac{2kN - p}{p} \right\rfloor = a_k - 1,$$

the transmitter $a_{k-1} + q$ must be in the same subnetwork with some transmitter in the set

$$\{x \mid a_{k-1} \leq x \leq a_{k-1} + q - 1\} \cup \{x \mid a_{k-1} + q + 1 \leq x \leq a_k - 1\}.$$

Therefore in any case, the transmitter $a_{k-1} + q$ is in the same subnetwork with some transmitter in the set

$$\{x \mid a_{k-1} \leq x \leq a_{k-1} + q - 1\} \cup \{x \mid a_{k-1} + q + 1 \leq x \leq a_k - 1\}.$$

So the set of transmitters

$$\{x \mid a_{k-1} \leq x \leq a_k - 1\}$$

are in the same subnetwork.

If $k \bmod \frac{R}{\gcd(NT,R)} > 0$, then by Corollary 4.12 the transmitter a_k is in the same subnetwork with the transmitter $a_k - (q+1) \in \{x \mid a_{k-1} \leq x \leq a_k - 1\}$. So in this case, all the transmitters

$$\{x \mid a_{k-1} \leq x \leq a_k\}$$

are in the same subnetwork. ■

Now we give the main result in this subsection in the following theorem.

THEOREM 4.16 *Suppose that $NT \bmod p \neq 0$. Then for any $0 \leq m < \gcd(NT, R)$, all transmitters in X_m and all receivers in Y_m constitutes a subnetwork. Therefore,*

$$L(T,R) = \frac{NT}{\gcd(NT,R)},$$
$$W(T,R) = \gcd(NT,R).$$

Proof. By Corollary 4.9, we only need to prove that all transmitters in X_m are in the same subnetwork. We represent X_m as follows.

$$\begin{aligned} X_m &= \left\{x \mid m\frac{NT}{\gcd(NT,R)} \le x \le (m+1)\frac{NT}{\gcd(NT,R)} - 1\right\} \\ &= \left\{x \mid a_{m\frac{R}{\gcd(NT,R)}} \le x \le a_{(m+1)\frac{R}{\gcd(NT,R)}} - 1\right\} \\ &= \left(\cup_{m\frac{NT}{\gcd(NT,R)}+1 \le k \le (m+1)\frac{R}{\gcd(NT,R)}-1}\{x \mid a_{k-1} \le x \le a_k\}\right) \\ &\cup \left\{x \mid a_{(m+1)\frac{R}{\gcd(NT,R)}-1} \le x \le a_{(m+1)\frac{R}{\gcd(NT,R)}} - 1\right\}. \end{aligned}$$

From Lemma 4.15, for any

$$m\frac{NT}{\gcd(NT,R)} + 1 \le k \le (m+1)\frac{R}{\gcd(NT,R)} - 1,$$

the set of transmitters

$$\{x \mid a_{k-1} \le x \le a_k\}$$

are in the same subnetwork. Also from Lemma 4.15, the set of transmitters

$$\left\{x \mid a_{(m+1)\frac{R}{\gcd(NT,R)}-1} \le x \le a_{(m+1)\frac{R}{\gcd(NT,R)}} - 1\right\}$$

are also in the same subnetwork. Therefore, all transmitters in X_m are in the same subnetwork. Therefore, all transmitters in X_m and all receivers in Y_m constitutes a subnetwork. So we have

$$\begin{aligned} L(T,R) &= \frac{NT}{\gcd(NT,R)}, \\ W(T,R) &= \gcd(NT,R). \end{aligned}$$

■

5.3 COST-PERFORMANCE RELATIONS

If NT is a multiple of p, $L(T,R) = \frac{p}{R}$ does not depend on T at all, and $W(T,R) = \frac{NTR}{p}$ increases linearly with T and the network size N. Therefore when N is large and the number of available wavelength is relatively small, we can choose T to be the smallest value such that NT

is a multiple of p. In this way, both the network cost and the required wavelength channels are reduced while the number of transmitters in the subnetwork remains the same, In particular, if N is a multiple of p, then we can choose $T = 1$, without degrading the performance.

If NT is not a multiple of p, the performance is very poor. In fact the best transmission cycle length is $\frac{NT}{R}$ and the maximal transmission concurrence is R, which happens only when NT is a multiple of R. The worst performance happens when R and NT are relatively prime. In this case, it degenerates into a pure TDM network, and the number of transmitters increase linearly with T. Again, the increasing in the number of transmitters or receivers may lead to the decreasing in the performance. Therefore, the parameter must be carefully selected to avoid this situation.

6. PONS BASED ON GENERALIZED KAUTZ DIGRAPH

In a generalized Kautz digraph $K(N, p)$, for each vertex $0 \leq a \leq N-1$, its i-th outgoing link is

$$a \to ((N - a - 1)p + i) \bmod N,$$

and its i-th incoming link is

$$N - 1 - \left\lfloor \frac{iN + a}{p} \right\rfloor \to a,$$

where $0 \leq i \leq p - 1$. This link ordering is asymmetric. One property of such ordering is that if the link $a \to b$ is the i-th outgoing link at a and j-th incoming link at b, then

$$(N - a - 1)p + i = jN + b,$$

which implies that

$$i = \frac{jN + b}{a} \bmod p$$

and

$$j = \left\lfloor \frac{(N - a - 1)p + i}{N} \right\rfloor.$$

In the transmission graph of the uniform consecutive partition based on the above link ordering, each transmitter has $\frac{p}{T}$ outgoing links, and each receiver has $\frac{p}{R}$ incoming links. The set of receivers it connects to is

$$\left\{\left(\left((N-a-1)p+t\frac{p}{T}+i\right) \bmod N, \left\lfloor\frac{(N-a-1)p+t\frac{p}{T}+i}{N\frac{p}{R}}\right\rfloor\right)\right.$$
$$\left.\mid 0 \leq i < \frac{p}{T}\right\}.$$

It implies that for each receiver (b,r), the set of transmitters a transmitter (a,t) connects from is

$$\left\{\left(N-1-\left\lfloor\frac{(r\frac{p}{R}+j)N+b}{p}\right\rfloor, \left\lfloor\frac{\left(\left(r\frac{p}{R}+j\right)N+b\right) \bmod p}{\frac{p}{T}}\right\rfloor\right)\right.$$
$$\left.\mid 0 \leq j < \frac{p}{R}\right\}.$$

We relabel each transmitter (a,t) by $(N-a-1)T+t$, and each receiver (b,r) by $rN+b$. Then in the transmission graph the transmitter set is

$$\{x \mid 0 \leq x \leq NT-1\}$$

and the receiver set is

$$\{y \mid 0 \leq b \leq NR-1\}.$$

The set of receivers a transmitter x connects to is

$$\left\{N\left\lfloor\frac{x\frac{p}{T}+i}{N\frac{p}{R}}\right\rfloor+\left(x\frac{p}{T}+i\right) \bmod N \mid 0 \leq i \leq \frac{p}{T}-1\right\}.$$

It's easy to see that the set of receivers a receiver y connects from is

$$\left\{\left\lfloor\frac{\left(\lfloor\frac{y}{N}\rfloor\frac{p}{R}+j\right)N+y \bmod N}{p}\right\rfloor T+\right.$$
$$\left.\left\lfloor\frac{\left(\left(\lfloor\frac{y}{N}\rfloor\frac{p}{R}+j\right)N+y \bmod N\right) \bmod p}{\frac{p}{T}}\right\rfloor \mid 0 \leq j \leq \frac{p}{R}-1\right\}.$$

By numbering the transmitters and receivers at each access node in the generalized Kautz digraph in this way, we get the similar transmission graph as that for generalized de Bruijin digraph. So the subnetwork structures and the cost-performance relations can be discussed in the same way.

7. PONS BASED ON HYPERCUBE

We order the links at each node in a n-dimensional hypercube Q_n, or n-cube in short, as follows. For each node $0 \leq a < N$, its i-th outgoing link is

$$a \rightarrow a \oplus 2^i,$$

and its i-th incoming link is

$$a \oplus 2^i \rightarrow a,$$

where $0 \leq i < n$ and the operator $\oplus$ is the parity operator (bit-wise exclusive or). This ordering is symmetric. In the transmission graph of the uniform consecutive partition based on the above link ordering, the set of receivers a transmitter (a, t) connects to is

$$\left\{ \left(a \oplus 2^i, \left\lfloor \frac{i}{\frac{n}{R}} \right\rfloor \right) \mid t\frac{n}{T} \leq i < (t+1)\frac{n}{T} \right\},$$

and the set of transmitters a receiver (b, r) connects from is

$$\left\{ \left(b \oplus 2^j, \left\lfloor \frac{j}{\frac{n}{T}} \right\rfloor \right) \mid r\frac{n}{R} \leq j < \frac{n}{R} - 1 \right\}$$

7.1 SUBNETWORK STRUCTURES

Let m be the least common multiple of $\frac{n}{T}$ and $\frac{n}{R}$ and

$$T' = \frac{m}{\frac{n}{T}} = \frac{T}{\frac{n}{m}},$$
$$R' = \frac{m}{\frac{n}{R}} = \frac{R}{\frac{n}{m}}.$$

As the link ordering is symmetric, from Corollary 4.3, for any subnetwork, there exists a unique integer $0 \leq k < \frac{n}{m}$ such that in this subnetwork the set of local indices of all transmitters is

$$\{t \mid kT' \leq t < (k+1)T'\},$$

and the set of local indices of all receivers is

$$\{r \mid kR' \leq r < (k+1)R'\}.$$

Furthermore, from Lemma 4.2, for any $kT' \leq t' < t < (k+1)T'$, the two transmitters (a, t) and $\left(a \oplus \sum_{i=t'+1}^{t}\left(2^{i\frac{n}{T}} \oplus 2^{i\frac{n}{T}-1}\right), t'\right)$ are in the same subnetwork; for any $kR' \leq r' < r < (k+1)R'$, the two receivers (a, r) and $\left(a \oplus \sum_{i=r'+1}^{r}\left(2^{i\frac{n}{R}} \oplus 2^{i\frac{n}{R}-1}\right), r'\right)$ are in the same subnetwork.

In the next we will study the structure of the set of node indices of all transmitters (or receivers) which are in the same subnetwork and have the same local index. We first introduce some definitions. The *weight* of any binary number is defined to be the number of its none-zero bits. The *Hamming distance* between two binary numbers is defined to be the number of different bits of them. Two binary numbers are said to have the same *parity* if they have even Hamming distance. It's easy to verify that two binary numbers have the same parity if and only if their weight are either both even or both odd. The set of n-bit binary numbers whose weights are even are closed under the parity operation $\oplus$. It's easy to prove by induction that in any subnetwork of a transmission graph,

- the node indices of all transmitters have the same parity,
- the node indices of all receivers have the same parity,
- the node index of any transmitter and the node index of any receiver have different parities.

Because the local indices of all transmitters and receivers are

$$\{t \mid kT' \leq t < (k+1)T'\}$$

and

$$\{r \mid kR' \leq r < (k+1)R'\}$$

for some $0 \leq k < \frac{n}{m}$ respectfully, the node indices of all transceivers in this subnetwork have the same bit at any position other than the positions $km, km+1, \cdots, (k+1)m-1$. Thus we can restrict our attention only to the positions $km, km+1, \cdots, (k+1)m-1$. For any n-bit binary number

$$x = x_{n-1} \cdots x_1 x_0,$$

we will call the m-bit binary number

$$x_{(k+1)m-1} \cdots x_{km+1} x_{km}$$

the *k-segment* of x. Let

$$A_k = \left\{ km + i\frac{n}{T} \mid 1 \le i \le T' - 1 \right\},$$
$$B_k = \left\{ km + i\frac{n}{R} \mid 1 \le i \le R' - 1 \right\},$$

Then

$$A_k \cap B_k = \emptyset,$$

since m is the least common multiple of $\frac{n}{T}$ and $\frac{n}{R}$. Let

$$C_k = (A_k \cup B_k) \cup \{km, (k+1)m - 1\} = \left\{ c_{k,0}, c_{k,1}, \cdots, c_{k,T'+R'-1} \right\},$$

where

$$km = c_{k,0} < c_{k,1} < \cdots < c_{k,T'+R'-1} = (k+1)m - 1.$$

For any $0 \le i < T' + R' - 1$, we will call the binary number

$$x_{c_{k,i+1}-1} \cdots x_{c_{k,i}+1} x_{c_{k,i}}$$

the (k,i)-section of x. Figure 4.5 gives an example of segments and sections for $n = 24, T = 6$ and $R = 8$.

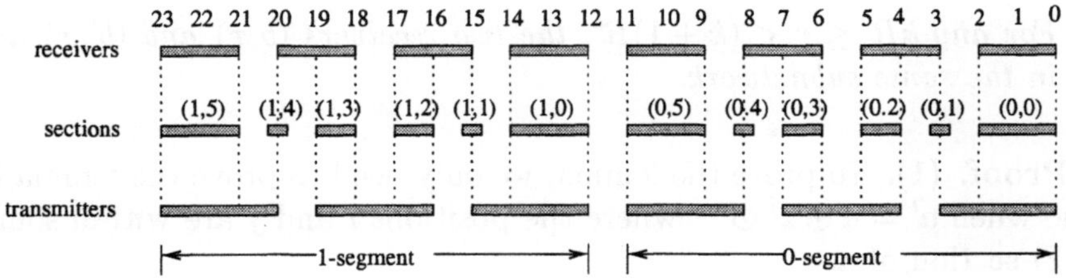

Figure 4.5. The definitions of segments and sections. In this example, $n = 24, T = 6$ and $R = 8$. There are two segments, and each segment has six sections.

For any $0 \le k < \frac{n}{m}$, we define a binary relation $\cong_k$ between two n-bit binary numbers as follows. For any two n-bit binary numbers x and y, $x \cong_k y$ if and only if

- x and y are only different at their k-segments;
- for each $0 \le i < T' + R' - 1$, the (k,i)-sections of x and y have the same parity;

It's easy to see that the binary relation $\cong_k$ is an equivalent relation. The next lemma gives the size of any equivalent class.

LEMMA 4.17 *For any $0 \leq k < \frac{n}{m}$ The size of any equivalent class under the binary equivalent relation $\cong_k$ is $2^{m-(T'+R')-1}$.*

Proof. For any $0 \leq i < T' + R' - 1$, let $\ell_i = c_{k,i+1} - c_{k,i}$, i.e., the length of any (k, i)-section of any binary number. Then the size of any equivalent class under the binary equivalent relation $\cong_k$ is

$$\prod_{i=0}^{T'+R'-2} 2^{\ell_i - 1} = 2^{\sum_{i=0}^{T'+R'-2}(\ell_i - 1)}$$
$$= 2^{\left(\sum_{i=0}^{T'+R'-2} \ell_i\right) - (T'+R'-1)} = 2^{m-(T'+R'-1)}.$$

■

The next lemma gives a sufficient condition for two transmitters/receivers which have the same local indices to be in the same subnetwork.

LEMMA 4.18 *Suppose that $a \cong_k a'$.*

1. *For any $kT' \leq t < (k+1)T'$, the two transmitters (a, t) and (a', t) are in the same subnetwork.*
2. *For any $kR' \leq r < (k+1)R'$, the two receivers (b, r) and (b', r) are in the same subnetwork.*

Proof. (1). To prove the lemma, we only need to prove the lemma is true when $a' = a \oplus 2^i \oplus 2^j$ where the positions i and j are within some (k, ℓ)-section of a.

Let

$$t' = \left\lfloor \frac{i}{\frac{n}{T}} \right\rfloor = \left\lfloor \frac{j}{\frac{n}{T}} \right\rfloor,$$
$$r' = \left\lfloor \frac{i}{\frac{n}{R}} \right\rfloor = \left\lfloor \frac{j}{\frac{n}{R}} \right\rfloor.$$

If $t = t'$, the following path

$$\begin{array}{ll} (a, t) \searrow & \\ & (a \oplus 2^i, r') \\ (a \oplus 2^i \oplus 2^j, t) \nearrow & \end{array}$$

implies that the two transmitters (a,t) and $(a \oplus 2^i \oplus 2^j, t)$ are in the same subnetwork.

If $t' > t$, then we have the following path

$$
\begin{aligned}
(a,t) &\rightsquigarrow \left(a \oplus \sum_{k=t}^{t'-1} \left(2^{(k+1)\frac{n}{T}-1} \oplus 2^{(k+1)\frac{n}{T}}\right), t'\right) \\
&\rightsquigarrow \left(a \oplus \sum_{k=t}^{t'-1} \left(2^{(k+1)\frac{n}{T}-1} \oplus 2^{(k+1)\frac{n}{T}}\right) \oplus 2^i \oplus 2^j, t'\right) \\
&= \left((a \oplus 2^i \oplus 2^j) \oplus \sum_{k=t}^{t'-1} \left(2^{(k+1)\frac{n}{T}-1} \oplus 2^{(k+1)\frac{n}{T}}\right), t'\right) \\
&\rightsquigarrow (a \oplus 2^i \oplus 2^j, t).
\end{aligned}
$$

So the lemma is also true when $t' > t$.

If $t' < t$, consider the following path

$$
\begin{aligned}
(a,t) &\rightsquigarrow \left(a \oplus \sum_{k=t'}^{t-1} \left(2^{(k+1)\frac{n}{T}-1} \oplus 2^{(k+1)\frac{n}{T}}\right), t'\right) \\
&\rightsquigarrow \left(a \oplus \sum_{k=t'}^{t-1} \left(2^{(k+1)\frac{n}{T}-1} \oplus 2^{(k+1)\frac{n}{T}}\right) \oplus 2^i \oplus 2^j, t'\right) \\
&= \left((a \oplus 2^i \oplus 2^j) \oplus \sum_{k=t'}^{t-1} \left(2^{(k+1)\frac{n}{T}-1} \oplus 2^{(k+1)\frac{n}{T}}\right), t'\right) \\
&\rightsquigarrow (a \oplus 2^i \oplus 2^j, t).
\end{aligned}
$$

Figure 4.6 illustrates the above path. Therefore the lemma is true in any case.

(2). The proof is similar to (1). ■

The next lemma says that the reverse of the above lemma is also true, i.e., the sufficient condition is also necessary.

LEMMA 4.19 *Suppose that* $0 \leq k < \frac{n-1}{m}$.

1. *For any* $kT' \leq t < (k+1)T'$, *if the two transmitters* (a,t) *and* (a',t) *are in the subnetwork, then* $a \cong_k a'$.

2. *For any* $kR' \leq r < (k+1)R'$, *if the two receivers* (b,r) *and* (b',r) *are in the same subnetwork, then* $b \cong_k b'$.

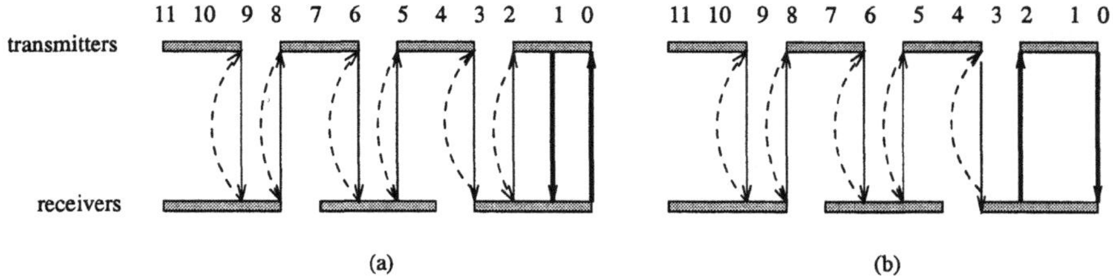

Figure 4.6. The two transmitters (a, t) and $(a \oplus 2^i \oplus 2^j, t)$ are in the same subnetwork if i and j are within some $(\lfloor \frac{t}{T'} \rfloor, \ell)$-section of a. In this example, $n = 12, T = 4, R = 3$ and $t = 3$. The thick solid lines represent the position of i and j. The thin solid lines represent the first half of the path, and the dash lines represent the second half of the path.

Proof. We consider three cases.

case 1. $T = R$. In this case $T' = R' = 1$ and each segment has only one section. So for (1) what we need to prove is that a and a' are only different at the t-segments and their t-segments have the same parity. Therefore, $a \cong_t a'$. This can be easily proved by induction on the length of the path between the two transmitters (a, t) and (a', t). (2) can follows from the similar argument.

Case 2. $T < R$. For convenience of the description, We first introduce some terminologies. We treat the local index of a transmitter as its *level.* Then if $T < R$, two transmitters which share a common receiver must be either at the same level or at two consecutive levels. Let (a, t) and (a', t') be such two transmitters, and p be the path from (a, t) to (a', t'). If $t = t'$, we call p be a *wandering* at level t, otherwise we call p be a *jump* between level t and level t'. In particular, if $t' = t + 1$, we call p a *jump-up* from level t to level $t + 1$. If $t' = t - 1$, we call p a *jump-down* from level t to level $t - 1$. Figure 4.7 illustrates these concepts.

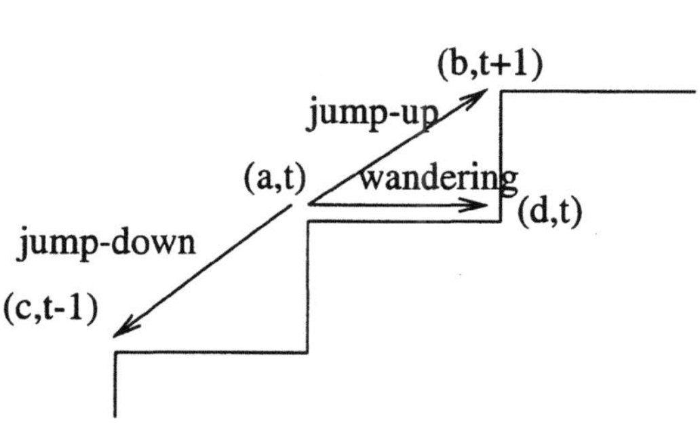

Figure 4.7. Wandering, jump-up and jump-down.

The section which ends with the position $(t+1)\frac{n}{T}-1$ and the section which begins with the position $(t+1)\frac{n}{T}$ in the $\lfloor\frac{t}{T'}\rfloor$-th segment are called the two *jump sections* between level t and level $t+1$. Figure 4.8 indicates all jump sections in the example in Figure 4.5.

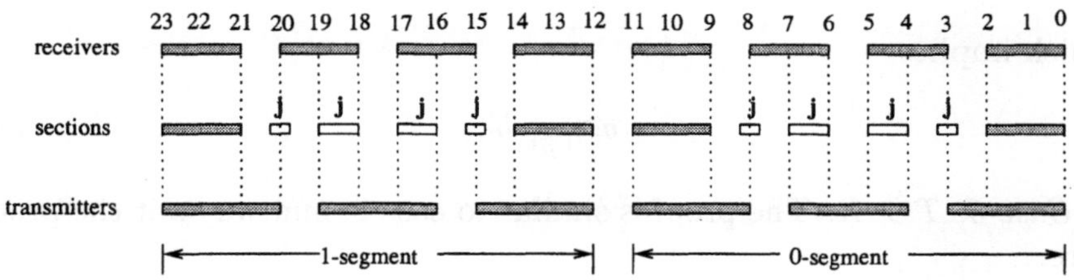

Figure 4.8. The jump sections, indicated by j, in the example in Figure 4.5.

If p is a wandering, then

$$a \cong_{\lfloor\frac{t}{T'}\rfloor} a'.$$

If p is a jump, then the two jump sections between level t and level $t+1$ of a and a' have odd Hamming distance respectively, and all other sections are equal respectively.

Now we give the proof for (1). Let p be any path from (a,t) to (a',t). Within the path p, there possibly exist jumps. Only the jump sections of a and a' could possibly have odd Hamming distance. However, notice that whenever there is a jump-up from some level t' to level $t'+1$, there are must be a jump-down from the level $t'+1$ to level t', and vice versa. Therefore, for any t', the number of jump-up's from level t' to level $t'+1$ is equal to the number of jump-down's from level $t'+1$ to level t'. So, after even number of jumps between level t' and level $t'+1$, the jump sections between level t' and level $t'+1$ also have even Hamming distance. Therefore,

$$a \cong_{\lfloor\frac{t}{T'}\rfloor} a'.$$

Next we give the proof for (2). The following two links

$$\left(b \oplus 2^{r\frac{n}{R}}, \left\lfloor \frac{r\frac{n}{R}}{\frac{n}{T}} \right\rfloor\right) \to (b,r),$$
$$\left(b' \oplus 2^{r\frac{n}{R}}, \left\lfloor \frac{r\frac{n}{R}}{\frac{n}{T}} \right\rfloor\right) \to (b',r),$$

imply that the two transmitters $\left(b \oplus 2^{r\frac{n}{R}}, \left\lfloor \frac{r\frac{n}{R}}{\frac{n}{T}} \right\rfloor\right)$ and $\left(b' \oplus 2^{r\frac{n}{R}}, \left\lfloor \frac{r\frac{n}{R}}{\frac{n}{T}} \right\rfloor\right)$ are in the same subnetwork. Therefore, from (1),

$$b \oplus 2^{r\frac{n}{R}} \cong_{\lfloor \frac{r}{R'} \rfloor} b' \oplus 2^{r\frac{n}{R}},$$

which implies

$$b \cong_{\lfloor \frac{r}{R'} \rfloor} b'.$$

Case 3. $T > R$. The proof is similar to case 2, and we omit the proof here.

Therefore in either case the lemma is true. ■

Now we are ready to present the main result in the this section.

THEOREM 4.20 *Suppose that* $0 \le k < \frac{n}{m}$.

1. *For any* $kT' \le t < (k+1)T'$, *the two transmitters* (a, t) *and* (a', t) *are in the same subnetwork if and only if* $a \cong_k a'$.

2. *For any* $kR' \le r < (k+1)R'$, *the two receivers* (b, r) *and* (b', r) *are in the same subnetwork if and only if* $b \cong_k b'$.

Therefore,

$$L(T, R) = T' 2^{m-(T'+R')+1},$$
$$W(T, R) = \frac{n}{m} 2^{n-m+T'+R'-1}.$$

Proof. The necessary and sufficient conditions stated in the theorem follows from Lemma 4.18 and Lemma 4.19. From Lemma 4.17, in any subnetwork the number of transmitters with the same local indices is $2^{m-(T'+R')+1}$. So the total number of transmitters in any subnetwork is

$$L(T, R) = T' 2^{m-(T'+R')+1}.$$

Thus

$$W(T, R) = \frac{T 2^n}{L(T, R)} = \frac{n}{m} 2^{n-m+T'+R'-1}.$$

■

7.2 COST-PERFORMANCE RELATIONS

First of all, swapping the number of transmitters and the number of receivers does not change the number of subnetworks. However, it does change the number of transmitters in each subnetwork. In fact, $W(T, R) = W(R, T)$ and if $T \leq R$ then $L(T, R) \leq L(R, T)$.

Now let's look at the impact of increasing the number of transmitters and/or receivers. We first consider the configurations in which T and R are equal. When $T = R$, we have $T' = R' = 1$ and $m = \frac{n}{T}$. Thus

$$L(T, T) = 2^{\frac{n}{T}-1},$$
$$W(T, T) = T2^{n-\frac{n}{T}+1};$$

and

$$L(kT, kT) = 2^{\frac{n}{kT}-1} = \frac{1}{2^{(1-\frac{1}{k})\frac{n}{T}}} L(T, T),$$
$$W(kT, kT) = kT2^{n-\frac{n}{kT}+1} = k2^{(1-\frac{1}{k})\frac{n}{T}} W(T, T).$$

If each access node has the same number of transmitters and receivers, increasing the number of transmitters and receivers simultaneously by the same amount at each node exponentially decreases the number of transmitters in each subnetwork but exponentially increases the number of subnetworks. As an example, Table 4.1 lists the values of $L(T, R)$ and $W(T, R)$ when $n = 8$. Such network contains $2^8 = 256$ access nodes. If $T = R = 1$, then $L(1, 1) = 128$ and $W(1, 1) = 2$. If we double the number of transmitters and receivers, then $L(2, 2) = 8$ and $W(2, 2) = 64$. If we double the transmitters and receivers once more, then $L(4, 4) = 2$ and $W(4, 4) = 512$.

$T \backslash R$	1	2	4
1	128	64	16
2	128	8	4
4	64	8	2

(a). $L(T, R)$

$T \backslash R$	1	2	4
1	2	4	16
2	4	64	128
4	16	128	512

(b). $W(T, R)$

Table 4.1. The values of $L(T, R)$ and $W(T, R)$ when $n = 8$.

When $R = kT$, we have $T' = 1$, $R' = k$ and $m = \frac{n}{T}$. Thus

$$L(T, kT) = 2^{\frac{n}{T}-k} = \frac{1}{2^{k-1}} L(T, T),$$
$$W(T, kT) = T2^{n-\frac{n}{T}+k} = 2^{k-1} W(T, T).$$

If the number of receivers is a multiple of the number of transmitters, then increasing the number of receivers alone while fixing the number of transmitters also exponentially decreases the number of transmitters in each subnetwork but exponentially increases the number of subnetworks. Again let's consider a network of size $2^8 = 256$. Suppose each access node has only one transmitter. If each access node has only one receiver, then $L(1,1) = 128$ and $W(1,1) = 2$. If we double the receivers at each access node, then $L(1,2) = 64$ and $W(1,2) = 4$. If we double the transmitters once more, then $L(1,4) = 16$ and $W(1,4) = 16$.

When $T = kR$, we have $T' = k$, $R' = 1$ and $m = \frac{n}{R}$. Thus

$$L(kR,R) = k2^{\frac{n}{R}-k} = \frac{k}{2^{k-1}}L(R,R) = kL(R,kR),$$
$$W(kR,R) = R2^{n-\frac{n}{R}+k} = 2^{k-1}W(R,R) = W(R,kR).$$

Notice that $L(2R,R) = L(R,R)$ but $W(2R,R) = 2W(R,R)$. This means if the each access node has the same number of transmitters, then doubling the number of transmitters alone does not change the number of transmitters in each subnetwork although the number of subnetworks is doubled. However, if we continue increasing the number of transmitters as multiples of the number of receivers, then the number of transmitters in each subnetwork will be decreased while the number of subnetworks will be exponentially decreased.

In general, it's easy to verify that if $T > R$, then

$$\frac{L(T,R)}{L(R,R)} = T'2^{\left(\frac{n}{R}-T'\right)+(R'-2)\left(\frac{n}{R}-1\right)},$$
$$\frac{W(R,R)}{W(T,R)} = \frac{R}{T}\frac{L(T,R)}{L(R,R)} = R'2^{\left(\frac{n}{R}-T'\right)+(R'-2)\left(\frac{n}{R}-1\right)};$$

and if $T < R$, then

$$\frac{L(T,R)}{L(T,T)} = T'2^{\left(\frac{n}{T}-R'\right)+(T'-2)\left(\frac{n}{T}-1\right)},$$
$$\frac{W(R,R)}{W(T,R)} = \frac{R}{T}\frac{L(T,R)}{L(R,R)} = R'2^{\left(\frac{n}{T}-R'\right)+(T'-2)\left(\frac{n}{T}-1\right)}.$$

Since m is the least common multiple of $\frac{n}{T}$ and $\frac{n}{R}$, $m \le \frac{n}{T}\cdot\frac{n}{R}$. Therefore,

$$T' = \frac{m}{\frac{n}{T}} \le \frac{n}{R},$$
$$R' = \frac{m}{\frac{n}{R}} \le \frac{n}{T}$$

and both equality holds if and only if $m = \frac{n}{T} \cdot \frac{n}{R}$, or equivalently, $\frac{n}{T}$ and $\frac{n}{R}$ are relatively prime.

In the next, we consider the case that neither of T and R is a multiple of each other.

LEMMA 4.21 *Suppose $T > R$ and T is not a multiple of R. Then*

- $L(T, R) \geq 3L(R, R)$, *and the equality holds if and only if* $n = 2T = 3R$;
- $W(R, R) \geq 2W(T, R)$, *and the equality holds if and only if* $n = 2T = kR$ *for some odd* $k > 1$.

Proof. Suppose that $T > R$ and T is not a multiple of R. Then $T' > R' \geq 2$ and $\frac{n}{R} > 2$. Thus,

$$T'2^{\left(\frac{n}{R}-T'\right)+(R'-2)\left(\frac{n}{R}-1\right)} \geq T' \geq 3$$

and the equation holds if and only if $T' = \frac{n}{R}$, $T' = 3$,and $R' = 2$, which is equivalent to $n = 2T = 3R$. Similarly

$$R'2^{\left(\frac{n}{R}-T'\right)+(R'-2)\left(\frac{n}{R}-1\right)} \geq R' \geq 2$$

and the equation holds if and only if $T' = \frac{n}{R}$ and $R' = 2$, which is equivalent to $n = 2T = kR$ for some odd $k > 1$. ∎

Similarly, we can prove the following lemma.

LEMMA 4.22 *Suppose $R > T$ and R is not a multiple of T. Then*

- $L(T, R) \geq 2L(R, R)$, *and the equality holds if and only if* $n = kT = 2R$ *for some odd* $k > 1$;
- $W(R, R) \geq 3W(T, R)$, *and the equality holds if and only if* $n = 3T = 2R$.

From the above two lemmas, if neither of T and R is a multiple of the other, then decreasing both T and R to $\min(T, R)$ would result in much less transmitters. This means that if there are sufficient large number of wavelength channels, the more number of transmitters or receivers could possibly leads to the degrade of the performance. As an example, Table 4.2 lists the values of $L(T, R)$ and $W(T, R)$ when $n = 6$. The size of this

$T\backslash R$	1	2	3
1	32	16	8
2	32	4	8
3	16	12	2

(a). $L(T, R)$

$T\backslash R$	1	2	3
1	2	4	8
2	4	32	16
3	8	16	96

(b). $W(T, R)$

Table 4.2. The values of $L(T, R)$ and $W(T, R)$ when $n = 6$.

network is $2^6 = 64$. If $T = R = 2$, then the performance is $L(2,2) = 4$. If we increase the transmitters by 1, then the performance becomes $L(3,2) = 12$. If we increase the receivers by 1, then the performance is $L(2,3) = 8$.

8. PONS BASED ON STAR GRAPH

We only consider the right star graph, but the left star graph can be treated similarly. In the right star graph, the vertices are permutations and the product of two permutations α and β, denoted by $\alpha\beta$, is traditionally given by

$$(\alpha\beta)\,(i) = \beta(\alpha(i))$$

The links at each node in a n-star are ordered as follows. For each $0 \le i \le n-2$, the i-th outgoing link at a node a is

$$a \to as_{i+2}$$

and the i-th incoming link at a node a is

$$as_{i+2} \to a,$$

where $s_k = (1, k)$, called a *transposition*. The permutation as_k can be interpreted as the one obtained from a by swapping the symbol 1 and the symbol k (not necessarily the symbols at position 0 and position k).

It's obvious the above link ordering is symmetric. In the transmission graph of the uniform consecutive partition based on such link ordering, the set of receivers a transmitter (a, t) connects to is

$$\left\{ \left(as_{i+2}, \left\lfloor \frac{i}{\frac{n-1}{R}} \right\rfloor \right) \mid t\frac{n-1}{T} \le i < (t+1)\frac{n-1}{T} \right\},$$

and the set of transmitters a receiver (b, r) connects from is

$$\left\{ \left(bs_{j+2}, \left\lfloor \frac{j}{\frac{n-1}{T}} \right\rfloor \right) \mid r\frac{n-1}{R} \leq j < (r+1)\frac{n-1}{R} \right\}.$$

8.1 SUBNETWORK STRUCTURE

Let

$$T' = \frac{m}{\frac{n-1}{T}} = \frac{T}{\frac{n-1}{m}},$$
$$R' = \frac{m}{\frac{n-1}{R}} = \frac{R}{\frac{n-1}{m}}$$

in which m is the least common multiple of $\frac{n-1}{T}$ and $\frac{n-1}{R}$. As the link ordering is symmetric, from Corollary 4.3, for any subnetwork, there exists a unique integer $0 \leq k < \frac{n-1}{m}$ such that in this subnetwork the set of local indices of all transmitters is

$$\{t \mid kT' \leq t < (k+1)T'\},$$

and the set of local indices of all receivers is

$$\{r \mid kR' \leq r < (k+1)R'\}.$$

Furthermore, from Lemma 4.2, for any $kT' \leq t' < t < (k+1)T'$, the two transmitters (a, t) and $\left(a \prod_{i=t'+1}^{t} s_{i\frac{n-1}{T}+2} s_{i\frac{n-1}{T}+1}, t'\right)$ are in the same subnetwork; for any $kR' \leq r' < r < (k+1)R'$, the two receivers (a, r) and $\left(a \prod_{i=r'+1}^{r} s_{i\frac{n-1}{R}+2} s_{i\frac{n-1}{R}+1}, r'\right)$ are in the same subnetwork.

In the next we will study the structure of the set of node indices of all transmitters (or receivers) which are in the same subnetwork and have the same local index. For convenience of description, we introduce some new terminologies. Let π be any permutation in P_n. A symbol i is called *invariant* under π if $\pi(i) = i$. A set S of symbols is called *closed* under π if for any symbol $i \in S$, $\pi(i) \in S$. Given a set S of symbols, π is called *local* to S if A is closed under π and other symbol not in S is invariant under π.

The node indices of all transmitters and receivers which are in the same subnetwork have the following obvious properties:

- the node indices of all transmitters have the same parity,
- the node indices of all receivers have the same parity,
- the node index of any transmitter and the node index of any receiver have different parity.

In addition, for any subnetwork there exists a unique integer $0 \le k < \frac{n}{m}$ such that the node indices of all transceivers in this subnetwork are local to the set of symbols

$$\{1\} \cup \{i \mid km + 2 \le i < (k+1)m + 2\}.$$

Before we further explore more properties of the node indices, we first give a decomposition of identity permutation e into a product of four transpositions.

LEMMA 4.23 *Let $0 \le i, j, k, \ell \le n - 2$ be any four different integers. Then*

$$e = s_i s_j s_k s_\ell s_i s_k s_j s_i s_\ell s_k.$$

The proof of this lemma is straightforward and we omit it here. The following lemma is based on such decomposition of identity.

LEMMA 4.24 *Let a be any permutation in P_n.*

1. *If $\left(t\frac{n-1}{T} + 1\right) \bmod \frac{n-1}{R} \ge 3$, then the two transmitters (a, t) and $(a, t-1)$ are in the same subnetwork.*
2. *If $\left(r\frac{n-1}{R} + 1\right) \bmod \frac{n-1}{T} \ge 3$, then the two receivers (a, r) and $(a, r-1)$ are in the same subnetwork.*

Proof. (1). The condition in the lemma implies that $\frac{n-1}{R} \ge 4$. Let

$$r = \left\lfloor \frac{t\frac{n-1}{T} + 1}{\frac{n-1}{R}} \right\rfloor.$$

Then from the condition in the lemma,

$$\left\lfloor \frac{t\frac{n-1}{T} - 2}{\frac{n-1}{R}} \right\rfloor = \left\lfloor \frac{t\frac{n-1}{T} - 1}{\frac{n-1}{R}} \right\rfloor = \left\lfloor \frac{t\frac{n-1}{T}}{\frac{n-1}{R}} \right\rfloor = r.$$

So

$$r = \left\lfloor \frac{tR}{T} \right\rfloor .$$

Let

$$\begin{aligned} i &= t\frac{n-1}{T} + 3, \\ j &= t\frac{n-1}{t} + 1, \\ k &= t\frac{n-1}{t}, \\ \ell &= t\frac{n-1}{T} + 2. \end{aligned}$$

So we have the following path.

$$\begin{array}{rl} (a,t) & \\ & \searrow (as_i, r) \\ (as_is_j, t-1) \nearrow & \\ & \searrow (as_is_js_k, r) \\ (as_is_js_ks_\ell, t) \nearrow & \\ & \searrow (as_is_js_ks_\ell s_i, r) \\ (as_is_js_ks_\ell s_is_k, t-1) \nearrow & \\ & \searrow (as_is_js_ks_\ell s_is_ks_j, r) \\ (as_is_js_ks_\ell s_is_ks_js_i, t) \nearrow & \\ & \searrow (as_is_js_ks_\ell s_is_ks_js_is_\ell, r) \\ (as_is_js_ks_\ell s_is_ks_js_is_\ell s_k, t-1) \nearrow & \end{array}$$

By Lemma 4.23,

$$(as_is_js_ks_\ell s_is_ks_js_is_\ell s_k, t-1) = (a, t-1).$$

Therefore, the two transmitters (a,t) and $(a,t-1)$ are in the same subnetwork.

(2). The proof is similar to (1). ■

8.1.1 SUBNETWORK STRUCTURE WHEN GCD$\left(\frac{N-1}{T}, \frac{N-1}{R}\right) > 1$

When $\gcd\left(\frac{n-1}{T}, \frac{n-1}{R}\right) > 1$, the conditions in Lemma 4.24 are met according to the next lemma.

LEMMA 4.25 *Suppose that* $\gcd\left(\frac{n-1}{T}, \frac{n-1}{R}\right) > 1$.

1. *If* $t \bmod T' > 0$, *then* $\left(t\frac{n-1}{T} + 1\right) \bmod \frac{n-1}{R} \geq 3$.

2. *If* $r \bmod R' > 0$, *then* $\left(r\frac{n-1}{R} + 1\right) \bmod \frac{n-1}{T} \geq 3$.

Proof. (1). The condition in this lemma implies that $T' > 1$. We first show that $T' > 1$ implies that $\frac{n-1}{R} \geq 4$. Suppose on the contrary that $\frac{n-1}{R} \leq 3$, i.e., $\frac{n-1}{R} = 2$ or 3. Then $\frac{n-1}{T}$ is a multiple of $\frac{n-1}{R}$ since $\gcd\left(\frac{n-1}{T}, \frac{n-1}{R}\right) > 1$. So $m = \frac{n-1}{T}$ and $T' = 1$ which is a contradiction. Therefore $\frac{n-1}{R} \geq 4$.

Let

$$p = \left(t\frac{n-1}{T} + 1\right) \bmod \frac{n-1}{R}.$$

Then

$$t\frac{n-1}{T} + 1 = q\frac{n-1}{R} + p$$

for some integer q. This implies that $\gcd\left(\frac{n-1}{T}, \frac{n-1}{R}\right)$ is a factor of $p-1$. Since $\gcd\left(\frac{n-1}{T}, \frac{n-1}{R}\right) > 1$, p can be neither 0 nor 2. Furthermore, $p \neq 1$ because $t \bmod T' > 0$. Therefore $p \geq 3$.

(2). The proof is similar to (1). ∎

From Lemma 4.24 and Lemma 4.25, we can immediately have the following corollary.

COROLLARY 4.26 *Suppose that* $\gcd\left(\frac{n-1}{T}, \frac{n-1}{R}\right) > 1$. *Let a be any permutation and* $0 \leq k < \frac{n-1}{m}$ *be any integer. Then*

1. *the* T' *transmitters* $\{(a,t) \mid kT' \leq t < (k+1)T'\}$ *are in the same subnetwork.*

2. *the* R' *transmitters* $\{(a,r) \mid kR' \leq r < (k+1)R'\}$ *are in the same subnetwork.*

Now we can completely determine the structure of the set of node indices of all transmitters/receivers which are in the same subnetwork and have the same local indices. We use Π_k to denote the all even permutations that are local to the set of symbols

$$\{1\} \cup \{i \mid km + 2 \leq i < (k+1)m + 2\}.$$

LEMMA 4.27 *Suppose that* $\gcd\left(\frac{n-1}{T}, \frac{n-1}{R}\right) > 1$ *and* $0 \le k < \frac{n-1}{m}$.

1. *If* $kT' \le t < (k+1)T'$, *two transmitters* (a,t) *and* (a',t) *are in the same subnetwork if and only if* $a' = a\pi$ *for some* $\pi \in \Pi_k$.

2. *If* $kR' \le r < (k+1)R'$, *two receivers* (b,r) *and* (b',r) *are in the same subnetwork if and only if* $b' = b\pi$ *for some* $\pi \in \Pi_k$.

Proof. (1). The necessary part is obvious. So we only prove the sufficient part. We prove it by induction on the distance between a and a', denoted by $d(a,a')$. Suppose that $d(a,a') = 2$, i.e., $a' = as_is_j$ for some two integers $km + 2 \le i, j < (k+1)m + 2$. Let

$$t' = \left\lfloor \frac{i}{\frac{n-1}{T}} \right\rfloor, t'' = \left\lfloor \frac{j}{\frac{n-1}{T}} \right\rfloor,$$

$$r' = \left\lfloor \frac{i}{\frac{n-1}{R}} \right\rfloor, r'' = \left\lfloor \frac{j}{\frac{n-1}{T}} \right\rfloor.$$

From Corollary 4.26, we have the following path:

$$\begin{array}{l}
(a,t) \\
\rightsquigarrow (a,t') \\
\searrow \\
\qquad (as_i, r') \rightsquigarrow \\
\qquad (as_i, r'') \\
\nearrow \\
(as_is_j, t'') \\
\rightsquigarrow (as_is_j, t).
\end{array}$$

Therefore, the two transmitters (a,t) and (as_is_j, t) must be in the same subnetwork.

Now we assume that $d(a,a') = 2\ell$ for some $\ell > 1$. Suppose that

$$a' = as_{i_1}s_{i_2}\cdots s_{i_{2\ell}}$$

where

$$km + 2 \le i_1, i_2, \cdots, i_{2\ell} < (k+1)m + 2.$$

Let $a'' = as_{i_1}s_{i_2}$. Then $d(a,a'') = 2$ and $d(a',a'') = 2(\ell - 1)$. So by the induction hypothesis, the two transmitters (a,t) and (a',t) are in the same subnetwork, so are the two transmitters (a', t) and (a'', t). Thus the two transmitters (a, t) and (a'', t) are in the same subnetwork. So by induction the lemma is true.

(2). The proof is similar to (1). ■

Now we are able to determine the structure of the subnetworks.

THEOREM 4.28 *Suppose that* $\gcd(\frac{n-1}{T}, \frac{n-1}{R}) > 1$ *and* $0 \leq k < \frac{n-1}{m}$. *Let* a *be any permutation in* P_n. *The set of transmitters*

$$\{(a\pi, t) \mid \pi \in \Pi_k, kT' \leq t < (k+1)T'\}$$

and the set of receivers

$$\{(as_{km+2}\pi, r) \mid \pi \in \Pi_k, kR' \leq r < (k+1)R'\}$$

constitute a subnetwork. Therefore,

$$L(T, R) = \frac{T'(m+1)!}{2},$$
$$W(T, R) = \frac{n-1}{m} \frac{2n!}{(m+1)!}.$$

Proof. The first part of the theorem follows from Corollary 4.26 and Lemma 4.27. As

$$|\Pi_k| = (m+1)!/2,$$

the total number of transmitters in each subnetwork is

$$L(T, R) = \frac{T'(m+1)!}{2}$$

and the total number of receivers in each subnetwork is $\frac{R'(m+1)!}{2}$. Thus,

$$W(T, R) = \frac{Tn!}{L(T, R)} = \frac{n-1}{m} \frac{2n!}{(m+1)!}.$$

■

8.1.2 SUBNETWORK STRUCTURE WHEN $\mathbf{GCD}\left(\frac{N-1}{T}, \frac{N-1}{R}\right) = 1$

When $\gcd\left(\frac{n-1}{T}, \frac{n-1}{R}\right) = 1$, we have

$$\begin{aligned} T' &= \frac{n-1}{R}, \\ R' &= \frac{n-1}{T}, \\ m &= \frac{n-1}{T}\frac{n-1}{R} = T'R'. \end{aligned}$$

Let x and y be the unique pair of integers satisfying that

$$xR' - yT' = 1, 0 < x < T', 0 < y < R'.$$

Then $(R' - y)T' - (T' - x)R' = 1$. Let

$$\begin{aligned} \ell &= \min(xR', (R' - y)T'), \\ u &= \min(x, T' - x), \\ v &= \min(y, R' - y). \end{aligned}$$

LEMMA 4.29 *Suppose that* $\gcd\left(\frac{n-1}{T}, \frac{n-1}{R}\right) = 1$. *Then*

1. $m \geq 2\ell$ *and the equality holds if and only if either* $T' = 2$ *or* $R' = 2$.
2. $T' \geq 2u$ *and the equality holds if and only if* $T' = 2$.
3. $R' \geq 2v$ *and the equality holds if and only if* $R' = 2$.

Proof. (1). First we show that $xR' \neq (R' - y)T'$. Suppose to the contrary that $xR' = (R' - y)T'$. Then xR' would be a common multiple of T' and R'. Since $x < T'$, this contradicts to that $m = T'R'$ is the least common multiple of T' and R'. Therefore $xR' \neq (R' - y)T'$. This implies that

$$xR' + (R' - y)T' \geq 2\ell + 1.$$

On the other hand,

$$xR' + (R' - y)T' = R'T' + xR' - yT' = m + 1.$$

So we have $m + 1 \geq 2\ell + 1$ or $m \geq 2\ell$.

In the next we show that $m = 2\ell$ if and only if either $T' = 2$ or $R' = 2$. The sufficiency of the condition is easy to verify. So we only need to prove that the condition is also necessary. We consider two cases.

Case 1. $\ell = xR'$. Then $m = T'R' = 2xR'$.So $T' = 2x$. Since $xR' - yT' = 1$ we have $xR' - 2xy = 1$ which implies that $x = 1$. Therefore $T' = 2$.

Case 2. $\ell = (R' - y)T'$. Then $m = T'R' = 2(R' - y)T'$. So $R' = 2(R' - y)$ or $R' = 2y$. Since $xR' - yT' = 1$ we have $2xy - yT' = 1$ which implies that $y = 1$. Therefore $R' = 2$.

Thus $m \geq 2\ell$ and the equality holds if and only if either $T' = 2$ or $R' = 2$.

(2). $T' \geq 2u$ is obvious. The equality holds if and only if $x = T' - x$. Since

$$xR' - yT' = 1,$$
$$(R' - y)T' - (T' - x)R' = 1,$$

by adding the two equations we get $(R' - 2y)T' = 2$. Since $T' > 1$, we must have $R' - 2y = 1$ and $T' = 2$.

(3). The proof is similar to (2). ∎

The following lemma is the analogy to Lemma 4.25.

LEMMA 4.30 *Suppose that* $\gcd\left(\frac{n-1}{T}, \frac{n-1}{R}\right) = 1$.

1. *if* $T' \geq 4$ *and* $t \bmod T' \neq 0, u, T' - u$, *then* $\left(t\frac{n-1}{T} + 1\right) \bmod \frac{n-1}{R} \geq 3$;
2. *if* $R' \geq 4$ *and* $r \bmod R' \neq 0, v, R' - v$, *then* $\left(r\frac{n-1}{R} + 1\right) \bmod \frac{n-1}{T} \geq 3$.

Proof. (1) Since $T' = \frac{n-1}{R}$ and $T' \geq 4$, $\frac{n-1}{R} \geq 4$. Let

$$p = \left(t\frac{n-1}{T} + 1\right) \bmod \frac{n-1}{R}.$$

Then

$$t\frac{n-1}{T} + 1 = q\frac{n-1}{R} + p$$

for some integer q.

If $p = 0$, then $q\frac{n-1}{R} - t\frac{n-1}{T} = 1$, which implies that either $t \bmod T' = u$ or $t \bmod T' = T' - u$. This contradicts to the condition in the lemma.

If $p = 1$, then $t\frac{n-1}{T} = q\frac{n-1}{R}$ which implies $t \bmod T' = 0$. This also contradicts to the condition in the lemma.

If $p = 2$, then $t\frac{n-1}{T} - q\frac{n-1}{R} = 1$, which implies that either $t \bmod T' = u$ or $t \bmod T' = T' - u$. This contradicts to the condition in the lemma too.

Therefore, $p \geq 3$.

(2). The proof is similar to (1). ■

The following corollary is the analogy of Corollary 4.26.

COROLLARY 4.31 *Suppose that* $\gcd\left(\frac{n-1}{T}, \frac{n-1}{R}\right) = 1$ *and* $0 \leq k < \frac{n-1}{m}$.

1. *For any permutation* $a \in P_n$, *the set of transmitters in each of the following three sets are in the same subnetwork.*

$$\{(a,t) \mid kT' \leq t \leq kT' + u - 1\},$$
$$\{(a,t) \mid kT' + u \leq t \leq (k+1)T' - u - 1\},$$
$$\{(a,t) \mid (k+1)T' - u \leq t \leq (k+1)T' - 1\}.$$

2. *For any permutation* $b \in P_n$, *the set of receivers in each of the following three sets are in the same subnetwork.*

$$\{(b,r) \mid kR' \leq r \leq kR' + v - 1\},$$
$$\{(b,r) \mid kR' + v \leq r \leq (k+1)R' - v - 1\},$$
$$\{(b,r) \mid (k+1)R' - v \leq r \leq (k+1)R' - 1\}.$$

Proof. (1). If $T' < 4$, then $u = 1$ by Lemma 4.29. So each of the three sets in the lemma either contains only one transmitter or is empty. Therefore, the lemma is trivially true. If $T' \geq 4$, the lemma can directly follow from Lemma 4.24 and Lemma 4.30.

(2) The proof is similar. ■

For any $0 \leq k < \frac{n-1}{m}$, the two symbols $km + \ell + 1$ and $(k+1)m - \ell + 2$ separate the m symbols

$$\{i \mid km + 2 \leq i \leq (k+1)m + 1\}$$

into three subsets

$$S_{k,1} = \{i \mid km + 2 \leq i \leq km + \ell\},$$
$$S_{k,2} = \{i \mid km + \ell + 2 \leq i \leq (k+1)m - \ell + 1\},$$
$$S_{k,3} = \{i \mid (k+1)m - \ell + 3 \leq i \leq (k+1)m + 1\}.$$

By the similar argument as in Lemma 4.27, we can show that

LEMMA 4.32 *Suppose that* $\gcd\left(\frac{n-1}{T}, \frac{n-1}{R}\right) = 1$. *Let* a *and* π *be any two permutations in* P_n. *Then for any* $kT' \le t \le (k+1)T' - 1$, *the two transmitters* (a, t) *and* $(a\pi, t)$ *are in the same subnetwork if* π *is an even permutation and one of the following condition is satisfied:*

1. $kT' \le t \le kT' + u - 1$ *and* π *is local to* $\{1\} \cup S_{k,1}$;
2. $kT' + u \le t \le (k+1)T' - u - 1$ *and* π *is local to* $\{1\} \cup S_{k,2}$;
3. $(k+1)T' - u \le t \le (k+1)T' - 1$ *and* π *is local to* $\{1\} \cup S_{k,3}$.

Similarly, for any $kR' \le r \le (k+1)R' - 1$, *the two receivers* (a, r) *and* $(a\pi, r)$ *are in the same subnetwork if* π *is an even permutation and one of the following condition is satisfied:*

1. $kR' \le r \le (k+1)R' - 1$ *and* π *is local to* $\{1\} \cup S_{k,1}$;
2. $kR' + v \le r \le (k+1)R' - v - 1$ *and* π *is local to* $\{1\} \cup S_{k,2}$;
3. $(k+1)R' - v \le r \le (k+1)R' - 1$ *and* π *is local to* $\{1\} \cup S_{k,3}$.

For any $0 \le k < \frac{n-1}{m}$, let

$$\begin{aligned}
\Pi_{k,1} = \{(\pi_1, \pi_2, \pi_3) \mid & \ \pi_1, \pi_2, \pi_3 \text{ are even and local to } \{1\} \cup S_{k,1}, \\
& \{km + \ell + 1\} \cup S_{k,2}, \{(k+1)m - \ell + 2\} \cup S_{k,3} \text{ resepctively}\} \\
\Pi_{k,2} = \{(\pi_1, \pi_2, \pi_3) \mid & \ \pi_1, \pi_2, \pi_3 \text{ are even and local to } \{km + \ell + 1\} \cup S_{k,1}, \\
& \{1\} \cup S_{k,2}, \{(k+1)m - \ell + 2\} \cup S_{k,3} \text{ resepctively}\} \\
\Pi_{k,3} = \{(\pi_1, \pi_2, \pi_3) \mid & \ \pi_1, \pi_2, \pi_3 \text{ are even and local to } \{km + \ell + 1\} \cup S_{k,1}, \\
& \{(k+1)m - \ell + 2\} \cup S_{k,2}, \{1\} \cup S_{k,3} \text{ resepctively}\}
\end{aligned}$$

The next lemma gives a sufficient condition for two transmitters/receivers which have the same local index to be in the same subnetwork.

LEMMA 4.33 *Suppose that* $\gcd\left(\frac{n-1}{T}, \frac{n-1}{R}\right) = 1$. *Let* a, π_1, π_2 *and* π_3 *be any permutations in* P_n. *Then for any* $kT' \le t \le (k+1)T' - 1$, *the two transmitters* (a, t) *and* $(a\pi_1\pi_2\pi_3, t)$ *are in the same subnetwork if* π_1, π_2, π_3 *are even permutations and one of the following condition is satisfied:*

1. $kT' \le t \le kT' + u - 1$ *and* $(\pi_1, \pi_2, \pi_3) \in \Pi_{k,1}$;

2. $kT' + u \leq t \leq (k+1)T' - u - 1$ *and* $(\pi_1, \pi_2, \pi_3) \in \Pi_{k,2}$*;*

3. $(k+1)T' - u \leq t \leq (k+1)T' - 1$ *and* $(\pi_1, \pi_2, \pi_3) \in \Pi_{k,3}$*.*

Similarly, for any $kR' \leq r \leq (k+1)R' - 1$*, the two receivers* (a, r) *and* $(a\pi_1\pi_2\pi_3, r)$ *are in the same subnetwork if* π_1, π_2, π_3 *are even permutations and one of the following condition is satisfied:*

1. $kR' \leq r \leq (k+1)R' - 1$ *and* $(\pi_1, \pi_2, \pi_3) \in \Pi_{k,1}$*;*

2. $kR' + v \leq r \leq (k+1)R' - v - 1$ *and* $(\pi_1, \pi_2, \pi_3) \in \Pi_{k,2}$*;*

3. $(k+1)R' - v \leq r \leq (k+1)R' - 1$ *and* $(\pi_1, \pi_2, \pi_3) \in \Pi_{k,3}$*.*

Proof. We only consider the transmitter part of the lemma. The receiver part follows from the same argument. In addition, we only prove the sufficiency of the first condition. The sufficiency of the other two conditions can be proved in the similar manner. To prove the sufficiency of the first condition, we only need to prove that

a. for any $km + 2 \leq i < j \leq km + \ell$, the two transmitters (a, t) and $(a(1, i, j), t)$ are in the same subnetwork;

b. for any $km + \ell + 2 \leq i < j \leq (k+1)m - \ell + 1$, the two transmitters (a, t) and $(a(km + \ell + 1, i, j), t)$ are in the same subnetwork;

c. for any $(k+1)m - \ell + 3 \leq i < j \leq (k+1)m + 1$, the two transmitters (a, t) and $(a((k+1)m - \ell + 2, i, j), t)$ are in the same subnetwork;

(a) is true from Lemma 4.32. Now we prove (b). We consider two different cases.

Case 1. $\ell \bmod R' = 0$. In this case, we have the following path.

$$
\begin{array}{l}
(a, t) \rightsquigarrow \\
(a, kT' + u - 1) \\
\searrow \\
\qquad (as_{km+\ell+1}, kR' + v) \rightsquigarrow \\
\qquad (as_{km+\ell+1}s_is_j, kR' + v) \\
\nearrow \\
(as_{km+\ell+1}s_is_js_{km+\ell+1}, kT' + u - 1) = \\
(a(km + \ell + 1, i, j), kT' + u - 1) \rightsquigarrow \\
(a(km + \ell + 1, i, j), t).
\end{array}
$$

Case 2. $\ell \bmod T' = 0$. In this case, we have the following path.

$$
\begin{aligned}
&(a,t) \leadsto \\
&(a, kT' + u - 1) \\
&\searrow \\
&\qquad (as_{km+\ell}, kR' + v - 1) \\
&\nearrow \\
&(as_{km+\ell}s_{km+\ell+1}, kT' + u) \leadsto \\
&(as_{km+\ell}s_{km+\ell+1}s_i s_j, kT' + u) \\
&\searrow \\
&\qquad (as_{km+\ell}s_{km+\ell+1}s_i s_j s_{km+\ell+1}, kR' + v - 1) = \\
&\qquad (as_{km+\ell}(km + \ell + 1, i, j), kR' + v - 1) \\
&\nearrow \\
&(as_{km+\ell}(km + \ell + 2, i, j)s_{km+\ell}, kT' + u - 1) = \\
&(a(km + \ell + 1, i, j), kT' + u - 1) \leadsto \\
&(a(km + \ell + 1, i, j), t).
\end{aligned}
$$

Therefore, (b) is true.

Now we prove (c). We also consider two different cases.

Case 1. $\ell \bmod R' = 0$. In this case, we have the following path.

$$
\begin{aligned}
&(a,t) \leadsto \\
&(a, kT' + u - 1) \\
&\searrow \\
&\qquad (as_{km+\ell+1}, kR' + v) \leadsto \\
&\qquad (as_{km+\ell+1}, (k+1)R' - v - 1) \\
&\nearrow \\
&(as_{km+\ell+1}s_{(k+1)m-\ell+2}, (k+1)T' - u) \leadsto \\
&(as_{km+\ell+1}s_{(k+1)m-\ell+2}s_i s_j, (k+1)T' - u) \\
&\searrow \\
&\qquad (as_{km+\ell+1}s_{(k+1)m-\ell+2}s_i s_j s_{(k+1)m-\ell+2}, (k+1)R' - v + 1) = \\
&\qquad (as_{km+\ell+1}((k+1)m - \ell + 2, i, j), (k+1)R' - v + 1) \leadsto \\
&\qquad (as_{km+\ell+1}(k+1)m - \ell + 2, i, j), kR' + v) \\
&\nearrow \\
&(as_{km+\ell+1}((k+1)m - \ell + 2, i, j)s_{km+\ell+1}, kT' + u - 1) = \\
&(a((k+1)m - \ell + 2, i, j), kT' + u - 1) \leadsto \\
&(a((k+1)m - \ell + 2, i, j), t).
\end{aligned}
$$

Case 2. $\ell \bmod T' = 0$. In this case, we have the following path.

$$
\begin{array}{l}
(a,t) \rightsquigarrow \\
(a, kT' + u - 1) \\
\searrow \\
\qquad (as_{km+\ell}, kR' + v - 1) \\
\nearrow \\
(as_{km+\ell}s_{km+\ell+1}, kT' + u) \rightsquigarrow \\
(as_{km+\ell}s_{km+\ell+1}, (k+1)T' - u - 1) \\
\searrow \\
\qquad (as_{km+\ell}s_{km+\ell+1}s_{(k+1)m-\ell+2}, (k+1)R' - v) \rightsquigarrow \\
\qquad (as_{km+\ell}s_{km+\ell+1}s_{(k+1)m-\ell+2}s_is_j, (k+1)R' - v) \\
\nearrow \\
(as_{km+\ell}s_{km+\ell+1}s_{(k+1)m-\ell+2}s_is_js_{(k+1)m-\ell+2}, (k+1)T' - u - 1) = \\
(as_{km+\ell}s_{km+\ell+1}((k+1)m - \ell + 2, i, j), (k+1)T' - u - 1) \rightsquigarrow \\
(as_{km+\ell}s_{km+\ell+1}((k+1)m - \ell + 2, i, j), kT' + u) \\
\searrow \\
\qquad (as_{km+\ell}s_{km+\ell+1}((k+1)m - \ell + 2, i, j)s_{km+\ell+1}, kR' + v - 1) = \\
\qquad (as_{km+\ell}((k+1)m - \ell + 2, i, j), kR' + v - 1) \\
\nearrow \\
(as_{km+\ell}(k+1)m - \ell + 2, i, j)s_{km+\ell}, kT' + u - 1) = \\
(a((k+1)m - \ell + 2, i, j), kT' + u - 1) \rightsquigarrow \\
(a((k+1)m - \ell + 2, i, j), t).
\end{array}
$$

Therefore (c) is also true. This completes the proof. ∎

In the next we prove that the conditions in the above lemma are also necessary. We first begin with the following lemma.

LEMMA 4.34 *Suppose that* $\gcd\left(\frac{n-1}{T}, \frac{n-1}{R}\right) = 1$. *If the two transmitters* (a,t) *and* (a',t) *(or the two receivers* (a,r) *and* (a',r) *respectively) are in the same subnetwork, then* $a' = a\pi_1\pi_2\pi_3$ *for some even permutations* π_1, π_2 *and* π_3 *in* P_n *which satisfies one of the following conditions:*

1. $(\pi_1, \pi_2, \pi_3) \in \Pi_{k,1}$ *if* $kT' \le t \le kT' + u - 1$ *(or* $kR' \le r \le (k+1)R' - 1$ *respectively);*
2. $(\pi_1, \pi_2, \pi_3) \in \Pi_{k,2}$ *if* $kT' + u \le t \le (k+1)T' - u - 1$ *(or* $kR' + v \le r \le (k+1)R' - v - 1$ *respectively);*

3. $(\pi_1, \pi_2, \pi_3) \in \Pi_{k,3}$ *if* $(k+1)T'-u \leq t \leq (k+1)T'-1$ *(or* $(k+1)R'-v \leq r \leq (k+1)R'-1$ *respectively).*

Proof. Again we only prove the transmitter part and for the transmitter part we only prove the necessity of the first condition. The others can be proved similarly.

We first prove that for any $kT'+u \leq t, t' \leq (k+1)T'-u-1$, if the two transmitters (a,t) and (a',t') are connected by a path in which the local indices of all other transmitters are between (including) $(k+1)T'-u$ and $(k+1)T'-1$, then $a' = a\pi_2\pi_3$ for some even permutations $\pi_2$and π_3 which are local to $\{1\} \cup S_{k,2}$ and $\{(k+1)m-\ell+2\} \cup S_{k,3}$ irrespectively. It's easy to see that the local indices of the two transmitters which are closest to the two transmitters (a,t) and (a',t') in the path must be equal to $(k+1)T'-u$. Let $(b,(k+1)T'-u)$ be the closest transmitter to transmitter (a,t) in the path, and $(b',(k+1)T'-u)$ be the closest transmitter to transmitter (a',t') in the path. Then within the path between the two transmitters $(b,(k+1)T'-u)$ and $(b',(k+1)T'-u)$, no transpositions can be taken at the position $(k+1)m-\ell+2$. So $b' = b\alpha$ for some even permutation α which is local to $S_{k,3}$. We consider two cases.

Case 1. $\ell \bmod T' = 0$. In this case,

$$b = as_{(k+1)m-\ell+2}s_i$$

for some $(k+1)m-\ell+3 \leq i \leq (k+1)m+1$ and

$$a' = b's_js_{(k+1)m-\ell+2} = b\alpha s_js_{(k+1)m-\ell+2}$$

for some $(k+1)m-\ell+3 \leq j \leq (k+1)m+1$. Therefore,

$$a' = as_{(k+1)m-\ell+2}s_i\alpha s_js_{(k+1)m-\ell+2}.$$

Let

$$\pi_3 = s_{(k+1)m-\ell+2}s_i\alpha s_js_{(k+1)m-\ell+2}.$$

Then π_3 is even and $a' = a\pi_3$. Since $\pi_3(1) = 1$ and only the symbols 1 and $(k+1)m-\ell+2, (k+1)m-\ell+2, \cdots, (k+1)m+1$ are possibly transposed in π_3, π_3 is local to $\{(k+1)m-\ell+2\} \cup S_{k,3}$.

Case 2. $\ell \bmod R' = 0$. In this case,

$$b = as_is_{(k+1)m-\ell+2}$$

for some $km+\ell+2 \leq i \leq (k+1)m-\ell+1$ and

$$a' = b's_{(k+1)m-\ell+2}s_j = b\alpha s_{(k+1)m-\ell+2}s_j$$

for some $km+\ell+2 \leq j \leq (k+1)m-\ell+1$. Therefore,

$$a' = as_j s_{(k+1)m-\ell+2} \alpha s_{(k+1)m-\ell+2} s_i.$$

Let

$$\pi_3 = s_{(k+1)m-\ell+2} \alpha s_{(k+1)m-\ell+2}.$$

Then π_3 is even. It's easy to verify that $\pi_3(1) = 1$ and thus π_3 is local to $\{(k+1)m-\ell+2\} \cup S_{k,3}$. So

$$a' = as_j \pi_3 s_i = as_j s_i \pi_3.$$

Let $\pi_2 = s_j s_i$. Then π_2 is an even permutation local to $\{1\} \cup S_{k,2}$ and $a' = a\pi_2\pi_3$.

Based on the above result, we can easily prove by induction that for any $kT'+u \leq t, t' \leq (k+1)T'-u-1$, if the two transmitters (a,t) and (a',t') are connected by a path in which the local indices of all transmitters are between (including) $kT'+u$ and $(k+1)T'-1$, then $a' = a\pi_2\pi_3$ for some even permutations $\pi_2$and π_3 which are local to $\{1\} \cup S_{k,2}$ and $\{(k+1)m-\ell+2\} \cup S_{k,3}$ respectively.

Following the similar arguments, we can prove that for any $kT' \leq t, t' \leq kT'+u-1$, if the two transmitters (a,t) and (a',t') are connected by a path in which the local indices of all other transmitters are between (including) $kT'+u$ and $(k+1)T'-1$, then $a' = a\pi_1\pi_2\pi_3$ for some even permutations $\pi_1, \pi_2$and π_3 which are local to $\{1\}\cup S_{k,1}$, $\{km+\ell+1\}\cup S_{k,2}$ and $\{(k+1)m-\ell+2\} \cup S_{k,3}$ respectively.

Once again we can prove by induction that for any $kT' \leq t, t' \leq kT'+u-1$, if the two transmitters (a,t) and (a',t') are in the same subnetwork, then $a' = a\pi_1\pi_2\pi_3$ for some even permutations $\pi_1, \pi_2$and π_3 which are local to $\{1\}\cup S_{k,1}$, $\{km+\ell+1\}\cup S_{k,2}$ and $\{(k+1)m-\ell+2\}\cup S_{k,3}$ respectively. ■

Now we are able to determine the structure of the subnetworks from Corollary 4.31, Lemma 4.33 and Lemma 4.34. Suppose that $0 \leq k < \frac{n-1}{m}$. Let a be any permutation in P_n. If $\ell \bmod T' = 0$, then the set of transmitters

$$\begin{aligned}&\{(as_{km+\ell+1}s_{km+2}\pi_1\pi_2\pi_3, t) \mid (\pi_1,\pi_2,\pi_3) \in \Pi_{k,1}, kT' \leq t < kT'+u\} \cup\\ &\{(a\pi_1\pi_2\pi_3, t) \mid (\pi_1,\pi_2,\pi_3) \in \Pi_{k,2}, kT'+u \leq t < (k+1)T'-u\} \cup\\ &\{(as_{(k+1)m-\ell+2}s_{(k+1)m+1}\pi_1\pi_2\pi_3, t) \mid (\pi_1,\pi_2,\pi_3) \in \Pi_{k,3},\\ &(k+1)T'-u \leq t < (k+1)T'\}\end{aligned}$$

and the set of receivers

$$\{(as_{km+\ell+1}\pi_1\pi_2\pi_3, r) \mid (\pi_1,\pi_2,\pi_3) \in \Pi_{k,1}, kR' \le r < kR' + v\} \cup$$
$$\{(as_{km+\ell+2}\pi_1\pi_2\pi_3, r) \mid (\pi_1,\pi_2,\pi_3) \in \Pi_{k,2}, kR' + v \le r < (k+1)R' - v\} \cup$$
$$\{(as_{(k+1)m-\ell+2}\pi_1\pi_2\pi_3, r) \mid (\pi_1,\pi_2,\pi_3) \in \Pi_{k,3},$$
$$(k+1)R' - v \le r < (k+1)R'\}$$

constitute a subnetwork. If $\ell \bmod R' = 0$, then the set of transmitters

$$\{(as_{km+\ell+1}\pi_1\pi_2\pi_3, t) \mid (\pi_1,\pi_2,\pi_3) \in \Pi_{k,1}, kT' \le t < kT' + u\} \cup$$
$$\{(as_{km+\ell+2}\pi_1\pi_2\pi_3, t) \mid (\pi_1,\pi_2,\pi_3) \in \Pi_{k,2}, kT' + u \le t < (k+1)T' - u\} \cup$$
$$\{(as_{(k+1)m-\ell+2}\pi_1\pi_2\pi_3, t) \mid (\pi_1,\pi_2,\pi_3) \in \Pi_{k,3},$$
$$(k+1)T' - u \le t < (k+1)T'\}$$

and the set of receivers

$$\{(as_{km+\ell+1}s_{km+2}\pi_1\pi_2\pi_3, r) \mid (\pi_1,\pi_2,\pi_3) \in \Pi_{k,1}, kR' \le r < kR' + v\} \cup$$
$$\{(a\pi_1\pi_2\pi_3, r) \mid (\pi_1,\pi_2,\pi_3) \in \Pi_{k,2}, kR' + v \le r < (k+1)R' - v\} \cup$$
$$\{(as_{(k+1)m-\ell+2}s_{(k+1)m+1}\pi_1\pi_2\pi_3, r) \mid (\pi_1,\pi_2,\pi_3) \in \Pi_{k,3},$$
$$(k+1)R' - v \le r < (k+1)R'\}$$

constitute a subnetwork. Therefore, we have the following theorem.

THEOREM 4.35 *Suppose that* $\gcd(\frac{n-1}{T}, \frac{n-1}{R}) = 1$. *Then*

$$L(T,R) = \begin{cases} T'\frac{(\ell!)^2(m+1-2\ell)!}{8} & \text{if } T' > 2 \text{ and } R' > 2, \\ T'\frac{(\ell!)^2}{4} & \text{if } T' = 2 \text{ or } R' = 2; \end{cases}$$

$$W(T,R) = \begin{cases} \frac{n-1}{m}\frac{8n!}{(\ell!)^2(m+1-2\ell)!} & \text{if } T' > 2 \text{ and } R' > 2, \\ \frac{n-1}{m}\frac{4n!}{(\ell!)^2} & \text{if } T' = 2 \text{ or } R' = 2. \end{cases}$$

Proof. From Lemma 4.29, $m \ge 2\ell$ and $m = 2\ell$ if and only if $T' = 2$ or $R' = 2$ then $m > 2\ell$ and the size of $Q_{k,2}$ is $(m+1-2\ell)!/2$. Thus, if $T' > 2$ and $R' = 2$,

$$|\Pi_{k,1}| = |\Pi_{k,2}| = |\Pi_{k,2}| = \frac{\ell!}{2}\frac{(m+1-2\ell)!}{2}\frac{\ell!}{2} = \frac{(\ell!)^2\,(m+1-2\ell)!}{8}.$$

Therefore,

$$L(T,R) = T'\frac{(\ell!)^2\,(m+1-2\ell)!}{8}$$
$$W(T,R) = \frac{n!T}{L(T,R)} = \frac{n-1}{m}\frac{8n!}{(\ell!)^2\,(m+1-2\ell)!}.$$

If $T' = 2$ or $R' = 2$,

$$|\Pi_{k,1}| = |\Pi_{k,2}| = |\Pi_{k,2}| = \frac{\ell!}{2}\frac{\ell!}{2} = \frac{(\ell!)^2}{4}.$$

Hence

$$L(T,R) = T'\frac{(\ell!)^2}{4}$$
$$W(T,R) = \frac{n-1}{m}\frac{4n!}{(\ell!)^2}.$$

■

8.2 COST-PERFORMANCE RELATIONS

When $\gcd\left(\frac{n-1}{T}, \frac{n-1}{R}\right) > 1$, from Theorem 4.28, if an access node has one transmitter (or receiver) in a subnetwork, then it has T' (or R') transmitters (or receivers) in this subnetwork. Since these transmitters (or receivers) have the same wavelength channel, they can not work concurrently, and therefore are useless duplications. So we can replace them by a single transmitter (or receiver), and the number of subnetworks is still the same but the number of transmitters and receivers are reduced. Note that the resulting network has the same number of transmitters and receivers at each access node. Thus we only need to consider the case that $T = R$. When $T = R$, we have $T' = R' = 1$ and $m = \frac{n-1}{T}$. Thus

$$L(T,T) = \frac{\left(\frac{n-1}{T}+1\right)!}{2},$$
$$W(T,T) = T\frac{2n!}{\left(\frac{n-1}{T}+1\right)!}.$$

$L(T,T)$ and $W(T,T)$ changes dramatically with T. Consider a network of size of $5! = 120$. If each access node has only one transmitter and only one receiver, then $L(1,1) = 60$ and $W(1,1) = 2$. If we double the number of transmitters and receivers at each access node, then $L(2,2) = 3$ and $W(2,2) = 80$.

Now let's look at the case that $\gcd\left(\frac{n-1}{T}, \frac{n-1}{R}\right) = 1$ with both $\frac{n-1}{T}$ and $\frac{n-1}{R}$ more than two. From 4.35, for any access node a,

- the set of transmitters $\{(a,t) \mid kT' \leq t \leq kT' + u - 1\}$ can be replaced by one transmitter;
- the set of transmitters $\{(a,t) \mid kT' + u \leq t \leq (k+1)T' - u - 1\}$ can be replaced by one transmitter;
- the set of transmitters $\{(a,t) \mid (k+1)T' - u \leq t \leq (k+1)T' - 1\}$ can be replaced by one transmitter;
- the set of receivers $\{(a,r) \mid kR' \leq r \leq kR' + v - 1\}$ can be replaced by one transmitter;
- the set of receivers $\{(a,r) \mid kR' + v \leq r \leq (k+1)R' - v - 1\}$ can be replaced by one transmitter;
- the set of receivers $\{(a,r) \mid (k+1)R' - v \leq r \leq (k+1)R' - 1\}$ can be replaced by one transmitter.

After this replacement, each access node has $3\frac{n-1}{m}$ transmitters and $3\frac{n-1}{m}$ receivers. Furthermore each access node has at most one transmitter and at most one receiver in each subnetwork. Thus the number of transmitters in each subnetwork is reduced to

$$3\frac{(\ell!)^2\,(m+1-2\ell)!}{8}$$

while the number of subnetworks remains the same. However, the transmitters at each access node might be responsible for *unequal* number of links, and so are the receivers. Therefore, the partition is no longer a uniform partition. But this does not matter. With $3\frac{n-1}{m}$ transmitters and $3\frac{n-1}{m}$ receivers available, we can further optimize the partition. For an example, we can partition the links by choosing $\ell = \left\lfloor \frac{m+1}{3} \right\rfloor$ as $\frac{(\ell!)^2(m+1-2\ell)!}{8}$ is minimized at $\ell = \left\lfloor \frac{m+1}{3} \right\rfloor$. With this partition, the number of transmitters in each subnetwork is then decreased, however the number of subnetworks is increased.

When $\gcd\left(\frac{n-1}{T}, \frac{n-1}{R}\right) = 1$ with one of $\frac{n-1}{T}$ and $\frac{n-1}{R}$ equal to two, we can perform similar replacement. If $\frac{n-1}{T} = 2$ and $\frac{n-1}{R}$ is odd, then for any access node a,

- the set of transmitters $\{(a,t) \mid kT' \leq t \leq kT' + u - 1\}$ can be replaced by one transmitter;
- the set of transmitters $\{(a,t) \mid kT' + u \leq t \leq (k+1)T' - u - 1\}$ can be replaced by one transmitter;

- the set of transmitters $\{(a,t) \mid (k+1)T' - u \leq t \leq (k+1)T' - 1\}$ c-an be replaced by one transmitter.

After this replacement, each access node has $3\frac{n-1}{m}$ transmitters and $2\frac{n-1}{m}$ receivers. Furthermore each access node has at most one transmitter and at most one receiver in each subnetwork. Thus the number of transmitters in each subnetwork is reduced to $\frac{3(\ell!)^2}{4}$ while the number of subnetworks remains the same.

If $\frac{n-1}{T} = 2$ and $\frac{n-1}{R}$ is odd, then for any access node a,

- the set of receivers $\{(a,r) \mid kR' \leq r \leq kR' + v - 1\}$ can be replaced by one transmitter;
- the set of receivers $\{(a,r) \mid kR' + v \leq r \leq (k+1)R' - v - 1\}$ can be replaced by one transmitter;
- the set of receivers $\{(a,r) \mid (k+1)R' - v \leq r \leq (k+1)R' - 1\}$ can be replaced by one transmitter.

After this replacement, each access node has $2\frac{n-1}{m}$ transmitters and $3\frac{n-1}{m}$ receivers. Furthermore each access node has at most one transmitter and at most one receiver in each subnetwork. Thus the number of transmitters in each subnetwork and the number of subnetworks remains the same.

9. PONS BASED ON ROTATOR DIGRAPHS

We only consider the left rotator digraph, but the right rotator digraph can be treated similarly. In the left rotator digraph, the vertices are permutations and the product of two permutations α and β, denoted by $\alpha\beta$, is the same as the functional composition, i.e.,

$$(\alpha\beta)(i) = \alpha(\beta(i))$$

For any node a and any $0 \leq i < n-1$, the link

$$x \to a\alpha_{i+2}$$

is called the i-th outgoing link of a, and the link

$$a\beta_{i+2} \to a$$

is called the i-th incoming link of a, where

$$\alpha_k = 23\cdots k1(k+1)\cdots n$$

and

$$\beta_k = k12\cdots(k-1)(k+1)\cdots n$$

for any $2 \leq k \leq n$. The permutation $a\alpha_k$ ($a\beta_k$) can be interpreted as the one obtained from a by rotating one place left (right) the k symbols at the first k positions (not necessarily the symbols 1 though k themselves).

It's obvious the above link ordering is symmetric. In the transmission graph of the uniform consecutive partition based on such link ordering, each transmitter has $\frac{n-1}{T}$ outgoing links, and each receiver has $\frac{n-1}{R}$ incoming links. The set of receivers a transmitter (a, t) connects to is the set of receivers a transmitter (a, t) connects to is

$$\left\{ \left(a\alpha_{i+2}, \left\lfloor \frac{i}{\frac{n-1}{R}} \right\rfloor \right) \mid t\frac{n-1}{T} \leq i < (t+1)\frac{n-1}{T} \right\},$$

The set of transmitters a receiver (b, r) connects from is

$$\left\{ \left(b\beta_{j+2}, \left\lfloor \frac{j}{\frac{n-1}{T}} \right\rfloor \right) \mid r\frac{n-1}{R} \leq j < (r+1)\frac{n-1}{R} \right\}.$$

The following lemma will be frequently used later in this section. We emphasize again that the product of the transposition is the functional composition.

LEMMA 4.36 *For any* $2 \leq i < j \leq n$,

$$\begin{aligned}
\alpha_i\beta_j &= (j, (j-1), \cdots, (i+1), i), \\
\alpha_j\beta_i &= (i, (i+1), \cdots, (j-1), j), \\
\beta_i\alpha_j &= (1, i+1, \cdots, j-1, j), \\
\beta_j\alpha_i &= (j, j-1, \cdots, i+1, 1).
\end{aligned}$$

The proof of this lemma is straightforward and we omit it here.

9.1 SUBNETWORK STRUCTURE

Let

$$\begin{aligned}
T' &= \frac{m}{\frac{n-1}{T}} = \frac{T}{\frac{n-1}{m}} \\
R' &= \frac{m}{\frac{n-1}{R}} = \frac{R}{\frac{n-1}{m}}
\end{aligned}$$

in which m is the least common multiple of $\frac{n-1}{T}$ and $\frac{n-1}{R}$. As the link ordering is symmetric, from Corollary 4.3, for any subnetwork, there exists a unique integer $0 \le k < \frac{n-1}{m}$ such that in this subnetwork the set of local indices of all transmitters is

$$\{t \mid kT' \le t < (k+1)T'\},$$

and the set of local indices of all receivers is

$$\{r \mid kR' \le r < (k+1)R'\}.$$

Furthermore, from Lemma 4.2, for any $kT' \le t' < t < (k+1)T'$, the two transmitters

$$(a,t), \left(a \prod_{i=t'+1}^{t} \left(i\frac{n-1}{T}+2, i\frac{n-1}{T}+1\right), r'\right)$$

are in the same subnetwork; for any $kR' \le r' < r < (k+1)R'$, the two receivers

$$(a,r), \left(a\left(r\frac{n-1}{R}+2, (r-1)\frac{n-1}{R}+2, \cdots, (r'+1)\frac{n-1}{R}+2, 1\right), r'\right)$$

are in the same subnetwork.

In the next we will study the structure of the set of node indices of all transmitters/receivers which are in the same subnetwork and have the same local indices. We first prove the following property of the subnetworks.

LEMMA 4.37 *For any be any permutation $\sigma \in P_n$,*

1. *if the two transmitters (a,t) and (a',t') are in the same subnetwork, so are the two transmitters $(\sigma a, t)$ and $(\sigma a', t')$;*

2. *if the two receivers (b,r) and (b',r') are in the same subnetwork, so are the two receivers $(\sigma b, r)$ and $(\sigma b', r')$.*

Proof. (1). Let p be any path between the two transmitters (a,t) and (a',t'). For each transceiver in p, we multiply from left the node index of each transceiver in p by σ, then the resulting path connects the two transmitters $(\sigma a, t)$ and $(\sigma a', t')$. Therefore, the two transmitters $(\sigma a, t)$ and $(\sigma a', t')$ are also in the same subnetwork.

(2) The proof is similar to (1). ∎

One immediate conclusion that can be drawn from the above lemma is that all subnetworks which have the same set of local indices of transmitters (or receivers) are isomorphic to each other. Let C_k be the subnetwork that contains the transmitter (e, kT'). In the following we will focus on the structure of the node indices of all transmitters/receivers in C_k. One obvious property of C_k is that the node indices of all transmitters/receivers in C_k are local to the set of symbols

$$\{i \mid km + 2 \leq i < (k+1)m + 2\}.$$

The following lemma gives an invariant property of the node indices of receivers in C_k.

LEMMA 4.38 *Suppose that* $0 \leq k < \frac{n-1}{m}$. *Then for any* $kT' < t < (k+1)T'$, *the symbol* $t\frac{n-1}{T} + 2$ *is invariant under the node index of any receiver in* C_k.

Proof. We first prove that if the two receivers (b, r) and (b', r') share a common transmitter, then $b' = b\pi$ for some permutation π under which for any $kT' \leq t < (k+1)T'$, the symbol $t\frac{n-1}{T} + 2$ is invariant. In fact if the two receivers (b, r) and (b', r') share a common transmitter, then there exist two integers $i, j \in \left\{t\frac{n-1}{T} + 2 \mid kT' \leq t < (k+1)T'\right\}$ such that $b' = b\beta_i\alpha_j$. Let $\pi = \beta_i\alpha_j$. From Lemma 4.36, if $i < j$ then

$$\pi = (1, i+1, \cdots, j-1, j),$$

and if $i > j$ then

$$\pi = (i, i-1, \cdots, j+1, 1).$$

In either case, for any $kT' \leq t < (k+1)T'$, the symbol $t\frac{n-1}{T} + 2$ is invariant under π.

Notice that the receiver (α_{km+2}, kR') is in C_k, and for any $kT' < t < (k+1)T'$ the symbol $t\frac{n-1}{T} + 2$ is invariant under α_{km+2}. Based on the proved result, we can prove by induction that for any $kT' < t < (k+1)T'$ the symbol $t\frac{n-1}{T} + 2$ is invariant under the node index of any receiver in C_k. ■

The following lemma gives a necessary condition for a permutation to be the node index of some receiver in C_k.

LEMMA 4.39 *Suppose that* $0 \leq k < \frac{n-1}{m}$. *Let* a *be the node index of any transmitter in* C_k. *Then* a *is local to the set of symbols*

$$\{i \mid km + 2 \leq i < (k+1)m + 2\},$$

and for any $kR' \leq r < (k+1)R'$ *the set of symbols*

$$\left\{i \mid r\frac{n-1}{R} + 2 \leq i \leq (r+1)\frac{n-1}{R} + 1\right\}$$

are closed under a. *If in addition the transmitter* (a, kT') *is in* C_k, *then the symbol* $t\frac{n-1}{T} + 2$ *is invariant under* a *for any* $kT' < t < (k+1)T'$.

Proof. We first prove that if the two transmitters (a, t) and (a', t') share a common receiver, then $a' = a\pi$ for some permutation π satisfying that

- π is local to the set of symbols

$$\{i \mid km + 2 \leq i < (k+1)m + 2\};$$

- for any $kR' \leq r < (k+1)R'$, the set of symbols

$$\left\{i \mid r\frac{n-1}{R} + 2 \leq i \leq (r+1)\frac{n-1}{R} + 1\right\}$$

  are closed under π.

Assume that the two transmitter (a, t) and (a', t') share a common receiver. Then there exist two integers

$$r\frac{n-1}{R} + 2 \leq i, j \leq (r+1)\frac{n-1}{R} + 1$$

such that $a' = a\alpha_i\beta_j$. Let $\pi = \alpha_i\beta_j$. From Lemma 4.36, if $i < j$ then

$$\pi = (j, j-1, \cdots, i+1, i),$$

and if $i > j$ then

$$\pi = (j, j+1, \cdots, i-1, i).$$

In either case, π is local to the set of symbols

$$\{i \mid km + 2 \leq i < (k+1)m + 2\},$$

and for any $kR' \leq r < (k+1)R'$, the set of symbols

$$\left\{i \mid r\frac{n-1}{R} + 2 \leq i \leq (r+1)\frac{n-1}{R} + 1\right\}$$

are closed under π.

Notice that the transmitter (e, kT') is in C_k, we can prove by induction that the first part of the lemma is true.

Now we assume that the transmitter (a, kT') is in C_k. Then the permutation $b = a\alpha_{km+2}$ is the node index of some receiver in C_k. Consider any $kT' < t < (k+1)T'$. From Lemma 4.38,

$$b(t\frac{n-1}{T} + 2) = t\frac{n-1}{T} + 2.$$

As $a = b\beta_{km+2}$,

$$\begin{aligned} &a\left(t\frac{n-1}{T} + 2\right) \\ &= b\left(\beta_{km+2}\left(t\frac{n-1}{T} + 2\right)\right) \\ &= b\left(t\frac{n-1}{T} + 2\right) \\ &= t\frac{n-1}{T} + 2. \end{aligned}$$

Thus the symbol $t\frac{n-1}{T} + 2$ is invariant under a. ■

The next lemma gives a sufficient condition for the transmitter (a, kT') to be in C_k.

LEMMA 4.40 *Suppose that* $0 \leq k < \frac{n-1}{m}$. *Then the transmitter* (a, kT') *is in* C_k *if*

1. *a is local to the set of symbols*

$$\{i \mid km + 2 \leq i < (k+1)m + 2\};$$

2. *for any* $kR' \leq r < (k+1)R'$, *the set of symbols*

$$\left\{i \mid r\frac{n-1}{R} + 2 \leq i \leq (r+1)\frac{n-1}{R} + 1\right\}$$

are closed under a;

3. for any $kT' < t < (k+1)T'$, the symbol $t\frac{n-1}{T}+2$ is invariant under a.

Proof. As the set of permutations

$$\{(i, i+1) \mid 1 \leq i \leq n-1\}$$

is a generator of P_n, we only need to prove that for any $r\frac{n-1}{R}+2 \leq i \leq (r+1)\frac{n-1}{R}$ and $(i-2) \bmod \frac{n-1}{T} > 0$,

a. if $(i-1) \bmod \frac{n-1}{T} > 0$, then the transmitter $((i, i+1), kT')$ is in C_k;

b. if $(i-1) \bmod \frac{n-1}{T} = 0$ and $i \leq (r+1)\frac{n-1}{R} - 1$, then the transmitter $((i, i+2), kT')$ is in C_k.

Let

$$t = \left\lfloor \frac{i-2}{\frac{n-1}{T}} \right\rfloor.$$

(a). If $t = kT'$, then (a) is obviously true. So we assume that $t > kT'$. Then we have the following path

$$\begin{aligned}
&(e, kT') \\
&\rightsquigarrow \left(\prod_{j=kT'+1}^{t} \left(j\frac{n-1}{T}+1, j\frac{n-1}{T}+2 \right), t \right) \\
&\rightsquigarrow \left(\left(\prod_{j=kT'+1}^{t} \left(j\frac{n-1}{T}+1, j\frac{n-1}{T}+2 \right) \right) (i, i+1), t \right) \\
&= \left((i, i+1) \prod_{j=kT'+1}^{t} \left(j\frac{n-1}{T}+1, j\frac{n-1}{T}+2 \right), t \right) \\
&\rightsquigarrow ((i, i+1), kT').
\end{aligned}$$

So (a) is also true.

(b). If $t = kT'$, then $i = km + \frac{n-1}{T} + 1$ and we the following path

$$\begin{aligned}
&(e, kT') \\
&\rightsquigarrow (\alpha_i \beta_{i+2}, kT'+1) \\
&\rightsquigarrow (\alpha_i \beta_{i+2} \alpha_{i+1} \beta_i, kT') \\
&= ((i+2, i+1, i)(i+1, i), kT') \\
&= ((i, i+2), kT').
\end{aligned}$$

If $t > kT'$, we have the following path

$$\begin{aligned}
&(e, kT') \\
&\rightsquigarrow \left(\prod_{j=kT'+1}^{t} \left(j\frac{n-1}{T} + 1, j\frac{n-1}{T} + 2 \right), t \right) \\
&\rightsquigarrow \left(\left(\prod_{j=kT'+1}^{t} \left(j\frac{n-1}{T} + 1, j\frac{n-1}{T} + 2 \right) \right) \alpha_i \beta_{i+2}, t+1 \right) \\
&\rightsquigarrow \left(\left(\prod_{j=kT'+1}^{t} \left(j\frac{n-1}{T} + 1, j\frac{n-1}{T} + 2 \right) \right) \alpha_i \beta_{i+2} \alpha_{i+1} \beta_i, t \right) \\
&= \left(\left(\prod_{j=kT'+1}^{t} \left(j\frac{n-1}{T} + 1, j\frac{n-1}{T} + 2 \right) \right) (i, i+2), t \right) \\
&= \left((i, i+2) \prod_{j=kT'+1}^{t} \left(j\frac{n-1}{T} + 1, j\frac{n-1}{T} + 2 \right), t \right) \\
&\rightsquigarrow ((i, i+2), kT').
\end{aligned}$$

So (b) is also true in this case. ■

From Lemma 4.39 and Lemma 4.40, we can fully determine the node indices of all transmitters in C_k whose local indices are kT'.

LEMMA 4.41 *Suppose that* $0 \le k < \frac{n-1}{m}$. *Then the transmitter* (a, kT') *is in* C_k *if and only if*

1. *a is local to the set of symbols*

$$\{i \mid km + 2 \le i < (k+1)m + 2\};$$

2. *for any* $kR' \le r < (k+1)R'$, *the set of symbols*

$$\left\{ i \mid r\frac{n-1}{R} + 2 \le i \le (r+1)\frac{n-1}{R} + 1 \right\}$$

are closed under a;

3. *for any* $kT' < t < (k+1)T'$, *the symbol* $t\frac{n-1}{T} + 2$ *is invariant under a.*

To count the number of transmitters in C_k with local indices kT', for any $0 \leq r < R$, we define

$$N_r = \left\{ i+2 \mid r\frac{n-1}{R} \leq i < (r+1)\frac{n-1}{R}, i \bmod \frac{n-1}{T} > 0 \right\}.$$

Then Lemma 4.41 can be restated as follows: for any $0 \leq k < \frac{n-1}{m}$, the transmitter (a, kT') is in C_k if and only if

1. a is local to the set $\{km+2\} \cup \left(\cup_{r=kR'}^{(k+1)R'-1} N_r \right)$;
2. the set $\{km+2\} \cup N_{kR'}$ is closed under a;
3. for any $kR' < r < (k+1)R'$, the set N_r is closed under a.

Thus the number of transmitters with local indices kT' in C_k is

$$(|N_{kR'}|+1)! \prod_{r=kR'+1}^{(k+1)R'-1} |N_r|!.$$

In the following, we will give the size of N_r. If T is a multiple of R, then $R' = 1$ and for any $0 \leq r < R$,

$$|N_r| = \frac{n-1}{R} - \frac{T}{R} = \frac{n-1-T}{R}.$$

If T is not a multiple of R, then $R' > 1$ and the calculation of $|N_r|$ will need the following lemma.

LEMMA 4.42 *Suppose that x and y are two positive integers with $x < y$. Then for any $0 \leq i < y$,*

$$\left\lceil \frac{(i+1)x}{y} \right\rceil - \left\lceil \frac{ix}{y} \right\rceil \leq 1$$

and the equality holds if and only if $i = \left\lfloor \frac{jy}{x} \right\rfloor$ for some $0 \leq j < x$.

Proof. For any $i \geq 0$, as $x < y$,

$$\left\lceil \frac{(i+1)x}{y} \right\rceil - \left\lceil \frac{ix}{y} \right\rceil \leq 1 + \frac{(i+1)x}{y} - \frac{ix}{y} = 1 + \frac{x}{y} < 2.$$

Since $\left\lceil \frac{(i+1)x}{y} \right\rceil - \left\lceil \frac{ix}{y} \right\rceil$ is an integer,

$$\left\lceil \frac{(i+1)x}{y} \right\rceil - \left\lceil \frac{ix}{y} \right\rceil \leq 1.$$

Now suppose that $i = \left\lfloor \frac{jy}{x} \right\rfloor$ for some $0 \leq j < x$. Then

$$i \leq \frac{jy}{x} < i + 1,$$

which implies that

$$\frac{ix}{y} \leq j < \frac{(i+1)x}{y}.$$

Therefore,

$$\left\lceil \frac{ix}{y} \right\rceil \leq j < \left\lceil \frac{(i+1)x}{y} \right\rceil,$$

which means that

$$\left\lceil \frac{(i+1)x}{y} \right\rceil - \left\lceil \frac{ix}{y} \right\rceil \geq 1.$$

On the other hand,

$$\left\lceil \frac{(i+1)x}{y} \right\rceil - \left\lceil \frac{ix}{y} \right\rceil \leq 1.$$

Thus

$$\left\lceil \frac{(i+1)x}{y} \right\rceil - \left\lceil \frac{ix}{y} \right\rceil = 1.$$

Finally, suppose that $\left\lfloor \frac{jy}{x} \right\rfloor < i < \left\lfloor \frac{(j+1)y}{x} \right\rfloor$ for some $0 \leq j < x$. Then

$$\frac{jy}{x} < i < i + 1 \leq \frac{(j+1)\,y}{x},$$

which implies that

$$j < \frac{ix}{y} < \frac{(i+1)x}{y} \leq j + 1.$$

Thus

$$\left\lceil \frac{ix}{y} \right\rceil = \left\lceil \frac{(i+1)x}{y} \right\rceil = j,$$

or equivalently,

$$\left\lceil \frac{(i+1)x}{y} \right\rceil - \left\lceil \frac{ix}{y} \right\rceil = 0.$$

■

The next lemma gives the exact size of N_r for any $0 \leq r < R$ when T is not a multiple of R.

LEMMA 4.43 *Suppose that T is not a multiple of R and $0 \leq r < R$. If*

$$r \bmod R' = \left\lfloor \frac{tR'}{T' \bmod R'} \right\rfloor$$

for some $0 \leq t < T' \bmod R'$, then

$$|N_r| = \frac{n-1}{R} - \left\lfloor \frac{T}{R} \right\rfloor - 1;$$

otherwise

$$|N_r| = \frac{n-1}{R} - \left\lfloor \frac{T}{R} \right\rfloor.$$

Proof. For any $0 \leq r < R$,

$$\left| \left\{ i \mid 0 \leq i < r\frac{n-1}{R} \right\} \cap \left\{ t\frac{n-1}{T} \mid 0 \leq t \leq T \right\} \right|$$
$$= \left\lceil \frac{r\frac{n-1}{R}}{\frac{n-1}{T}} \right\rceil = \left\lceil \frac{rT}{R} \right\rceil.$$

By Lemma 4.42,

$$\left| \left\{ i \mid r\frac{n-1}{R} \leq i < (r+1)\frac{n-1}{R} \right\} \cap \left\{ t\frac{n-1}{T} \mid 0 \leq t \leq T \right\} \right|$$
$$= \left\lceil \frac{(r+1)\,T}{R} \right\rceil - \left\lceil \frac{rT}{R} \right\rceil$$
$$= \left\lceil \frac{(r+1)\,T'}{R'} \right\rceil - \left\lceil \frac{rT'}{R'} \right\rceil \quad \left(\text{as } \frac{T}{T'} = \frac{R}{R'} \right)$$
$$= \left\lfloor \frac{T}{R} \right\rfloor + \left\lceil \frac{(r \bmod R' + 1)\,(T' \bmod R')}{R'} \right\rceil - \left\lceil \frac{(r \bmod R')\,(T' \bmod R')}{R'} \right\rceil$$
$$= \begin{cases} \left\lfloor \frac{T}{R} \right\rfloor + 1 & \text{if } r \bmod R' = \left\lfloor \frac{tR'}{T' \bmod R'} \right\rfloor \text{ for some } 0 \leq t < T' \bmod R'; \\ \left\lfloor \frac{T}{R} \right\rfloor & \text{otherwise.} \end{cases}$$

Thus the lemma follows. ∎

Now we are able to calculate $L(T,R)$ and $W(T,R)$.

THEOREM 4.44 *If T is a multiple of R, then*

$$L(T,R) = \frac{T}{R}\left(\frac{n-1-T}{R}+1\right)!,$$
$$W(T,R) = R\frac{n!}{(\frac{n-1-T}{R}+1)!}.$$

If T is not a multiple of R, then

$$L(T,R) = T'\left(\left(\frac{n-1}{R}-\left\lfloor\frac{T}{R}\right\rfloor-1\right)!\right)^{T' \bmod R'-1}\left(\left(\frac{n-1}{R}-\left\lfloor\frac{T}{R}\right\rfloor\right)!\right)^{R'-T' \bmod R'+1},$$
$$W(T,R) = \frac{n-1}{m}\frac{n!}{((\frac{n-1}{R}-\lfloor\frac{T}{R}\rfloor-1)!)^{T' \bmod R'-1}((\frac{n-1}{R}-\lfloor\frac{T}{R}\rfloor)!)^{R'-T' \bmod R'+1}}.$$

Proof. If T is a multiple of R, then

$$|N_{kR'}| = \frac{n-1-T}{R}.$$

Thus The total number of transmitters with local indices kT' in C_k is

$$(|N_{kR'}|+1)!\prod_{r=kR'+1}^{(k+1)R'-1}|N_r|! = \left(\frac{n-1-T}{R}+1\right)!.$$

So the total number of transmitters in C_k is

$$L(T,R) = T'\left(\frac{n-1-T}{R}+1\right)! = \frac{T}{R}\left(\frac{n-1-T}{R}+1\right)!.$$

If T is not a multiple of R, then from Lemma 4.43, the total number of transmitters with local indices kT' in C_k is that

$$(|N_{kR'}|+1)!\prod_{r=kR'+1}^{(k+1)R'-1}|N_r|!$$
$$= \left(\left(\frac{n-1}{R}-\left\lfloor\frac{T}{R}\right\rfloor-1\right)!\right)^{T' \bmod R'-1}\left(\left(\frac{n-1}{R}-\left\lfloor\frac{T}{R}\right\rfloor\right)!\right)^{R'-T' \bmod R'+1}.$$

Thus the total number of transmitters in C_k is

$$L(T,R) = T'\left(\left(\frac{n-1}{R} - \left\lfloor\frac{T}{R}\right\rfloor - 1\right)!\right)^{T' \bmod R'-1}\left(\left(\frac{n-1}{R} - \left\lfloor\frac{T}{R}\right\rfloor\right)!\right)^{R'-T' \bmod R'+1}.$$

This implies that

$$W(T,R) = \frac{Tn!}{L(T,R)} = \frac{n-1}{m}\frac{n!}{\left(\left(\frac{n-1}{R} - \lfloor\frac{T}{R}\rfloor - 1\right)!\right)^{T' \bmod R'-1}\left(\left(\frac{n-1}{R} - \lfloor\frac{T}{R}\rfloor\right)!\right)^{R'-T' \bmod R'+1}}.$$

■

9.2 COST-PERFORMANCE RELATIONS

The expressions for $L(T,R)$ and $W(T,R)$ in Theorem 4.44 are very complex. They can be simplified for many special cases of T and R. We consider the following scenarios.

When $T < R$, $\lfloor\frac{T}{R}\rfloor = 0$ and $T' \bmod R' = T'$. Thus the expression for $L(T,R)$ and $W(T,R)$ can be further simplified as follows:

$$L(T,R) = T'\left(\left(\frac{n-1}{R} - 1\right)!\right)^{T'-1}\left(\left(\frac{n-1}{R}\right)!\right)^{R'-T'+1},$$

$$W(T,R) = \frac{n-1}{m}\frac{n!}{\left(\left(\frac{n-1}{R} - 1\right)!\right)^{T'-1}\left(\left(\frac{n-1}{R}\right)!\right)^{R'-T'+1}}.$$

When $R = kT$, $T' = 1$ and $R' = k$. In this case,

$$L(T,kT) = \left(\left(\frac{n-1}{kT}\right)!\right)^{k},$$

$$W(T,kT) = T\frac{n!}{\left(\left(\frac{n-1}{kT}\right)!\right)^{k}}.$$

When $T = R$, $T' = R' = 1$ and therefore

$$L(R,R) = \left(\frac{n-1}{R}\right)!,$$

$$W(R,R) = R\frac{n!}{\left(\frac{n-1}{R}\right)!}.$$

When $T = kR$, $T' = k$ and $R' = 1$. In this case,

$$L(kR, R) = k\left(\frac{n-1}{R} - k + 1\right)!,$$

$$W(kR, R) = R\frac{n!}{(\frac{n-1}{R} - k + 1)!}.$$

In general $L(T, R)$ and $W(T, R)$ change dramatically with T and R. As an example, Table 4.3 lists the values of $L(T, R)$ and $W(T, R)$ when $n = 5$.

$T \backslash R$	1	2
1	24	4
2	12	2

(a). $L(T, R)$

$T \backslash R$	1	2
1	5	30
2	20	120

(b). $W(T, R)$

Table 4.3. The values of $L(T, R)$ and $W(T, R)$ when $n = 5$.

10. EXTENSION TO ARBITRARY NUMBER OF TRANSCEIVERS

In the uniform consecutive partition, we assume that both T, the number of transmitters at each access node, and R, the number of receivers at each access node, must be factors of d, the degree of the regular interconnection topology to be realized. Such requirement restricts the network to a very limited number of configurations. For an example, if the degree d is a prime number, then T and R are limited to one or d. In this section, we briefly discuss how to relax such restriction to allow any number of transmitters and receivers to be used at each access node.

10.1 SINGLE-HOP PONS

For any $0 < R < N$, we partition the N access nodes into R subsets

$$A_0, A_1, \cdots, A_{R-1}$$

with

$$|A_r| = \begin{cases} \lceil \frac{N}{R} \rceil & \text{if } 0 \le r < N \bmod R, \\ \lfloor \frac{N}{R} \rfloor & \text{if } N \bmod R \le r < R. \end{cases}$$

Then the receiver (b, r) will be responsible for the incoming links at b from access nodes in A_r.

Similarly, for any $0 < T < N$, we partition the N access nodes into T subsets

$$B_0, B_1, \cdots, B_{T-1}$$

with

$$|B_t| = \begin{cases} \lceil \frac{N}{T} \rceil & \text{if } 0 \leq t < N \bmod T, \\ \lfloor \frac{N}{T} \rfloor & \text{if } N \bmod T \leq t < T. \end{cases}$$

Then the transmitter (a, t) will be responsible for the outgoing links at a to access nodes in B_t.

With this partition, it's easy to verify that for any $0 \leq t < T$ and $0 \leq r < R$, the set of transmitters

$$\{(a, t) \mid a \in A_r\}$$

and the set of receivers

$$\{(b, r) \mid b \in B_t\}$$

constitute a subnetwork. Thus the maximal number of transmitters in a subnetwork is

$$L(T, R) = \left\lceil \frac{N}{R} \right\rceil$$

and the number of subnetworks is

$$W(T, R) = TR.$$

10.2 PONS BASED ON GENERALIZED DE BRUIJIN DIGRAPH

Here we only consider those generalized de Bruijin digraphs whose degree p is a factor of the network size N. Let i be any integer with $0 \leq i < \frac{N}{p}$. For any $0 < R < N$, we partition the p access nodes

$$\{k\frac{N}{p} + i \mid 0 \leq k < p\}$$

into R subsets

$$A_{i,0}, A_{i,1}, \cdots, A_{i,R-1}$$

with

$$|A_{i,r}| = \begin{cases} \lceil \frac{p}{R} \rceil & \text{if } 0 \leq r < p \bmod R, \\ \lfloor \frac{p}{R} \rfloor & \text{if } p \bmod R \leq r < R. \end{cases}$$

Then for any access node b with $\left\lfloor \frac{b}{p} \right\rfloor = i$, the receiver (b, r) will be responsible for the incoming links at b from access nodes in $A_{i,r}$.

Similarly, for any $0 < T < N$, we partition the p access nodes

$$\{ip + k \mid 0 \leq k < p\}$$

into T subsets

$$B_{i,0}, B_{i,1}, \cdots, B_{i,T-1}$$

with

$$|B_{i,t}| = \begin{cases} \lceil \frac{p}{T} \rceil & \text{if } 0 \leq t < p \bmod T, \\ \lfloor \frac{p}{T} \rfloor & \text{if } p \bmod T \leq t < T. \end{cases}$$

Then for any access node a with $a \bmod \frac{N}{p} = i$, the transmitter (a, t) will be responsible for the outgoing links at a to access nodes in $B_{i,t}$.

With this partition, it's easy to verify that for any $0 \leq i < \frac{N}{p}$, $0 \leq t < T$ and $0 \leq r < R$, the set of transmitters

$$\{(a, t) \mid a \in A_{i,r}\}$$

and the set of receivers

$$\{(b, r) \mid b \in B_{i,t}\}$$

constitute a subnetwork. Thus the maximal number of transmitters in a subnetwork is

$$L(T, R) = \left\lceil \frac{N}{R} \right\rceil$$

and the number of subnetworks is

$$W(T, R) = \frac{N}{p} TR.$$

10.3 PONS BASED ON GENERALIZED KAUTZ DIGRAPH

Here we also only consider those generalized Kautz digraphs whose degree p is a factor of the network size N. For any T and R, we partition the N access nodes into

$$\{A_{i,r} \mid 0 \leq i < \frac{N}{p}, 0 \leq r < R\}$$

and into

$$\{B_{i,t} \mid 0 \leq i < \frac{N}{p}, 0 \leq t < T\}$$

the same way as in the previous subsection. Then for any access node b with $\left\lfloor \frac{b}{p} \right\rfloor = \frac{N}{p} - 1 - i$, the receiver (b, r) will be responsible for the incoming links at b from access nodes in $A_{i,r}$; for any access node a with $a \bmod \frac{N}{p} = i$, the transmitter (a, t) will be responsible for the outgoing links at a to access nodes in $B_{\frac{N}{p}-1-i,t}$.

With this partition, it's easy to verify that for any $0 \leq i < \frac{N}{p}$, $0 \leq t < T$ and $0 \leq r < R$, the set of transmitters

$$\{(a, t) \mid a \in A_{i,r}\}$$

and the set of receivers

$$\{(b, r) \mid b \in B_{\frac{N}{p}-1-i,t}\}$$

constitute a subnetwork. Thus the maximal number of transmitters in a subnetwork is

$$L(T, R) = \left\lceil \frac{N}{R} \right\rceil$$

and the number of subnetworks is

$$W(T, R) = \frac{N}{p} TR.$$

10.4 PONS BASED ON HYPERCUBE

For any

$$S \subseteq \{0, 1, \cdots, n-1\}$$

and any n-bit binary number a, we use $a|_S$ to denote the $|S|$-bit binary number consisting of the bits of a at positions in S, and $a|_{\overline{S}}$ to denote the $(n - |S|)$-bit binary number consisting of the bits of a at positions not in S.

We begin with the simplest case that $T = R$. We consecutively partition the n dimensions into T groups

$$D_0, D_1, \cdots, D_{T-1}$$

with

$$|D_t| = \begin{cases} \left\lceil \frac{n}{T} \right\rceil & \text{if } 0 \leq t < n \bmod T, \\ \left\lfloor \frac{n}{T} \right\rfloor & \text{if } n \bmod T \leq t < T. \end{cases}$$

Then the transmitter (a, t) will be responsible for the outgoing links at a along dimensions in D_t, and the receiver (a, r) will be responsible for the incoming links at a along dimensions in D_r. Then it can be shown that for any $0 \leq t < T$, the two transmitters (a, t) and (a', t) are in the same subnetwork if and only if $(a \oplus a')\,|_{D_r}$ is even and $(a \oplus a')\,|_{\overline{D_r}}$ is zero. Thus the maximal number of transmitters in each subnetwork is

$$L(T, R) = \max\left\{2^{|D_t|-1} : 0 \leq t < T\right\} = 2^{\left\lceil \frac{n}{T} \right\rceil - 1}$$

and the number of subnetworks is

$$\begin{aligned} & W(T, R) \\ &= \sum_{t=0}^{T-1} 2^{n-|D_t|+1} \\ &= (n \bmod T)\, 2^{n-\left\lceil \frac{n}{T} \right\rceil+1} + (T - n \bmod T)\, 2^{n-\left\lfloor \frac{n}{T} \right\rfloor+1} \\ &= T2^{n-\left\lfloor \frac{n}{T} \right\rfloor+1} - (n \bmod T)\left(2^{n-\left\lfloor \frac{n}{T} \right\rfloor+1} - 2^{n-\left\lceil \frac{n}{T} \right\rceil+1}\right) \\ &= T2^{n-\left\lfloor \frac{n}{T} \right\rfloor+1} - (n \bmod T)\, 2^{n-\left\lceil \frac{n}{T} \right\rceil+1}. \end{aligned}$$

Now we look at the case that $T = 1$ and $R > 1$. We consecutively partition the n dimensions into R groups

$$D_0, D_1, \cdots, D_{R-1}$$

of *arbitrary* sizes, and at each access node a let the receiver (a, r) be responsible for the incoming links at a along dimensions in D_r. Then it can be shown that the two transmitters $(a, 0)$ and $(a', 0)$ are in the same subnetwork if and only if for any $0 \leq r < R$, $(a \oplus a') |_{D_r}$ is even. Thus the number of transmitters in each subnetwork is

$$L(T, R) = \prod_{r=1}^{R-1} 2^{|D_r|-1} = 2^{\sum_{r=0}^{R-1} |D_t| - R} = 2^{n-R}$$

and the number of subnetworks is

$$W(T, R) = 2^R.$$

In the next we assume that $0 < T < R < n$ and R is a multiple of T. We consecutively partition the n dimensions into T groups

$$D_0, D_1, \cdots, D_{T-1}$$

with

$$|D_t| = \begin{cases} \lceil \frac{n}{T} \rceil & \text{if } 0 \leq t < n \bmod T, \\ \lfloor \frac{n}{T} \rfloor & \text{if } n \bmod T \leq t < T. \end{cases}$$

Then the transmitter (a, t) will be responsible for the outgoing links at a along dimensions in D_t. For each $0 \leq t < T$, we further consecutively partition D_t into $\frac{R}{T}$ subgroups

$$D_{t,0}, D_{t,1}, \cdots, D_{t,\frac{R}{T}-1}$$

of *arbitrary* sizes, and at each access node a let the receiver $(a, t\frac{R}{T} + i)$ be responsible for the incoming links at a along dimensions in $D_{t,i}$ for any $0 \leq i < \frac{R}{T}$. Then it can be shown that the two transmitters (a, t) and (a', t) are in the same subnetwork if and only if $(a \oplus a') |_{\overline{D_t}}$ is zero and for any $0 \leq i < \frac{R}{T}$, $(a \oplus a') |_{D_{t,i}}$ is even. Thus the maximal number of transmitters in each subnetwork is

$$L(T, R) = \max \left\{ 2^{|D_t| - \frac{R}{T}} : 0 \leq t < T \right\} = 2^{\lceil \frac{n}{T} \rceil - \frac{R}{T}}$$

and the number of subnetworks is

$$\begin{aligned}
W(T,R) &= \sum_{t=0}^{T-1} 2^{n-|D_t|} 2^{\frac{R}{T}} \\
&= \sum_{t=0}^{T-1} 2^{n-|D_t|+\frac{R}{T}} \\
&= (n \bmod T)\, 2^{n-\lceil \frac{n}{T} \rceil + \frac{R}{T}} + (T - n \bmod T)\, 2^{n-\lfloor \frac{n}{T} \rfloor + \frac{R}{T}} \\
&= T 2^{n-\lfloor \frac{n}{T} \rfloor + \frac{R}{T}} - (n \bmod T) \left(2^{n-\lfloor \frac{n}{T} \rfloor + \frac{R}{T}} - 2^{n-\lceil \frac{n}{T} \rceil + \frac{R}{T}} \right) \\
&= T 2^{n-\lfloor \frac{n}{T} \rfloor + \frac{R}{T}} - (n \bmod T)\, 2^{n-\lceil \frac{n}{T} \rceil + \frac{R}{T}}.
\end{aligned}$$

Finally we assume that $0 < T < R < n$ and R is not necessarily a multiple of T. We consecutively partition the n dimensions into T groups

$$D_0, D_1, \cdots, D_{T-1}$$

with

$$|D_t| = \begin{cases} \lceil \frac{n}{T} \rceil & \text{if } 0 \le t < n \bmod T, \\ \lfloor \frac{n}{T} \rfloor & \text{if } n \bmod T \le t < T. \end{cases}$$

Then the transmitter (a, t) will be responsible for the outgoing links at a along dimensions in D_t. For each $0 \le t < R \bmod T$, we further consecutively partition D_t into $\lceil \frac{R}{T} \rceil$ subgroups

$$D_{t,0}, D_{t,1}, \cdots, D_{t, \lceil \frac{R}{T} \rceil - 1}$$

of *arbitrary* sizes, and at each access node a let the receiver $(a, t\lceil \frac{R}{T} \rceil + i)$ be responsible for the incoming links at a along dimensions in $D_{t,i}$ for any $0 \le i < \lceil \frac{R}{T} \rceil$. For each $R \bmod T \le t < T$,we consecutively partition D_t into $\lfloor \frac{R}{T} \rfloor$ subgroups

$$D_{t,0}, D_{t,1}, \cdots, D_{t, \lfloor \frac{R}{T} \rfloor - 1}$$

of *arbitrary* sizes, and at each access node a let the receiver $(a, R \bmod T + t\lfloor \frac{R}{T} \rfloor + i)$ be responsible for the incoming links at a along dimensions in $D_{t,i}$ for any $0 \le i < \lfloor \frac{R}{T} \rfloor$. Then it can be shown that the two transmitters (a, t) and (a', t) are in the same subnetwork if and only if $(a \oplus a')\,|_{\overline{D_t}}$ is zero and $(a \oplus a')\,|_{D_{t,i}}$ is even for any $0 \le i < \lceil \frac{R}{T} \rceil$ if

$0 \leq t < R \bmod T$ or any $0 \leq i < \lfloor \frac{R}{T} \rfloor$ if $R \bmod T \leq t < T$. It's then routine to verify that

$$L(T,R) = 2^{\lceil \frac{n-R}{T} \rceil},$$
$$W(T,R) = (T + (R-n) \bmod T)\, 2^{n - \lceil \frac{n-R}{T} \rceil}.$$

When R is more than T, we can perform similar partition and analysis. The details are omit over here.

10.5 PONS BASED ON STAR GRAPH

From the analysis in Section 8., we only have to consider the configuration in which $T = R$ for any other configurations can be improved to this configuration without increasing any cost. We consecutively partition the $n-1$ dimensions into T groups

$$D_0, D_1, \cdots, D_{T-1}$$

with

$$|D_t| = \begin{cases} \lceil \frac{n-1}{T} \rceil & \text{if } 0 \leq t < (n-1) \bmod T, \\ \lfloor \frac{n-1}{T} \rfloor & \text{if } (n-1) \bmod T \leq t < T. \end{cases}$$

Then the transmitter (a,t) will be responsible for the outgoing links at a along dimensions in D_t, and the receiver (a,r) will be responsible for the incoming links at a along dimensions in D_r. Then it can be shown that for any $0 \leq t < T$, the two transmitters (a,t) and (a',t) are in the same subnetwork if and only if $a^{-1}a'$ is an even permutation which is local to the set of symbols

$$\{1\} \cup \{i+2 \mid i \in D_t\}.$$

Thus the maximal number of transmitters in each subnetwork is

$$\begin{aligned} &L(T,R) \\ &= \max\left\{ \frac{(|D_t|+1)!}{2} : 0 \leq t < T \right\} \\ &= \frac{(\lceil \frac{n-1}{T} \rceil + 1)!}{2} \end{aligned}$$

and the number of subnetworks is

$$\begin{aligned} & W(T,R) \\ & = \sum_{t=0}^{T-1} 2\frac{n!}{(|D_t|+1)!} \\ & = ((n-1) \bmod T)\frac{2n!}{(\lceil \frac{n-1}{T} \rceil + 1)!} + (T-(n-1) \bmod T)\frac{2n!}{(\lfloor \frac{n-1}{T} \rfloor + 1)!}. \end{aligned}$$

10.6 PONS BASED ON ROTATOR DIGRAPH

We begin with the simplest case that $T = R$. We consecutively partition the $n-1$ dimensions into T groups

$$D_0, D_1, \cdots, D_{T-1}$$

with

$$|D_t| = \begin{cases} \lceil \frac{n-1}{T} \rceil & \text{if } 0 \le t < (n-1) \bmod T, \\ \lfloor \frac{n-1}{T} \rfloor & \text{if } (n-1) \bmod T \le t < T. \end{cases}$$

Then the transmitter (a, t) will be responsible for the outgoing links at a along dimensions in D_t, and the receiver (a, r) will be responsible for the incoming links at a along dimensions in D_r. Then it can be shown that for any $0 \le t < T$, the two transmitters (a, t) and (a', t) are in the same subnetwork if and only if $a^{-1}a'$ is local to the set of symbols

$$\{i + 2 \mid i \in D_t\}.$$

Thus the maximal number of transmitters in each subnetwork is

$$\begin{aligned} & L(T,R) \\ & = \max\{|D_t|! : 0 \le t < T\} \\ & = \left\lceil \frac{n-1}{T} \right\rceil ! \end{aligned}$$

and the number of subnetworks is

$$\begin{aligned} & W(T,R) \\ & = \sum_{t=0}^{T-1} \frac{n!}{|D_t|!} \\ & = ((n-1) \bmod T)\frac{n!}{\lceil \frac{n-1}{T} \rceil !} + (T-(n-1) \bmod T)\frac{n!}{\lfloor \frac{n-1}{T} \rfloor !}. \end{aligned}$$

Now we look at the case that $0 < T < R < n-1$ and R is a multiple of T. We consecutively partition the $n-1$ dimensions into R groups

$$\left\{D_{t,i} : 0 \leq t < T, 0 \leq i < \frac{R}{T}\right\}$$

satisfying that

- for any $0 \leq t < T$ and $0 \leq i < \frac{R}{T}$,
$$\left\lfloor \frac{n-1}{R} \right\rfloor \leq |D_{t,i}| \leq \left\lceil \frac{n-1}{R} \right\rceil;$$
- for $0 \leq t < ((n-1) \bmod R) \bmod T$,
$$\left|\left\{D_{t,i} : |D_{t,i}| = \left\lceil \frac{n-1}{R} \right\rceil, 0 \leq i < \frac{R}{T}\right\}\right| = \left\lceil \frac{(n-1) \bmod R}{T} \right\rceil;$$
- for $((n-1) \bmod R) \bmod T \leq t < T$,
$$\left|\left\{D_{t,i} : |D_{t,i}| = \left\lceil \frac{n-1}{R} \right\rceil, 0 \leq i < \frac{R}{T}\right\}\right| = \left\lfloor \frac{(n-1) \bmod R}{T} \right\rfloor.$$

For any $0 \leq t < T$, let

$$D_t = \bigcup_{i=0}^{\frac{R}{T}-1} D_{t,i}.$$

At each access node a the transmitter (a, t) is responsible for the outgoing links at a along dimensions in D_t, and the receiver $(a, t\frac{R}{T} + i)$ is responsible for the incoming links at a along dimensions in $D_{t,i}$. Then it can be shown that the two transmitters (a, t) and (a', t) are in the same subnetwork if and only if for any $0 \leq i < \frac{R}{T}$, $a^{-1}a'$ is local to the set of symbols in $D_{t,i}$. Thus the number of transmitters in each subnetwork is

$$\begin{aligned}
&L(T, R) \\
&= \max\left\{\prod_{i=0}^{\frac{R}{T}-1} |D_{t,i}|! : 0 \leq t < T\right\} \\
&= \left(\left\lceil \frac{n-1}{R} \right\rceil !\right)^{\left\lceil \frac{(n-1) \bmod R}{T} \right\rceil} \left(\left\lfloor \frac{n-1}{R} \right\rfloor !\right)^{\frac{R}{T} - \left\lceil \frac{(n-1) \bmod R}{T} \right\rceil} \\
&= \left(\left\lfloor \frac{n-1}{R} \right\rfloor !\right)^{\frac{R}{T}} \left\lceil \frac{n-1}{R} \right\rceil^{\left\lceil \frac{(n-1) \bmod R}{T} \right\rceil}.
\end{aligned}$$

and the number of subnetworks is

$$W(T, R)$$
$$= \sum_{t=0}^{T-1} \frac{n!}{\prod_{i=0}^{\frac{R}{T}-1} |D_{t,i}|!}$$
$$= (((n-1) \bmod R) \bmod T) \frac{n!}{\left(\lfloor \frac{n-1}{R} \rfloor !\right)^{\frac{R}{T}} \lceil \frac{n-1}{R} \rceil^{\left\lceil \frac{(n-1) \bmod R}{T} \right\rceil}} +$$
$$(T - ((n-1) \bmod R) \bmod T) \frac{n!}{\left(\lfloor \frac{n-1}{R} \rfloor !\right)^{\frac{R}{T}} \lceil \frac{n-1}{R} \rceil^{\left\lfloor \frac{(n-1) \bmod R}{T} \right\rfloor}}.$$

Now we look at the case that $0 < R < T < n-1$ and T is a multiple of R. We consecutively partition the $n-1$ dimensions into R groups

$$\left\{ D_{t,i} : 0 \le t < T, 0 \le i < \frac{R}{T} \right\}$$

satisfying that

- for any $0 \le t < T$ and $0 \le i < \frac{R}{T}$,

$$\left\lfloor \frac{n-1}{R} \right\rfloor \le |D_{t,i}| \le \left\lceil \frac{n-1}{R} \right\rceil;$$

- for $0 \le t < ((n-1) \bmod R) \bmod T$,

$$\left| \left\{ D_{t,i} : |D_{t,i}| = \left\lceil \frac{n-1}{R} \right\rceil, 0 \le i < \frac{R}{T} \right\} \right| = \left\lceil \frac{(n-1) \bmod R}{T} \right\rceil;$$

- for $((n-1) \bmod R) \bmod T \le t < T$,

$$\left| \left\{ D_{t,i} : |D_{t,i}| = \left\lceil \frac{n-1}{R} \right\rceil, 0 \le i < \frac{R}{T} \right\} \right| = \left\lfloor \frac{(n-1) \bmod R}{T} \right\rfloor.$$

For any $0 \le t < T$, let

$$D_t = \bigcup_{i=0}^{\frac{R}{T}-1} D_{t,i}.$$

At each access node a the transmitter (a, t) is responsible for the outgoing links at a along dimensions in D_t, and the receiver $(a, t\frac{R}{T} + i)$ is responsible for the incoming links at a along dimensions in $D_{t,i}$. Then it can be shown that the two transmitters (a,t) and (a',t) are in the same subnetwork if and only if for any $0 \le i < \frac{R}{T}$, $a^{-1}a'$ is local to the set of symbols in $D_{t,i}$. Thus the number of transmitters in each subnetwork is

$$\begin{aligned} & L(T,R) \\ & = \max\left\{ \prod_{i=0}^{\frac{R}{T}-1} |D_{t,i}|! : 0 \le t < T \right\} \\ & = \left(\left\lfloor \frac{n-1}{R} \right\rfloor ! \right)^{\frac{R}{T}} \left\lceil \frac{n-1}{R} \right\rceil^{\left\lceil \frac{(n-1) \bmod R}{T} \right\rceil}. \end{aligned}$$

and the number of subnetworks is

$$\begin{aligned} & W(T,R) \\ & = \sum_{t=0}^{T-1} \frac{n!}{\prod_{i=0}^{\frac{R}{T}-1} |D_{t,i}|!} \\ & = (((n-1) \bmod R) \bmod T) \frac{n!}{\left(\lfloor \frac{n-1}{R} \rfloor !\right)^{\frac{R}{T}} \lceil \frac{n-1}{R} \rceil^{\left\lceil \frac{(n-1) \bmod R}{T} \right\rceil}} + \\ & (T - ((n-1) \bmod R) \bmod T) \frac{n!}{\left(\lfloor \frac{n-1}{R} \rfloor !\right)^{\frac{R}{T}} \lceil \frac{n-1}{R} \rceil^{\left\lfloor \frac{(n-1) \bmod R}{T} \right\rfloor}}. \end{aligned}$$

When $0 < T < R < n-1$ and R is not a multiple of T, the most balanced partition which leads to the least maximal number of transmitters in each subnetwork is unknown yet. We leave it as an *open* problem for further study.

In the next we assume that $0 < R < T < n$ and T is a multiple of R. We consecutively partition the $n-1$ dimensions into R groups

$$D_0, D_1, \cdots, D_{R-1}$$

with

$$|D_r| = \begin{cases} \left\lceil \frac{n-1}{R} \right\rceil & \text{if } 0 \le r < (n-1) \bmod R, \\ \left\lfloor \frac{n-1}{R} \right\rfloor & \text{if } (n-1) \bmod R \le r < R. \end{cases}$$

Then the receiver (a,r) will be responsible for the incoming links at a along dimensions in D_r. For each $0 \le r < R$, we further consecutively partition D_r into $\frac{T}{R}$ subgroups

$$D_{r,0}, D_{r,1}, \cdots, D_{r,\frac{T}{R}-1}$$

of *arbitrary* sizes. At each access node a let the transmitter $(a, r\frac{T}{R} + i)$ be responsible for the outgoing links at a along dimensions in $D_{r,i}$ for any $0 \le i < \frac{T}{R}$. Then it can be shown that the two transmitters $(a, r\frac{T}{R})$ and $(a', r\frac{T}{R})$ are in the same subnetwork if and only if $a^{-1}a'$ is local to the set of symbols in D_r except the minimal number in each $D_{r,i}$ for $1 \le i < \frac{T}{R}$. Thus the number of transmitters in each subnetwork is

$$
\begin{aligned}
&L(T, R) \\
&= \frac{T}{R} \max \left\{ \left(|D_r| - \frac{T}{R} + 1 \right)! : 0 \le r < R \right\} \\
&= \frac{T}{R} \left(\left\lceil \frac{n-1}{R} \right\rceil - \frac{T}{R} + 1 \right)!
\end{aligned}
$$

and the number of subnetworks is

$$
\begin{aligned}
&W(T, R) \\
&= \sum_{r=0}^{R-1} \frac{n!}{(|D_r| - \frac{T}{R} + 1)!} \\
&= \\
&= ((n-1) \bmod R) \frac{n!}{(\lceil \frac{n-1}{R} \rceil - \frac{T}{R} + 1)!} + \\
&(R - (n-1) \bmod R) \frac{n!}{(\lfloor \frac{n-1}{R} \rfloor - \frac{T}{R} + 1)!}.
\end{aligned}
$$

Finally we assume that $0 < R < T < n$ and T is not necessarily a multiple of R. We consecutively partition the $n-1$ dimensions into R groups

$$D_0, D_1, \cdots, D_{R-1}$$

with

$$
|D_r| = \begin{cases} \lceil \frac{n-1}{R} \rceil & \text{if } 0 \le r < (n-1) \bmod R, \\ \lfloor \frac{n-1}{R} \rfloor & \text{if } (n-1) \bmod R \le r < R. \end{cases}
$$

Then the receiver (a, r) will be responsible for the incoming links at a along dimensions in D_r. For each $0 \le r < T \bmod R$, we further consecutively partition D_r into $\lceil \frac{T}{R} \rceil$ subgroups

$$D_{r,0}, D_{r,1}, \cdots, D_{r, \lceil \frac{T}{R} \rceil - 1}$$

of *arbitrary* sizes. At each access node a let the transmitter $(a, r\lceil\frac{T}{R}\rceil + i)$ be responsible for the outgoing links at a along dimensions in $D_{r,i}$ for any $0 \le i < \lceil\frac{T}{R}\rceil$. For each $T \bmod R \le r < R$, we consecutively partition D_r into $\lfloor\frac{T}{R}\rfloor$ subgroups

$$D_{r,0}, D_{r,1}, \cdots, D_{r,\lfloor\frac{T}{R}\rfloor-1}$$

of *arbitrary* sizes. At each access node a let the transmitter $(a, T \bmod R + r\lfloor\frac{T}{R}\rfloor + i)$ be responsible for the outgoing links at a along dimensions in $D_{r,i}$ for any $0 \le i < \lfloor\frac{T}{R}\rfloor$.Then it can be shown that the two transmitters $(a, r\frac{T}{R})$ and $(a', r\frac{T}{R})$ are in the same subnetwork if and only if $a^{-1}a'$ is local to the set of symbols in D_r except the minimal number in each $D_{r,i}$ for $1 \le i < \lceil\frac{T}{R}\rceil$ if $0 \le r < T \bmod R$ or $1 \le i < \lfloor\frac{T}{R}\rfloor$ if $T \bmod R \le r < R$.

If $(n-1) \bmod R = T \bmod R$, then

$$\begin{aligned} &L(T,R)\\ &= \max\left\{\left\lceil\frac{T}{R}\right\rceil\left(\left\lceil\frac{n-1}{R}\right\rceil - \left\lceil\frac{T}{R}\right\rceil + 1\right)!, \left\lfloor\frac{T}{R}\right\rfloor\left(\left\lfloor\frac{n-1}{R}\right\rfloor - \left\lfloor\frac{T}{R}\right\rfloor + 1\right)!\right\}\\ &= \left\lceil\frac{T}{R}\right\rceil\left(\frac{n-1-T}{R} + 1\right)! \end{aligned}$$

and

$$\begin{aligned} &W(T,R)\\ &= ((n-1) \bmod R)\frac{n!}{(\lceil\frac{n-1}{R}\rceil - \lceil\frac{T}{R}\rceil + 1)!} +\\ &(R - (n-1) \bmod R)\frac{n!}{(\lfloor\frac{n-1}{R}\rfloor - \lfloor\frac{T}{R}\rfloor + 1)!}\\ &= R\frac{n!}{(\frac{n-1-T}{R} + 1)!}. \end{aligned}$$

If $(n-1) \bmod R < T \bmod R$, then

$$\begin{aligned} &L(T,R)\\ &= \max\left\{\left\lceil\frac{T}{R}\right\rceil\left(\left\lceil\frac{n-1}{R}\right\rceil - \left\lceil\frac{T}{R}\right\rceil + 1\right)!, \left\lfloor\frac{T}{R}\right\rfloor\left(\left\lfloor\frac{n-1}{R}\right\rfloor - \left\lfloor\frac{T}{R}\right\rfloor + 1\right)!\right\}\\ &= \begin{cases} \lceil\frac{T}{R}\rceil(\lceil\frac{n-1-T}{R}\rceil + 1)! & \text{if } (n-1) \bmod R > 0,\\ \lfloor\frac{T}{R}\rfloor(\lceil\frac{n-1-T}{R}\rceil + 1)! & \text{if } (n-1) \bmod R = 0. \end{cases} \end{aligned}$$

and

$$
\begin{aligned}
&W(T,R)\\
&= ((n-1) \bmod R) \frac{n!}{(\lceil \frac{n-1}{R} \rceil - \lceil \frac{T}{R} \rceil + 1)!}\\
&+ (T \bmod R - (n-1) \bmod R) \frac{n!}{(\lfloor \frac{n-1}{R} \rfloor - \lceil \frac{T}{R} \rceil + 1)!}\\
&+ (R - T \bmod R) \frac{n!}{(\lfloor \frac{n-1}{R} \rfloor - \lfloor \frac{T}{R} \rfloor + 1)!}\\
&= \left(R + (T \bmod R - (n-1) \bmod R) \left\lceil \frac{n-1-T}{R} \right\rceil\right) \frac{n!}{(\lceil \frac{n-1-T}{R} \rceil + 1)!}.
\end{aligned}
$$

If $(n-1) \bmod R > T \bmod R$, then

$$
\begin{aligned}
&L(T,R)\\
&= \max\left\{\left\lceil \frac{T}{R} \right\rceil \left(\left\lceil \frac{n-1}{R} \right\rceil - \left\lceil \frac{T}{R} \right\rceil + 1\right)!, \left\lfloor \frac{T}{R} \right\rfloor \left(\left\lceil \frac{n-1}{R} \right\rceil - \left\lfloor \frac{T}{R} \right\rfloor + 1\right)!\right\}\\
&= \left\lfloor \frac{T}{R} \right\rfloor \left(\left\lceil \frac{n-1-T}{R} \right\rceil + 1\right)!
\end{aligned}
$$

and

$$
\begin{aligned}
&W(T,R)\\
&= (T \bmod R) \frac{n!}{(\lceil \frac{n-1}{R} \rceil - \lceil \frac{T}{R} \rceil + 1)!}\\
&+ ((n-1) \bmod R - T \bmod R) \frac{n!}{(\lceil \frac{n-1}{R} \rceil - \lfloor \frac{T}{R} \rfloor + 1)!}\\
&+ (R - (n-1) \bmod R) \frac{n!}{(\lfloor \frac{n-1}{R} \rfloor - \lfloor \frac{T}{R} \rfloor + 1)!}\\
&= R \frac{n!}{(\lfloor \frac{n-1-T}{R} \rfloor + 1)!} - ((n-1-T) \bmod R) \left\lceil \frac{n-1-T}{R} \right\rceil \frac{n!}{(\lceil \frac{n-1-T}{R} \rceil + 1)!}.
\end{aligned}
$$

Chapter 5

LOGICAL TOPOLOGY AND OPTIMAL ROUTING

As discussed in Chapter 1, the routing in multi-hop passive optical networks is traditionally based on some embedded regular interconnection topology with desired structural properties such as simple routing, predictable path lengths, balanced loads and fault-tolerance among others. However, one important fact that has been ignored in the past is that the process to realize the regular interconnection topology actually creates some *by-products*, the additional (logical) links not present in the original regular interconnection topology. Because of the better connectivity, such *side-effect* can be exploited to reduce the distance in terms of the number of hops among stations. Before we move on to the details, let's first examine a simple example.

Consider a PON with eight stations into which a three-dimensional hypercube is embedded as follows. Each station has a single transmitter and a single receiver. The transmitters at stations $000, 011, 101, 110$ and the receivers at stations $001, 010, 100, 111$ are assigned with wavelength λ_0, while the receivers at stations $000, 011, 101, 110$ and the transmitters at stations $001, 010, 100, 111$ are assigned with wavelength λ_1. Such wavelength assignment realizes the 3-cube. Now we consider the routing from station 000 to station 111. If the routing is simply based on the routing in the 3-cube, then the shortest distance is three hops. However, notice that the transmitter at station 000 and the receiver at the station 111 have the same wavelength λ_0, the station 000 can talk to the station 111 directly, and therefore their distance is just one. A graph theoretic explanation to this improvement is the difference between the embedded 3-cube and the actual logical topology. Figure 5.1 shows the logical topology of the above wavelength assignment. It is the supergraph of 3-cube. In addition to the links in the 3-cube, four additional links are present in the logical topology: the link between 000 and 111, the link between 001 and 110, the link between 010 and 101, and the link between 100 and 011. It's easy to see that the diameter of this logical topology is two, in contrast to the fact that the diameter of the 3-cube is three.

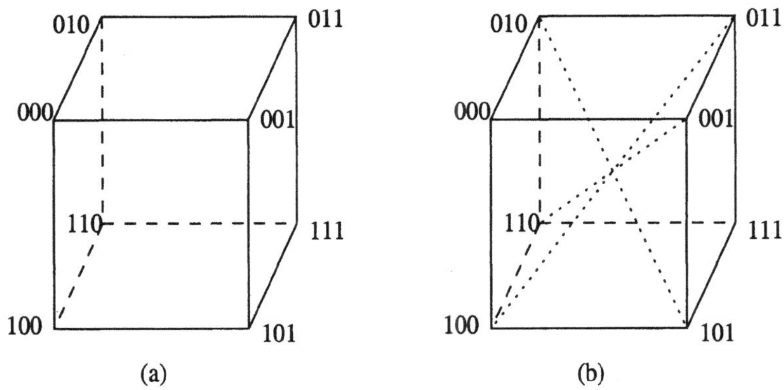

Figure 5.1. The logical topology is a supergraph of the embedded regular topology.

The above example raises the question on how better is the logical topology than the original regular interconnection topology in terms of the network properties such as routing, the load balancing and fault tolerance. To address this question, we have to first determine the structure of the logical topology. This is in general a very difficult task. However, with the results on the subnetwork structures obtained in the previous chapter, we are now ready to solve it. As will be turned out, the logical topology is also a regular topology in general. However, the optimal routing in these logical topology is sometimes hard to find.

This chapter will characterize the logical topologies of various passive optical networks studied in the previous chapter and discuss their routing properties. We first notice that for the PONs based on the generalized de Bruijn digraph whose nodal degree is a factor of the network size, the logical topology and the embedded generalized de Bruijn digraph are exactly the same. Thus therefor no improvements can be made. Similarly, the PONs based on generalized Kautz digraph whose nodal degree is a factor of the network size also have identical logical topology and the embedded regular topology. Therefore, we will only study the PONs based on hypercube, star graph and rotator digraph.

1. PONS BASED ON HYPERCUBE

For the simplicity of discussion, we first introduce some notations. For any

$$S \subseteq \{0, 1, \cdots, n-1\}$$

and any n-bit binary number a, we use $a|_S$ to denote the $|S|$-bit binary number consisting of the bits of a at positions in S, and $a|_{\overline{S}}$ to denote the $(n - |S|)$-bit binary number consisting of the bits of a at positions not in S. For any two stations a and b, we use $H(a, b)$ to denote their distance in terms the number of hops in the logical topology.

Due to the symmetry of the hypercube, swapping the number of transmitters and receivers does not change the connectivity of the logical topology. Thus we only consider the case that the number of transmitters is no more than the number of receivers. We will discuss the structural and routing properties of the logical topology from the simplest configuration to the most complicated yet most general configuration. For each configuration, we first determine the set of neighbors of each station and the nodal degree in the logical topology, which reflects the connectivity of the logical topology. After that we present the optimal routing algorithm in the logical topology. Finally we give the analytic expression for the diameter in terms of the number of hops in the logical topology.

1.1 $T = R$

1.1.1 LOGICAL TOPOLOGY

We begin with the simplest case that $T = R$. The n dimensions are consecutively partition into T groups $D_0, D_1, \cdots, D_{T-1}$ with

$$|D_t| = \begin{cases} \lceil \frac{n}{T} \rceil & \text{for } 0 \le t < n \bmod T, \\ \lfloor \frac{n}{T} \rfloor & \text{for } n \bmod T \le t < T. \end{cases}$$

The transmitter (a, t) is responsible for the outgoing links at a along dimensions in D_t, and the receiver (a, r) is responsible for the incoming links at a along dimensions in D_r. Then the subnetwork that the transmitter (a, t) belongs to consist of all transmitters

$$\left\{(b, t) : (a \oplus b)\,|_{D_t} \text{ is even and } (a \oplus b)\,|_{\overline{D_t}} = 0\right\}$$

and all receivers

$$\left\{(b,t) : (a \oplus b)\,|_{D_t} \text{ is odd and } (a \oplus b)\,|_{\overline{D_t}} = 0\right\}.$$

Thus in the logical topology, the adjacent stations of the station a is

$$\bigcup_{t=0}^{T-1} \left\{b : (a \oplus b)\,|_{D_t} \text{ is odd and } (a \oplus b)\,|_{\overline{D_t}} = 0\right\}.$$

So the nodal degree of each station in the logical topology is

$$\begin{aligned}
&\sum_{t=0}^{T-1} 2^{|D_t|-1} \\
&= (n \bmod T)\, 2^{\lceil \frac{n}{T} \rceil -1} + (T - n \bmod T)\, 2^{\lfloor \frac{n}{T} \rfloor -1} \\
&= (T + n \bmod T)\, 2^{\lfloor \frac{n}{T} \rfloor -1}.
\end{aligned}$$

As

$$(T + n \bmod T)\, 2^{\lfloor \frac{n}{T} \rfloor -1} = n$$

if and only if $T > \frac{n}{2}$, each station has more neighbors in the logical topology if $T \leq \frac{n}{2}$, and thus may reduce the distances between some nodes.

1.1.2 OPTIMAL ROUTING

Now let's look at the optimal routing in the logical topology. The routing from station a to station b is equivalent to changing the bits of a to the bits of b according to certain rules. In the logical topology with $T = R$, at each step any odd number of bits at positions in some D_t are allowed to be reversed simultaneously. Recall that in the original n-cube, only *one* bit can be changed at a time. Thus the distance in terms of the number of steps/hops to change a to b should be smaller.

Note that at each step the reversing of bits at positions in some D_t has no impact on the bits in other positions. Thus to change the bits of a at positions in D_t to the bits of b at positions in D_t, we only have to look at $(a \oplus b)\,|_{D_t}$. Suppose that $(a \oplus b)\,|_{\overline{D_t}} = 0$ and $(a \oplus b)\,|_{D_t} \neq 0$. Then

- if $(a \oplus b)\,|_{D_t}$ is odd, then a single hop is needed from station a to station b;

- if $(a \oplus b)\,|_{D_t}$ is even, then two hops are needed from station a to station b;

Therefore, the routing can be performed sequentially for each $0 \leq t < T$. The optimal routing given in Table 5.1 is very similar to the well-known Z-routing in the hypercube. It is given in the recursive format for the simplicity of description.

Algorithm Routing1(a, b)
if $a = b$, *stop;*
find the smallest t *with* $(a \oplus b)\,|_{D_t} \neq 0$;
if $(a \oplus b)\,|_{D_t}$ *is odd then*
pick any a' *satisfying that*
$(a' \oplus b)\,|_{D_t} = (a' \oplus a)\,|_{\overline{D_t}} = 0$
and $(a' \oplus a)\,|_{D_t}$ *is odd;*
a transmits to a' *via transmitter* (a, t);
Routing1(a', b);
else
pick any a' *satisfying that*
$(a' \oplus a)\,|_{\overline{D_t}} = 0$
and $(a' \oplus a)\,|_{D_t}$ *is odd;*
pick any a'' *satisfying that*
$(a'' \oplus b)\,|_{D_t} = (a'' \oplus a)\,|_{\overline{D_t}} = 0$;
a transmits to a' *via transmitter* (a, t);
a' *transmits to* a'' *via transmitter* (a', t);
Routing1(a'', b);
End-Algorithm

Table 5.1. Optimal routing from a to b when $T = R$.

1.1.3 DISTANCE AND DIAMETER

In the next we will study the diameter of the logical topology. For any binary number a and amy $0 \leq t < T$, we define $h_t(a)$ as follows.

- if $a|_{D_t} = 0$, then $h_t(a) = 0$;
- if $a|_{D_t} \neq 0$ and $a|_{D_t}$ is odd, then $h_t(a) = 1$;
- if $a|_{D_t} \neq 0$ and $a|_{D_t}$ is even, then $h_t(a) = 2$.

Then the distance between the station a and the station b is given by

$$H(a, b) = \sum_{t=1}^{T-1} h_t(a \oplus b).$$

It's easy to see that for any $0 \leq t < T$, the maxima of $h_t(a)$ is two if $|D_t| > 1$, and is one if $|D_t| = 1$. Therefore, if $n \geq 2T$, the diameter is equal to $2T$. Now we assume that $n < 2T$. For any $0 \leq t < n - T$, $|D_t| = 2$ and thus the maxima of $h_t(a)$ is two. For any $n - T \leq t < T$, $|D_t| = 1$ and thus the maxima of $h_t(a)$ is one. Therefore the diameter is

$$2(n - T) + (2T - n) = n.$$

In summary, when $T = R$, the diameter of the logical topology is $\min\{n, 2T\}$. Hence the fewer the number of transmitters/receivers, the shorter the diameter. However, the fewer number of transmitters may cause larger number of transmitters/receivers in each subnetwork and result in longer channel access delay. The worst number of time slots can be as large as

$$2T \cdot 2^{\lceil \frac{n}{T} \rceil - 1} = T 2^{\lceil \frac{n}{T} \rceil}$$

which might decrease as T increase.

1.2 $T < R$

1.2.1 LOGICAL TOPOLOGY

Now we consider the configuration with $0 < T < R < n$. The n dimensions are consecutively partitioned into T groups

$$D_0, D_1, \cdots, D_{T-1}$$

with

$$|D_t| = \begin{cases} \lceil \frac{n}{T} \rceil & \text{for } 0 \leq t < n \bmod T, \\ \lfloor \frac{n}{T} \rfloor & \text{for } n \bmod T \leq t < T. \end{cases}$$

The transmitter (a, t) is responsible for the outgoing links at a along dimensions in D_t. For each $0 \leq t < R \bmod T$, D_t is further consecutively partitioned into $\lceil \frac{R}{T} \rceil$ subgroups

$$D_{t,0}, D_{t,1}, \cdots, D_{t, \lceil \frac{R}{T} \rceil - 1}$$

of *arbitrary* sizes, and the receiver $(a, t\lceil \frac{R}{T} \rceil + i)$ is responsible for the incoming links at a along dimensions in $D_{t,i}$ for any $0 \leq i < \lceil \frac{R}{T} \rceil$. For each $R \bmod T \leq t < T, D_t$ is further consecutively partitioned into $\lfloor \frac{R}{T} \rfloor$ subgroups

$$D_{t,0}, D_{t,1}, \cdots, D_{t, \lfloor \frac{R}{T} \rfloor - 1}$$

of *arbitrary* sizes, and the receiver $(a, R \bmod T + t\lfloor \frac{R}{T} \rfloor + i)$ is responsible for the incoming links at a along dimensions in $D_{t,i}$ for any $0 \le i < \lfloor \frac{R}{T} \rfloor$. Then if $0 \le t < R \bmod T$ the subnetwork that the transmitter (a, t) belongs to consists of all transmitters

$$\left\{(b,t) : (a \oplus b)\mid_{\overline{D_t}} = 0, (a \oplus b)\mid_{D_{t,i}} \text{ is even for any } 0 \le i < \left\lceil \frac{R}{T} \right\rceil \right\}$$

and all receivers

$$\bigcup_{i=0}^{\lceil \frac{R}{T} \rceil - 1} \left\{(b, t\left\lceil \frac{R}{T} \right\rceil + i) : (a \oplus b)\mid_{\overline{D_t}} = 0, (a \oplus b)\mid_{D_{t,i}} \text{ is odd} \right.$$
$$\left. \text{and } (a \oplus b)\mid_{D_{t,j}} \text{ is even for any } j \ne i\right\}.$$

If $R \bmod T \le t < T$ the subnetwork that the transmitter (a, t) belongs to consists of all transmitters

$$\left\{(b,t) : (a \oplus b)\mid_{\overline{D_t}} = 0, (a \oplus b)\mid_{D_{t,i}} \text{ is even for any } 0 \le i < \left\lfloor \frac{R}{T} \right\rfloor \right\}$$

and all receivers

$$\bigcup_{i=0}^{\lfloor \frac{R}{T} \rfloor - 1} \left\{(b, R \bmod T + t\left\lfloor \frac{R}{T} \right\rfloor + i) : (a \oplus b)\mid_{\overline{D_t}} = 0, (a \oplus b)\mid_{D_{t,i}} \text{ is odd} \right.$$
$$\left. \text{and } (a \oplus b)\mid_{D_{t,j}} \text{ is even for any } j \ne i\right\}.$$

Thus in the logical topology, the adjacent stations of the station a is

$$\bigcup_{t=0}^{R\bmod T - 1} \bigcup_{i=0}^{\lceil \frac{R}{T} \rceil - 1} \{b : (a \oplus b)\mid_{\overline{D_t}} = 0, (a \oplus b)\mid_{D_{t,i}} \text{ is odd},$$
$$(a \oplus b)\mid_{D_{t,j}} \text{ is even } \forall j \ne i\} \cup$$
$$\bigcup_{t=R\bmod T}^{T-1} \bigcup_{i=0}^{\lceil \frac{R}{T} \rceil - 1} \{b : (a \oplus b)\mid_{\overline{D_t}} = 0, (a \oplus b)\mid_{D_{t,i}} \text{ is odd},$$
$$(a \oplus b)\mid_{D_{t,j}} \text{ is even } \forall j \ne i\}$$

So the nodal degree of each station in the super topology is

$$\sum_{t=0}^{R\bmod T - 1} \left\lceil \frac{R}{T} \right\rceil 2^{|D_t| - \lceil \frac{R}{T} \rceil} + \sum_{t=R\bmod T}^{T-1} \left\lfloor \frac{R}{T} \right\rfloor 2^{|D_t| - \lfloor \frac{R}{T} \rfloor}$$

which can be simplified as follows:

- If $n \bmod T = R \bmod T$, then the degree is

$$R2^{\frac{n-R}{T}}.$$

- If $n \bmod T < R \bmod T$, then the degree is

$$R2^{\lceil \frac{n-R}{T} \rceil} + (n \bmod T - R \bmod T) \left\lceil \frac{R}{T} \right\rceil 2^{\lfloor \frac{n-R}{T} \rfloor}.$$

- If $n \bmod T > R \bmod T$, then the degree is

$$\left(R + ((n-R) \bmod T) \left\lfloor \frac{R}{T} \right\rfloor \right) 2^{\lfloor \frac{n-R}{T} \rfloor}.$$

1.2.2 OPTIMAL ROUTING

Now let's look at the optimal routing in the logical topology. We again treat the routing from a to b as a number of steps to change the bits of a to the bits of b. For any $0 \leq t < T$, any odd number of bits at positions in some $D_{t,i}$ and any even number of bits at positions in any $D_{t,j}$ with $j \neq i$ are allowed to be reversed simultaneously within a single step. Note that at each step the reversing of bits at positions in some D_t has no impact on the bits in other positions. Thus to change the bits of a at positions in D_t to the bits of b at positions in D_t, we only have to look at $(a \oplus b)|_{D_t}$. Suppose that $(a \oplus b)|_{\overline{D_t}} = 0$ and $(a \oplus b)|_{D_t} \neq 0$. Then

- if $(a \oplus b)|_{D_{t,i}}$ is even for all i, then two hops are needed from station a to station b;

- otherwise, the minimal number of hops required from station a to station b is equal to the number of i's with odd $a|_{D_{t,i}}$.

Such procedure can be repeated sequentially for each $0 \leq t < T$. The recursive version of an optimal routing algorithm is given in Table 5.2.

Algorithm Routing2(a, b)
  if $a = b$, *stop;*
  find the smallest t *with* $(a \oplus b)|_{D_t} \neq 0$;
  find the set $S = \{i : (a \oplus b)|_{D_{t,i}} \text{ is odd}\}$;
  if $S = \emptyset$ *then*
    choose any a' *satisfying that*
      $(a' \oplus a)|_{\overline{D_t}} = 0$
      and $(a' \oplus a)|_{D_{t,i}}$ *is odd for some* i;
    choose any a'' *satisfying that*
      $(a'' \oplus b)|_{D_t} = (a'' \oplus a)|_{\overline{D_t}} = 0$;
    a transmits to a' *via transmitter* (a, t);
    a' *transmits to* a'' *via transmitter* (a', t);
    Routing2(a'', b);
  else
    for each $i \in S$ *in the increasing order do*
      if i *is the last one in* S *then*
        choose any a' *satisfying that*
          $(a' \oplus b)|_{D_t} = (a' \oplus a)|_{\overline{D_t}} = 0$
          and $(a' \oplus a)|_{D_{t,i}}$ *is odd;*
      else
        choose any a' *satisfying that*
          $(a' \oplus a)|_{\overline{D_t}} = 0$, $(a' \oplus a)|_{D_{t,i}}$ *is odd*
          and $(a' \oplus a)|_{D_{t,j}}$ *is even* $\forall\, j \neq i$;
      a transmits to a' *via transmitter* (a, t);
      replace a by a';
    Routing2(a, b);
End-Algorithm

Table 5.2. Optimal routing from a to b when $T < R$.

1.2.3 DISTANCE AND DIAMETER

In the next we determine the diameter of the logical topology. For any binary number a and any $0 \leq t < T$, we define $h_t(a)$ as follows.

- if $a|_{D_t} = 0$, then $h_t(a) = 0$;
- if $a|_{D_t} \neq 0$ and $a|_{D_{t,i}}$ is even for all i, then $h_t(a) = 2$;
- if $a|_{D_t} \neq 0$ and $a|_{D_{t,i}}$ is odd for some i, then $h_t(a)$ is equal the number of i's satisfying that $a|_{D_{t,i}}$ is odd.

Then the distance between the station a and the station b is given by

$$H(a, b) = \sum_{t=1}^{T-1} h_t(a \oplus b).$$

Thus the diameter is the sum of the maxima of $h_t(a)$ over all $0 \leq t < T$. In general, if the number of receivers at each station that are responsible for the links along dimensions in D_t is more than one, the maxima of $h_t(a)$ is equal to such number. If there is only one receiver at each station that is responsible for the links along dimensions in D_t, then the maxima of $h_t(a)$ is equal to two if $|D_t| > 1$ and equal to one if $|D_t| > 1$.

If $R \geq 2T$,

$$\left\lceil \frac{R}{T} \right\rceil \geq \left\lfloor \frac{R}{T} \right\rfloor \geq 2.$$

Thus for any $0 \leq t < T$, there are at least two receivers that are responsible for the links along dimensions in D_t. This implies that the maxima of $h_t(a)$ is equal to the number of receivers at each station that are responsible for the links along dimensions in D_t. So the diameter is equal to R, the total number of transmitters at each station.

If $T < R < 2T \leq n$,

$$\left\lceil \frac{n}{T} \right\rceil \geq \left\lfloor \frac{n}{T} \right\rfloor \geq 2.$$

For $0 \leq t < R - T$, there are exactly two receivers that are responsible for the links along dimensions in D_t, hence the maxima of $h_t(a)$ is 2. For $R - T \leq t < T$, $|D_t| \geq 2$ while there is only one receiver at each station that is responsible for the links along dimensions in D_t. So the maxima of $h_t(a)$ is also equal to two. Therefore the diameter is $2T$.

If $T < R < n < 2T$, we show that the diameter is n. In fact, for $0 \leq t < R - T$, $|D_t| = \lceil \frac{n}{T} \rceil = 2$ and there are exactly two receivers that are responsible for the two links along dimensions in D_t. Hence the maxima of $h_t(a)$ is 2. For $R - T \leq t < n - T$, $|D_t| = \lceil \frac{n}{T} \rceil = 2$ but there is only one receiver at each station that is responsible for the two links along dimensions in D_t. So the maxima of $h_t(a)$ is also equal to two. For $N - T \leq t < T - 1$, $|D_t| = \lfloor \frac{n}{T} \rfloor = 1$ and there is only one receiver at each station that is responsible for the link along dimension in D_t. So the maxima of $h_t(a)$ is equal to one. Therefore the diameter is

$$2(R - T) + 2(n - T) + (2T - n) = n.$$

In summary, when $0 < T < R < n$, the diameter is equal to

$$\min\{n, \max\{R, 2T\}\}.$$

2. PONS BASED ON STAR GRAPH

As discussed in previous chapter, any configuration with different T and R can be reduced to the configuration with $\min\{T, R\}$ transmitters and $\min\{T, R\}$ receivers without change the subnetwork structure. Thereby, we only consider the logical topology with equal T and R.

The $n-1$ dimensions are consecutively partitioned into T groups

$$D_0, D_1, \cdots, D_{T-1}$$

with

$$|D_t| = \begin{cases} \left\lceil \frac{n-1}{T} \right\rceil & \text{if } 0 \le t < (n-1) \bmod T, \\ \left\lfloor \frac{n-1}{T} \right\rfloor & \text{if } (n-1) \bmod T \le t < T. \end{cases}$$

The transmitter (a, t) is responsible for the outgoing links at a along dimensions in D_t, and the receiver (a, r) is responsible for the incoming links at a along dimensions in D_r. Then the subnetwork that the transmitter (a, t) belongs to consist of all transmitters

$$\left\{(b, t) : a^{-1}b \text{ is even and local to } \{1\} \cup \{i+2 \mid i \in D_t\}\right\}$$

and all receivers

$$\left\{(b, t) : a^{-1}b \text{ is odd and local to } \{1\} \cup \{i+2 \mid i \in D_t\}\right\}.$$

Thus in the logical topology, the adjacent stations of the station a is

$$\bigcup_{t=0}^{T-1} \left\{b : a^{-1}b \text{ is odd and local to } \{1\} \cup \{i+2 \mid i \in D_t\}\right\}.$$

So the nodal degree of each station in the logical topology is

$$\begin{aligned} &\sum_{t=0}^{T-1} \frac{(|D_t|+1)!}{2} \\ &= ((n-1) \bmod T) \frac{(\lceil \frac{n-1}{T} \rceil + 1)!}{2} + (T - (n-1) \bmod T) \frac{(\lfloor \frac{n-1}{T} \rfloor + 1)!}{2} \\ &= \left(T + ((n-1) \bmod T) \left\lceil \frac{n-1}{T} \right\rceil\right) \frac{(\lfloor \frac{n-1}{T} \rfloor + 1)!}{2}. \end{aligned}$$

It's easy to show that

$$\left(T + ((n-1) \bmod T) \left\lceil \frac{n-1}{T} \right\rceil\right) \frac{(\lfloor \frac{n-1}{T} \rfloor + 1)!}{2} > n-1$$

for any $0 < T < n-1$. So the logical topology provides more connectivity than the star graph, and thus may reduce the distances between some nodes.

Now let's look at the optimal routing in the logical topology. We only have to find the optimal routing from any permutation to the identity permutation. We first introduce the concept of length of a permutation. For each cycle permutation, or simply cycle, it's *length* is defined to be the minimum number of transpositions it can be decomposed into. Thus if it does not contain the symbol 1, it's length is equal to the number of symbols plus one. If it does contain the symbol 1, it's length is equal to the number of symbols minus one. The length of any non-identity permutation is then defined to be the sum of the lengths of all cycles it contains, and the length of the identity permutation is defined to be zero. Let a be any permutation which contains m symbols and consists of k cycles. If the symbol 1 is contained in a, then the length of a is

$$m - 1 + k - 1 = m + k - 2 \leq n + \left\lfloor \frac{n}{2} \right\rfloor - 2 = \left\lfloor \frac{3n}{2} \right\rfloor - 2.$$

If the symbol 1 is not contained in a, then the length of a is

$$m + k \leq (n-1) + \left\lfloor \frac{n-1}{2} \right\rfloor = \left\lfloor \frac{3n-1}{2} \right\rfloor - 1.$$

The concept of length plays an important role in the design of the routing algorithm in the star graphs. Actually it can be proved that the distance between any permutation a and the identity permutation is equal to the length of a. Thus any route from a to the identity permutation in which the length is decreased by one during each step is an optimal route. There are many options to decrease the length at each step. Here let's examine the effect on the change of the length caused by multiplying a cycle with a transposition. If the cycle contains the symbol 1, then the length is decreased by one if the transposition is between the symbol immediately before 1 in the cycle and the symbol 1. Otherwise, the length is increased by one. If the cycle does not contain the symbol 1, then the length is decreased by one for any transposition between some symbol in the cycle and the symbol 1, and is increase by one otherwise. In the logical topology, we will follow the similar ideas. Note that in the logical topology, any odd number of transpositions can be taken along the dimensions in some D_t. Therefore, the decreasing of the length by more than one is possible at each step in the logical topology. In the next we study how to maximize the decreasing of the length at each step.

Two symbols are called *neighboring* in a cycle if either no symbol or only symbol 1 is between them. It can be shown that if there exist two neighboring symbols which both belong to $\{i + 2 \mid i \in D_t\}$, then after an odd number of transpositions along dimensions in some D_t, we can come up with a permutation in which

- any two symbols in the set $\{i+2 \mid i \in D_t\}$ do not appear neighboring in any cycle;
- the symbol 1 appears in a cycle;
- all symbols in the set $\{i + 2 \mid i \in D_t\}$ appear in at most two cycles, and one must contain the symbol 1.

So in the first stage we try to remove the neighboring symbols. The implementation needs more subtle treatment. At the end of each step, we have to merge all, or all but one cycles containing symbols in the set $\{i+2 \mid i \in D_t\}$ to form a cycle containing the symbol 1. If some cycle can not be merged, the choice of such cycle is tricky. This cycle, if there is any, must contain a symbol in another set $\{i+2 \mid i \in D_{t'}\}$ such that some cycle containing two neighboring symbols in the set $\{i + 2 \mid i \in D_{t'}\}$. If such t' exists, then this t' will be used in the next step; otherwise, pick any cycle and find any t' such that there exist two symbols which both belong to $\{i+2 \mid i \in D_{t'}\}$ that are neighboring in some cycle. Note that each step in the first stage removes at least one symbol and therefore the length is decreased by at least one.

After the first stage, no two symbols which are both in some $\{i + 2 \mid i \in D_t\}$ appear neighboring in any cycle. However, the resulting permutation may consist of multiple cycles, one of which contains the symbol 1. In the second stage, we will merge those cycles which do not contain the symbol 1 into the cycle that contains the symbol 1. At each step, any odd number of cycles which contain at least one symbol from a common set $\{i + 2 \mid i \in D_t\}$ can be merged simultaneously into the cycle contains the symbol 1. However, the optimal merging is currently unsolved. Here we adopt a greedy merging procedure. At each step, we merge as many cycles as possible. This can be achieved by finding the number of cycles that contain symbol(s) from the each set $\{i+2 \mid i \in D_t\}$. This merging procedure is optimal locally at each step, but might not be optimal globally. We also note that each step decreases the length of the permutation by at most one.

At the end of the second stage, the permutation becomes a cycle permutation which contains one and have no neighboring symbols belonging to any set $\{i+2 \mid i \in D_t\}$. This permutation can be represented as a product of transpositions, the tonal number of which is equal to the length of the permutation. Therefore the number of steps required is at most the length of the permutation. We conjecture that this is also the minimal number of steps required.

3. PONS BASED ON ROTATOR DIGRAPH

We only consider the simplest case that $T = R$. The $n-1$ dimensions are consecutively partitioned into T groups

$$D_0, D_1, \cdots, D_{T-1}$$

with

$$|D_t| = \begin{cases} \left\lceil \frac{n-1}{T} \right\rceil & \text{if } 0 \le t < (n-1) \bmod T, \\ \left\lfloor \frac{n-1}{T} \right\rfloor & \text{if } (n-1) \bmod T \le t < T. \end{cases}$$

The transmitter (a, t) is responsible for the outgoing links at a along dimensions in D_t, and the receiver (a, r) will be responsible for the incoming links at a along dimensions in D_r. For any $0 \le t < T$, let

$$x_t = \min\{i+2 \mid i \in D_t\}$$
$$y_t = \max\{i+2 \mid i \in D_t\}.$$

Then the subnetwork that the transmitter (a, t) belongs to consists of all transmitters

$$\left\{(b, t) : a^{-1}b \text{ is local to } \{i+2 \mid i \in D_t\}\right\}$$

and all receivers

$$\left\{(b\alpha_{x_t}, t) : a^{-1}b \text{ is local to } \{i+2 \mid i \in D_t\}\right\}.$$

Thus in the logical topology, the adjacent stations of the station a is

$$\bigcup_{t=0}^{T-1} \left\{b\alpha_{x_t} : a^{-1}b \text{ is local to } \{i+2 \mid i \in D_t\}\right\}.$$

So the nodal degree of each station in the logical topology is

$$\begin{aligned}
&\sum_{t=0}^{T-1} |D_t|! \\
&= ((n-1) \bmod T)\left\lceil \frac{n-1}{T} \right\rceil ! + (T - (n-1) \bmod T)\left\lfloor \frac{n-1}{T} \right\rfloor ! \\
&= \left(T + ((n-1) \bmod T)\left\lfloor \frac{n-1}{T} \right\rfloor\right)\left\lfloor \frac{n-1}{T} \right\rfloor !.
\end{aligned}$$

It's easy to show that

$$\left(T + ((n-1) \bmod T)\left\lfloor \frac{n-1}{T} \right\rfloor\right)\left\lfloor \frac{n-1}{T} \right\rfloor ! \geq n-1$$

for any $0 < T < n-1$. So the logical topology provides more connectivity than the star graph, and thus may reduce the distances between some nodes.

Now let's look at the optimal routing in the logical topology. We only have to find the optimal routing from any permutation to the identity permutation. We also define the length of a permutation in the context of the rotator digraph. For any permutation a, the length of a is defined to be i if

$$a(i) > a(i+1) < a(i+2) < \cdots < a(n),$$

in other words, $a(i+1)a(i+2)\cdots a(n)$ is the longest increasingly sorted contiguous trailing region of a. It's obvious that the length of any permutation is at most $n-1$. The distance from a permutation a to the identity permutation in the rotator digraph is exactly the length of a. The optimal routing in rotator digraph is based on insertion sorting. At each step, the first symbol is inserted to the sorted trailing region such that the trailing sorted region of the new permutation is lengthened by one. As there is only one move that will lengthen the trailing sorted region, the shortest path from any permutation to the identity permutation is unique.

The routing in the logical topology follows the similar idea such that at each step, the length of the permutation is reduced by one or more. Let's examine how to maximize the decreasing of the length. Let a be the current permutation, and ℓ be the position that the first symbol of a, $a(1)$, should be inserted to according to the routing algorithm in the rotator digraph. Suppose that $\ell \in \{i+2 \mid i \in D_t\}$. If the length of

a is not in the $\{i+2 \mid i \in D_t\}$, then the station a sends data to the station $a\alpha_\ell$ via the transmitter (a,t). However, if the length of a is in the $\{i+2 \mid i \in D_t\}$, we can potentially do a little better. Let b be the unique permutation satisfying that

- $b(j) = a(j)$ for any $j \notin \{i+2 \mid i \in D_t\}$;
- over $\{i+2 \mid i \in D_t\}$, b is either monotonically increasing or *bitonic*, i.e., $b(x_t) > \cdots > b(\ell') < \cdots < b(y_t)$ for some $\ell' \in \{i+2 \mid i \in D_t\}$.

It's obvious that the length of b is no more than the length of a. As $a^{-1}b$ is local to $\{i+2 \mid i \in D_t\}$, the two transmitters (a,t) and (b,t) is in the same subnetwork. Thus the receiver $(b\alpha_\ell, t)$ is also in the subnetwork that the transmitter (a,t) belongs to. So we let the station a send data to the station $b\alpha_\ell$ via the transmitter (a,t). Note that the length of $b\alpha_\ell$ is one less then the length of b, which is no more than the length of a. Thus if the length of b is less than the length of a, the length decreasing is more than one.

While the exact value of the diameter of the logical topology is still unknown, one thing can be sure is that the diameter is less than $n-1$, the diameter of the rotator digraph. Any permutation other than the one in which the n symbols are monotonically decreasing, $n(n-1)\cdots 21$, has length of less than $n-1$, and thus its distance to the identity permutation is less than $n-1$. Therefore, we only have to show that the distance from $n(n-1)\cdots 21$ to the identity permutation is also less than $n-1$. We first show that its true when $n-1 \geq 2T$. The receiver selected in the first step by our routing algorithm is $(n-1)(n-2)\cdots(|D_{T-1}|+1)\,12\cdots|D_{T-1}|\,n$. As its length is $n-|D_{T-1}|-1$, the distance from $n(n-1)\cdots 21$ to the identity permutation is at most

$$1+(n-|D_{T-1}|-1) = n-|D_{T-1}| \leq n - \left\lfloor \frac{n-1}{T} \right\rfloor \leq n-2.$$

Now we assume that $T < n-1 < 2T$. In this case, $\lfloor \frac{n-1}{T} \rfloor = 1$ and $\lceil \frac{n-1}{T} \rceil = 2$. For $1 \leq i \leq 2T-(n-1)$, the receiver at step i is $(n-i)(n-i-1)\cdots 321(n-i+1)\cdots n$. However, at step $2T-(n-1)+1$, the receiver

$$(2\,(n-1-T))\,(2\,(n-1-T)-1)\cdots 312\,(2\,(n-1-T)+1)\cdots n$$

is selected in our routing algorithm. As it's length is $(2\,(n-1-T))-2$, the distance from $n(n-1)\cdots 21$ to the identity permutation is at most

$$(2T-(n-1))+1+((2\,(n-1-T))-2) = n-2.$$

Thus in any case, our routing algorithm can traverse in at most $n-2$ steps from $n(n-1)\cdots 21$ to the identity permutation. Thus the distance from $n(n-1)\cdots 21$ to the identity permutation is at most $n-2$. So the diameter of the logical topology is no more than $n-2$.

Chapter 6

MULTI-STAR CLUSTER INTERCONNECTION NETWORKS

1. INTRODUCTION

The previous two chapters have studied the PONs built upon single passive star coupler. The size of these networks are limited by the dimension of the passive star coupler and the transmission concurrence in these networks are limited by the number of available wavelength channels. The *cluster interconnection networks (CINs)* [11], described in Chapter 1, overcome these limitations by using multiple passive star couplers and sharing wavelengths in the additional spatial domain. The key design issue is the minimal channel set assignment. Since two output clusters can share a common channel set if and only if they are not adjacent to any common input star coupler, the number of output star couplers adjacent to any input star coupler gives a lower bound on the minimal number of channel sets. In this chapter, we will determine the suboptimal or optimal upper bounds on the minimal number of channel sets. The study relies on the formulation of the optimal channel set assignment problem into vertex coloring problem as described below.

Let G be the regular interconnection topology that the CIN is used to connect the couplers. When G is an undirected graph such as hypercube and star graph, each edge in G is treated as two unidirectional fiber links in opposite directions. If the CIN does not provide self-cluster links, then two output star couplers share a common input star coupler if and only if the distances of the corresponding two vertices in G is two. If we regard each channel set as a color, then a channel set assignment corresponds to a vertex coloring of G such that no two vertices with distance two have the same color. If the CIN does provide self-cluster links, then two output star couplers share a common input star coupler if and only if the distances of the corresponding two vertices in G is either one or two. If we also regard each channel set again as a color, then a channel set assignment at this time corresponds to a vertex coloring of G such that no two vertices with distance of at most two have the same color. Based on these observations, we define a *2-vertex-coloring* of an undirected graph G, denoted by 2-*VC*, to be any vertex coloring scheme such that no two vertices of distance two share the same color. Similarly, we define a $\overline{2}$*-vertex-coloring* of an undirected graph G, denoted by $\overline{2}$-*VC*, to be any vertex coloring scheme such that no two vertices of distance at most two share the same color.

When G which is a directed graph such as rotator digraph, each link corresponds to one unidirectional fiber link. If the cluster interconnection topology does not provide self-cluster links, then two output star couplers share a common input star coupler if and only if the corresponding two vertices in G are both immediate predecessors of some common vertex. Thus if we regard each channel set as a color, then a channel set assignment corresponds to a vertex coloring of G such that the immediate predecessors of any vertex have distinct colors. If the cluster interconnection topology does provide self-cluster links, then two output star couplers share a common input star coupler if and only if the corresponding two vertices in G either are adjacent or are both immediate predecessors of some common vertex. Based on these observations, we define a *2-vertex-coloring* of a directed graph G, denoted by 2-*VC*, to be any vertex coloring scheme such that the immediate predecessors of any vertex have distinct colors. Similarly, we define a $\overline{2}$*-vertex-coloring* of a directed graph G, denoted by $\overline{2}$-*VC*, to be any vertex coloring scheme such that any vertex and its immediate predecessors have distinct colors.

For any graph G, undirected or directed, we use $\chi_2(G)$ and $\chi_{\overline{2}}(G)$ to denote the minimal number of colors required by any 2-VC and $\overline{2}$-VC respectively. We will assume that G has no self-loops. It's obvious that the nodal degree of G is a lower bound on $\chi_2(G)$, and one plus the nodal degree of G is a lower bound on $\chi_{\overline{2}}(G)$. In the next section, we will present tight upper bounds on $\chi_2(G)$ and $\chi_{\overline{2}}(G)$ when G is the hypercube. In the last two sections, we will give the exact values of $\chi_2(G)$ and $\chi_{\overline{2}}(G)$ when G is a star graph or a rotator digraph.

The study on the minimal number of channel sets has been studied for various cluster interconnection topologies in [10]. In this chapter, we will extend their study on some other cluster interconnection topologies based on hypercubes, star graphs and rotator digraphs [155, 156, 158]. The optimal/suboptimal vertex coloring scheme for hypercubes, the star graph and the rotator digraphs will be presented in Section 2., Section 3. and Section 4. respectively.

2. VERTEX COLORING OF HYPERCUBE

The coloring schemes in hypercubes shall use a special class of matrices called *binary representation matrices.* The binary representation matrix of order k, denoted by BRM_k is a $\{0,1\}$-matrix consisting of k rows and $\lceil \log_2(k+1) \rceil$ columns. For any $1 \le i \le k$, the i-th row vector of BRM_k is the binary $\lceil \log_2(k+1) \rceil$-vector of the number i. For example, the matrix

$$BRM_5 = \begin{pmatrix} 0 & 0 & 1 \\ 0 & 1 & 0 \\ 0 & 1 & 1 \\ 1 & 0 & 0 \\ 1 & 0 & 1 \end{pmatrix}$$

is the binary representation matrix of order 5.

For simplicity of description, we will use C_n to denote the n-dimensional hypercube. For any binary n-vector

$$v = (v_1, v_2, \cdots, v_{n-1}, v_n),$$

we use $\hat{v}$ to denote the binary $(n-1)$-vector obtained from v by removing the last entry of v, i.e.,

$$\hat{v} = (v_1, v_2, \cdots, v_{n-1}).$$

In the next we will present a 2-VC in Subsection 2.1, and then a $\overline{2}$-VC in Subsection 2.2. Our coloring uses binary vectors to represent colors.

2.1 2-VC OF HYPERCUBE

Consider the following vertex coloring scheme:

Cube_VC_2 For any vertex v, coloring v with the color $\hat{v}BRM_{n-1}$.

The next theorem states that **Cube_VC_2** is a 2-VC of C_n.

THEOREM 6.1 *The coloring **Cube_VC_2** is a 2-VC of C_n.*

Proof. Let u and v be any two vertices with distance of two. Suppose that $u_i \neq v_i$ and $u_j \neq v_j$ for some $1 \leq i < j \leq n$. We consider two cases.

Case 1. $j = n$. In this case,

$$\begin{aligned} &\hat{u}BRM_{n-1} - \hat{v}BRM_{n-1} \\ &= (\hat{u} - \hat{v})BRM_{n-1} \\ &= (BRM_{n-1})_i \\ &\neq 0 \end{aligned}$$

as $(BRM_{n-1})_i$ is the binary $\lceil \log_2 n \rceil$-vector of the number i. Therefore

$$\hat{u}BRM_{n-1} \neq \hat{v}BRM_{n-1},$$

which means u and v have different colors.

Case 2. $j < n$. In this case,

$$\begin{aligned} &\hat{u}BRM_{n-1} - \hat{v}BRM_{n-1} \\ &= (\hat{u} - \hat{v})BRM_{n-1} \\ &= (BRM_{n-1})_i + (BRM_{n-1})_j \\ &\neq 0 \end{aligned}$$

as $i \neq j$. This implies that u and v have different colors.

So in either case, u and v have different colors. This completes the proof of the lemma. ∎

A vertex coloring scheme is said to be *balanced* if each color is assigned to the same number of vertices. The next lemma shows that **Cube_VC_2** has this nice property.

LEMMA 6.2 *The coloring **Cube_VC_2** uses $2^{\lceil \log_2 n \rceil}$ colors in total, and each color is assigned to $2^{n-\lceil \log_2 n \rceil}$ vertices.*

Proof. It's easy to see that **Cube_VC_2** uses $2^{\lceil \log_2 n \rceil}$ colors. So we only need to prove that each color is assigned to $2^{n-\lceil \log_2 n \rceil}$ vertices. Since the rank of the matrix BRM_{n-1} is equal to $\lceil \log_2 n \rceil$, the dimension of the null space of BRM_{n-1} is equal to $n-1-\lceil \log_2 n \rceil$. Notice that each binary $(n-1)$-vector $\hat{v}$ corresponds to two binary n-vectors. Therefore any color is assigned to $2 \cdot 2^{n-1-\lceil \log_2 n \rceil} = 2^{n-\lceil \log_2 n \rceil}$ vertices of C_n. ■

When n is a power of 2, **Cube_VC_2** uses n colors, which is the lower-bound for $\chi_2(C_n)$. Thus in this case, **Cube_VC_2** is an optimal 2-VC of C_n. If n is not a power of 2, then

$$\chi_2(C_n) \leq 2^{\lceil \log_2 n \rceil} < 2^{1+\log_2 n} = 2n \leq 2\chi_2(C_n).$$

Thus the number of colors used by **Cube_VC_2** is less than twice the lower bound.

2.2 $\overline{2}$-VC OF HYPERCUBE

The vertex coloring given below is very similar to **Cube_VC_2**:

Cube_VC_$\overline{2}$ For any vertex v, coloring v with the color $vBRM_n$.

The next theorem states that **Cube_VC_$\overline{2}$** is a $\overline{2}$-VC of C_n.

THEOREM 6.3 *The coloring **Cube_VC_$\overline{2}$** is a $\overline{2}$-VC of C_n.*

Proof. Let u and v be any two vertices whose distance is at most two. We consider two cases.

*Case 1.*The distance between u and v is one. Suppose that $u_i \neq v_i$ for some $1 \leq i \leq n$. Then

$$\begin{aligned} &uBRM_n - vBRM_n \\ &= (u-v)BRM_n \\ &= (BRM_n)_i \neq 0 \end{aligned}$$

as $(BRM_n)_i$ is the binary $\lceil \log_2(n+1) \rceil$-vector of the number i. Therefore

$$uBRM_n \neq vBRM_n$$

which means u and v have different colors.

Case 2. The distance between u and v is two. Suppose that $u_i \neq v_i$ and $u_j \neq v_j$for some $1 \leq i < j \leq n$. Then

$$\begin{aligned} &uBRM_n - vBRM_n \\ &= (u-v)BRM_n \\ &= (BRM_n)_i + (BRM_n)_j \neq 0 \end{aligned}$$

since $i \neq j$. This implies that u and v have different colors.

So in either case, u and v have different colors. This completes the proof of the lemma. ■

Similar to **Cube_VC_2**, **Cube_VC_$\overline{2}$** is also a balanced vertex coloring scheme.

LEMMA 6.4 *The coloring* ***Cube_VC_$\overline{2}$*** *uses* $2^{\lceil \log_2(n+1) \rceil}$ *colors and each color is assigned to* $2^{n-\lceil \log_2(n+1) \rceil}$ *vertices.*

The proof is similar to the proof for Lemma 6.2 and we omit it here.

When $n+1$ is a power of 2, the **Cube_VC_$\overline{2}$** uses $n+1$ colors, which is the lower-bound for $\chi_{\overline{2}}(C_n)$. Thus in this case, **Cube_VC_$\overline{2}$** is an optimal $\overline{2}$-VC of C_n. When $n+1$ is not a power of 2, then

$$\chi_{\overline{2}}(C_n) \leq 2^{\lceil \log_2(n+1) \rceil} < 2^{1+\log_2(n+1)} = 2(n+1) \leq 2\chi_{\overline{2}}(C_n),$$

Thus the number of colors used by **Cube_VC_$\overline{2}$** is less than twice the lower bound.

2.3 CONJECTURE ON THE OPTIMALITY

From Theorem 6.1 and Theorem 6.3, we have the following bounds on $\chi_2(C_n)$ and $\chi_{\overline{2}}(C_n)$:

$$n \leq \chi_2(C_n) \leq 2^{\lceil \log_2 n \rceil},$$
$$n+1 \leq \chi_{\overline{2}}(C_n) \leq 2^{\lceil \log_2(n+1) \rceil}.$$

The discussions above indicate that the lower-bound can be achieved in some networks of special sizes. One question is whether the lower-bound is always achievable. The following two examples give a negative answer.

Example 1. $\chi_2(C_3) = 2^{\lceil \log_2 3 \rceil} = 4$. This can be proved in the following way. Without loss of generality, assume that vertex 000 has color 1, vertex 011 has color 2 and vertex 101 has color 3. We consider the color of vertex 110. Since the vertex 110 has distance of 2 from each of the three vertices 000, 011 and 101, it must have some color different from color $1, 2$ and 3. Therefore, $\chi_2(C_3) \geq 4$. On the other hand, 4 is an upper bound for $\chi_2(C_3)$. So $\chi_2(C_3) = 4$.

Example 2. $\chi_{\overline{2}}(C_2) = 2^{\lceil \log_2 3 \rceil} = 4$. The proof is straightforward, since in C_2 the distance between any pair of vertices is at most two.

The above two examples indicates the optimality of our vertex coloring scheme. As we have not been able to find any hypercubes in which our vertex coloring scheme are not optimal, we raise the following conjecture.

CONJECTURE 6.5 *For any positive integer n,*

$$\chi_2(C_n) = 2^{\lceil \log_2 n \rceil},$$
$$\chi_{\overline{2}}(C_n) = 2^{\lceil \log_2 (n+1) \rceil}.$$

If this conjecture is true, then **Cube_VC_2** is an optimal 2-VC of C_n and **Cube_VC_$\overline{2}$** is an optimal $\overline{2}$-VC of C_n.

3. VERTEX COLORING OF STAR GRAPH

In this section, both left star graphs and right star graphs will be considered. A left star graph of order n is denoted by LS_n, and a right star graph of order n is denoted by RS_n. We will present the optimal $\overline{2}$-VCs of both left star graphs and right star graphs. In particular, we will prove that $\chi_{\overline{2}}(LS_n) = \chi_{\overline{2}}(RS_n) = n$.

Before we move on to the details on the optimal $\overline{2}$-VCs, let's look at the implication of them on the optimal 2-VCs. Note that any $\overline{2}$-VC of a graph is also a 2-VC of the same graph. Thus the optimal $\overline{2}$-VCs of a graph impose an upper bound on the colors used by the optimal 2-VC of the same graph. As $\chi_2(LS_n)$ and $\chi_2(RS_n)$ are both bounded from below by $n-1$, which is the nodal degree of LS_n and RS_n. Thus

$$n - 1 \le \chi_2(LS_n) \le \chi_{\overline{2}}(LS_n) = n,$$
$$n - 1 \le \chi_2(RS_n) \le \chi_{\overline{2}}(RS_n) = n.$$

This means that our optimal $\overline{2}$-VC uses at most one more color than any optimal 2-VC. However, any optimal 2-VC might have to use the same number of colors as any optimal $\overline{2}$-VC does. For an example, consider the star graph of order 3. Both LS_3 and RS_3 are a cycle consisting of six vertices. In such a cycle, any three alternating vertices of the cycle must have different colors, and therefore any optimal 2-VC of LS_3 or RS_3 must use three colors, the same number of colors used by any optimal $\overline{2}$-VC of LS_3 or RS_3. In general we conjecture that $\chi_2(LS_n) = \chi_2(RS_n) = n$.

In the next, we present the optimal $\overline{2}$-VCs of both left star graphs and right star graphs.

3.1 $\overline{2}$-VC OF LEFT STAR GRAPH

In our vertex coloring, the colors are indexed by positive integers. Our vertex coloring is described as follows.

LS_VC_$\overline{2}$ Coloring the vertex α with color $\alpha^{-1}(1)$.

We first prove that **LS_VC_$\overline{2}$** is a $\overline{2}$-VC of LS_n.

THEOREM 6.6 *The coloring* ***LS_VC_$\overline{2}$*** *an optimal $\overline{2}$-VC of LS_n, and therefore $\chi_{\overline{2}}(LS_n) = n$.*

Proof. It's obvious that **LS_VC_$\overline{2}$** uses n colors, which is a lower bound on $\chi_{\overline{2}}(LS_n)$. Thus to prove the theorem, we only have to prove that **LS_VC_$\overline{2}$** is a $\overline{2}$-VC of LS_n. Let α and β be two vertices in LS_n that have distance of at most two. We consider two cases.

case 1. The distance between α and β is one. Then there exist some integer $2 \leq i \leq n$ such that $\beta = \alpha \cdot s_i$. As $s_i^{-1} = s_i$,

$$\beta^{-1} = (\alpha \cdot s_i)^{-1} = s_i^{-1} \cdot \alpha^{-1} = s_i \cdot \alpha^{-1}.$$

So we have

$$\begin{aligned}\beta^{-1}(1) &= (s_i \cdot \alpha^{-1})(1) \\ &= \alpha^{-1}(s_i(1)) \\ &= \alpha^{-1}(i) \\ &\neq \alpha^{-1}(1).\end{aligned}$$

This implies that α and β have different colors.

case 2. The distance between α and β is two. Then there are two different integers $2 \leq i < j \leq n$ such that $\beta = \alpha \cdot s_i \cdot s_j$. In this case,

$$\beta^{-1} = (\alpha \cdot s_i \cdot s_j)^{-1} = s_j^{-1} \cdot s_i^{-1} \cdot \alpha^{-1} = s_j \cdot s_i \cdot \alpha^{-1}.$$

So we have

$$\begin{aligned}\beta^{-1}(1) &= (s_j \cdot s_i \cdot \alpha^{-1})(1)\\ &= \alpha^{-1}(s_i(s_j(1)))\\ &= \alpha^{-1}(s_i(j))\\ &= \alpha^{-1}(j)\\ &\neq \alpha^{-1}(1).\end{aligned}$$

This also implies that α and β have different colors.

Thus in both cases α and β have different colors. Therefore, **LS_VC_$\overline{2}$** is a $\overline{2}$-VC of LS_n. ■

The next lemma indicates that **LS_VC_$\overline{2}$** is also a balanced vertex coloring.

LEMMA 6.7 *The coloring **LS_VC_$\overline{2}$** assigns each color to exactly* $(n-1)!$ *vertices.*

Proof. It's easy to see that for any $1 \leq i \leq n$, all the $(n-1)!$ vertices which map i to 1 have the color i. ■

3.2 $\overline{2}$-VC OF RIGHT STAR GRAPH

Our vertex coloring of the right star graphs is given below.

RS_VC_$\overline{2}$ Coloring the vertex α with color $\alpha(1)$.

The next theorem states that **RS_VC_$\overline{2}$** is a $\overline{2}$-VC of RS_n.

THEOREM 6.8 *The coloring **RS_VC_$\overline{2}$** is an optimal $\overline{2}$-VC of RS_n, and therefore $\chi_{\overline{2}}(RS_n) = n$.*

Proof. It's obvious that **RS_VC_$\overline{2}$** uses n colors, which is a lower bound on $\chi_{\overline{2}}(RS_n)$. Thus to prove the theorem, we only have to prove that **RS_VC_$\overline{2}$** is a $\overline{2}$-VC of RS_n. Let α and β be two vertices in RS_n that have distance of at most two. We consider two cases.

case 1. The distance between α and β is one. Then there exist some integer $2 \le i \le n$ such that $\beta = \alpha \circ s_i$. Thus

$$\begin{aligned}\beta(1) &= (\alpha \circ s_i)(1)\\ &= \alpha(s_i(1))\\ &= \alpha(i)\\ &\neq \alpha(1).\end{aligned}$$

This implies that α and β have different colors.

case 2. The distance between α and β is two. Then there are two different integers $2 \le i < j \le n$ such that $\beta = \alpha \circ s_i \circ s_j$. Thus,

$$\begin{aligned}\beta(1) &= (\alpha \circ s_i \circ s_j)(1)\\ &= \alpha(s_i(s_j(1)))\\ &= \alpha(s_i(j))\\ &= \alpha(j)\\ &\neq \alpha(1).\end{aligned}$$

This also implies that α and β have different colors.

So in both cases α and β have different colors. Therefore, **RS_VC_$\overline{2}$** is a $\overline{2}$-VC of RS_n. ∎

RS_VC_$\overline{2}$ also shares the balance property with **LS_VC_$\overline{2}$**.

LEMMA 6.9 *The coloring **RS_VC_$\overline{2}$** assigns each color to exactly* $(n-1)!$ *vertices.*

Proof. It's easy to see that for any $1 \le i \le n$, all the $(n-1)!$ vertices which map 1 to i have the color i. ∎

3.3 2-VC OF STAR GRAPH

The lower bound on $\chi_2(LS_n)$ and $\chi_2(RS_n)$ is $n-1$, which is the nodal degree of LS_n and RS_n. On the hand, the $\overline{2}$-VC is also a 2-VC. Thus

$$\begin{aligned}n-1 \le \chi_2(LS_n) \le \chi_{\overline{2}}(LS_n) = n,\\ n-1 \le \chi_2(RS_n) \le \chi_{\overline{2}}(RS_n) = n.\end{aligned}$$

This means that the optimal $\overline{2}$-VC uses at most one more color than the optimal 2-VC. However, the optimal 2-VC might have to use the same number of colors as any optimal $\overline{2}$-VC does. For an example, consider the star graph of order 3. Both the left star graph of order 3 and the right star graph of order 3 are a cycle consisting of six vertices. In such a cycle, each vertex has two neighbors. However, any three alternating vertices of the cycle must have different colors, and therefore any optimal 2-VC of LS_3 or RS_3 must use three colors, the same number of colors used by any optimal $\overline{2}$-VC of LS_3 or RS_3. In general we conjecture that $\chi_2(LS_n = \chi_2(RS_n) = n$.

4. VERTEX COLORING OF ROTATOR DIGRAPH

In this section, both left rotator digraphs and right rotator digraphs will be considered. A left rotator digraph of order n is denoted by LR_n, and a right rotator digraph of order n is denoted by RR_n. We will present the optimal $\overline{2}$-VCs of both left rotator digraphs and right rotator digraphs. In particular, we will prove that

$$\chi_{\overline{2}}(LR_n) = \chi_{\overline{2}}(RR_n) = n.$$

This will imply that

$$n - 1 \leq \chi_2(LR_n) \leq \chi_{\overline{2}}(LR_n) = n,$$
$$n - 1 \leq \chi_2(RR_n) \leq \chi_{\overline{2}}(RR_n) = n,$$

as any $\overline{2}$-VC of a graph is also a 2-VC of the same graph. This means that our optimal $\overline{2}$-VC uses at most one more color than any optimal 2-VC. We notice that $\chi_2(LS_n)$ and $\chi_2(RS_n)$ might be equal to $n - 1$. For an example, consider the rotator digraph of order 3. The vertex coloring which assigns one color to the three vertices 123, 231, and 312, and another color to the remaining three vertices 213, 132, and 321 is a 2-VC of both LS_3 and RS_3. Thus $\chi_2(LS_3) = \chi_2(RS_3) = 2$. In general we believe that $\chi_2(LS_n) = \chi_2(RS_n) = n - 1$.

In the next, we present the optimal $\overline{2}$-VCs of both left star graphs and right star graphs.

4.1 $\overline{2}$-VC OF LEFT ROTATOR DIGRAPH

In our vertex coloring, the colors are indexed by positive integers. Our vertex coloring is described as follows.

LR_VC_$\overline{2}$ Coloring the vertex π with color $\pi(1)$.

We first prove that **LR_VC_$\overline{2}$** is a $\overline{2}$-VC of LR_n.

THEOREM 6.10 *The coloring **LR_VC_$\overline{2}$** is an optimal $\overline{2}$-VC of LR_n, and therefore $\chi_{\overline{2}}(LR_n) = n$.*

Proof. It's obvious that **LR_VC_$\overline{2}$** uses n colors, which is a lower bound on $\chi_{\overline{2}}(LR_n)$. Thus to prove the theorem, we only have to prove that **LR_VC_$\overline{2}$** is a $\overline{2}$-VC of LR_n. Let π and σ be two vertices in LR_n. We consider two cases.

case 1. π is an immediate successor of σ. Then there exist some integer $2 \leq i \leq n$ such that $\pi = \sigma \circ \alpha_i$. Thus,

$$\begin{aligned}\pi(1) &= (\sigma \circ \alpha_i)(1)\\ &= \sigma(\alpha_i(1))\\ &= \sigma(i)\\ &\neq \sigma(1).\end{aligned}$$

This implies that π and σ have different colors.

case 2. σ and π have a common immediate successor. Then there are two different integers $2 \leq i < j \leq n$ such that $\pi = \sigma \circ \alpha_i \circ \beta_j$. Thus

$$\begin{aligned}\pi(1) &= (\sigma \circ \alpha_i \circ \beta_j)(1)\\ &= \sigma(\alpha_i(\beta_j(1)))\\ &= \sigma(\alpha_i(j))\\ &\neq \sigma(\alpha_i(i))\\ &= \sigma(1).\end{aligned}$$

This also implies that α and β have different colors.

Thus in both cases α and β have different colors. This completes the proof of the lemma. ■

The next lemma shows that **LR_VC_$\overline{2}$** is a balanced vertex coloring.

LEMMA 6.11 *The coloring* ***LR_VC_$\overline{2}$*** *assigns each color to exactly* $(n-1)!$ *vertices.*

Proof. It's easy to see that for any $1 \leq i \leq n$, all the $(n-1)!$ vertices which map 1 to i have the color i. ■

Another interesting property is that the $n-1$ immediate successors of any node have the same color.

LEMMA 6.12 *The coloring* ***LR_VC_$\overline{2}$*** *assigns the* $n-1$ *immediate successors of any node in* LR_n *the same color.*

Proof. We only need to prove that for any vertex π and any $2 \leq i \leq n$, the value of $(\pi \circ \alpha_i)(1)$ depends only on π rather than on i. In fact,

$$(\pi \circ \alpha_i)(1) = \pi(\alpha_i(1)) = \pi(2).$$

Therefore, the lemma is true. ■

4.2 $\overline{2}$-VC OF RIGHT ROTATOR DIGRAPH

In this section, we will present an optimal $\overline{2}$-VC of RR_n. The vertex coloring is similar to **LR_VC_$\overline{2}$** and is given below.

RR_VC_$\overline{2}$ Coloring the vertex π with color $\pi(1)$.

The next theorem proves that **RR_VC_$\overline{2}$** is an optimal $\overline{2}$-VC of RR_n.

THEOREM 6.13 *The coloring* ***RR_VC_$\overline{2}$*** *is an optimal* $\overline{2}$*-VC of* RR_n*, and therefore* $\chi_{\overline{2}}(RR_n) = n$.

Proof. It's obvious that **RR_VC_$\overline{2}$** uses n colors, which is a lower bound on $\chi_{\overline{2}}(RR_n)$. Thus to prove the theorem, we only have to prove that **RR_VC_$\overline{2}$** is a $\overline{2}$-VC of RR_n. Let π and σ be two vertices in RR_n. We consider two cases.

case 1. σ is an immediate successor of π. Then there exist some integer $2 \leq i \leq n$ such that $\sigma = \pi \cdot \beta_i$. Thus,

$$\sigma^{-1} = \alpha_i \cdot \pi^{-1}.$$

So we have

$$\begin{aligned} \sigma^{-1}(1) &= (\alpha_i \cdot \pi^{-1})(1) \\ &= \pi^{-1}(\alpha_i(1)) \\ &= \pi^{-1}(i) \\ &\neq \pi^{-1}(1). \end{aligned}$$

This implies that α and β have different colors.

case 2. σ and π have a common immediate successor. Then there are two different integers $2 \leq i < j \leq n$ such that $\sigma = \pi \cdot \beta_i \cdot \alpha_j$. Thus,

$$\sigma^{-1} = \beta_j \cdot \alpha_i \cdot \pi^{-1}.$$

Therefore,

$$\begin{aligned} \sigma^{-1}(1) &= (\beta_j \cdot \alpha_i \cdot \pi^{-1})(1) \\ &= \pi^{-1}(\alpha_i(\beta_j(1))) \\ &= \pi^{-1}(\alpha_i(j)) \\ &\neq \pi^{-1}(\alpha_i(i)) \\ &= \pi^{-1}(1). \end{aligned}$$

This also implies that α and β have different colors.

Thus in both cases α and β have different colors. Therefore, the coloring scheme **RR_VC_$\overline{2}$** is a $\overline{2}$-VC of RR_n. ∎

The next lemma shows that **RR_VC_$\overline{2}$** is a balanced vertex coloring.

LEMMA 6.14 *The coloring* ***RR_VC_$\overline{2}$*** *assigns each color to exactly* $(n-1)!$ *vertices.*

Proof. It's easy to see that for any $1 \leq i \leq n$, all the $(n-1)!$ vertices which map i to 1 have the color i. ∎

Similar to **LR_VC_$\bar{2}$** of LR_n, the **RR_VC_$\bar{2}$** of RR_n also assigns the same color to the $n-1$ immediate successors of any node.

LEMMA 6.15 *The coloring **RR_VC_$\bar{2}$** assigns the $n-1$ immediate successors of any node in LR_n the same color.*

Proof. We only need to prove that for any vertex π and any $2 \leq i \leq n$, the value of $(\pi \cdot \beta_i)^{-1}(1)$ depends only on π rather than on i. In fact,

$$(\pi \cdot \beta_i)^{-1}(1) = (\alpha_i \cdot \pi^{-1})(1) = \pi^{-1}(\alpha(1)) = \pi^{-1}(2).$$

Therefore, the lemma is true. ∎

III

WAVELENGTH-SWITCHED OPTICAL NETWORKS

Chapter 7

LIGHTPATH COLORING IN RINGS AND TREES OF RINGS

1. INTRODUCTION

Wavelength-switched optical networks are the underlying network architecture for optical transport networks. In this chapter, we assume that the optical switches have no wavelength conversion functions. In these networks, an end-to-end connection is established as a lightpath. At any time, two lightpaths sharing a common link must have different wavelengths. A basic design problem is to choose a set of lightpaths with minimal number of wavelengths to connect a given set of communication pairs. This optimization problem is known as the *route and wavelength assignment problem* if the routings of all communication pairs are not prespecified, or as the *path-coloring problem* if the routings of all communication pairs are given in advance. The latter is equivalent to a graph-coloring problem on the *intersecting graph* of the set of paths–the graph whose vertices are the given paths, with an edge between each pair of paths that share a link in the network. Note that if the underlying network has a linear or tree topology, the routing between any pair of nodes is unique and thus these two optimization problems become identical. This chapter will address only the path-coloring problem in rings and trees of rings, which are of most practical interest due to their easy support for fault protection.

Given a set of paths $\mathcal{P}$ in a graph, we use ℓ, ω and χ to denote its load, clique number and chromatic number respectively. An obvious relation among them is that $\ell \leq \omega \leq \chi$. If the underlying network has a linear topology, the path-coloring problem is the well-known *interval coloring* problem, which can be solved by a simple greedy algorithm in polynomial time. In addition, for any set of interval, $\ell = \omega = \chi$.

If the underlying network is a ring, the path-coloring problem is the well-known *circular-arc coloring* problem [63] [151]. The clique number ω of a set of circular arcs can be calculated in polynomial time [24] [174], and the following relation is known in the literature:

$$\chi \leq \min\left\{2\ell - 1, \left\lfloor \frac{3\omega}{2} \right\rfloor\right\}.$$

The relation $\chi \leq 2\ell - 1$ follows from Tucker's algorithm [151]: pick up any node which is an endpoint of some arc, assign each path across this node with a unique color, and then apply any algorithm for interval coloring to color the remaining paths. The relation $\chi \leq \lfloor \frac{3\omega}{2} \rfloor$ follows from Karapetian's algorithm [90], which will be described later in Section 2.. For certain instances, the equality in the above two relations can hold. Consider the following instance of $2k+1$ paths in a ring of $4k+2$ nodes for any integer $k \geq 1$ (see Figure 7.1 (a)): the set of paths consists of the arcs from node[1] $4i$ to $4i+3$ and the arcs from node $4i+2$ to $4i+5$ for $1 \leq i \leq k$. The intersecting graph of these paths is a circuit of $2k+1$ vertices (see Figure 7.1 (b)). Thus $\ell = \omega = 2$ and $\chi = 3$. So for this instance, $\chi = 2\ell - 1 = \lfloor \frac{3\omega}{2} \rfloor$.

For circular-arc coloring, both χ and ω can be as large as $2\ell - 1$. Consider the following instance of $2k+1$ paths in a ring of $2k+1$ nodes for any integer $k \geq 1$ (see Figure 7.2 (a)): the set of paths consists of the arcs from node i to $i+k+1$ for $0 \leq i \leq 2k$. The intersecting graph of these paths is a clique of $2k+1$ vertices (see Figure 7.2 (b)). Thus $\chi = \omega = 2k+1$ and $\ell = k$. So for this instance, $\chi = \omega = 2\ell - 1$. In this chapter, we will present yet another heuristic similar to but simpler than the one in [90], which uses at most $\min\{2\ell - 1, \omega + \lfloor \frac{\ell}{2} \rfloor\}$ colors. This implies the following stronger relation

$$\chi \leq \min\left\{2\ell - 1, \omega + \left\lfloor \frac{\ell}{2} \right\rfloor\right\}.$$

A tree of rings, also known as a *cactus*, is a connected graph whose biconnected components are all rings (see Figure 7.3), i.e., a pair of

[1]All arithematic involoving node indices in a ring will be implicitly done using operation modulo the ring size.

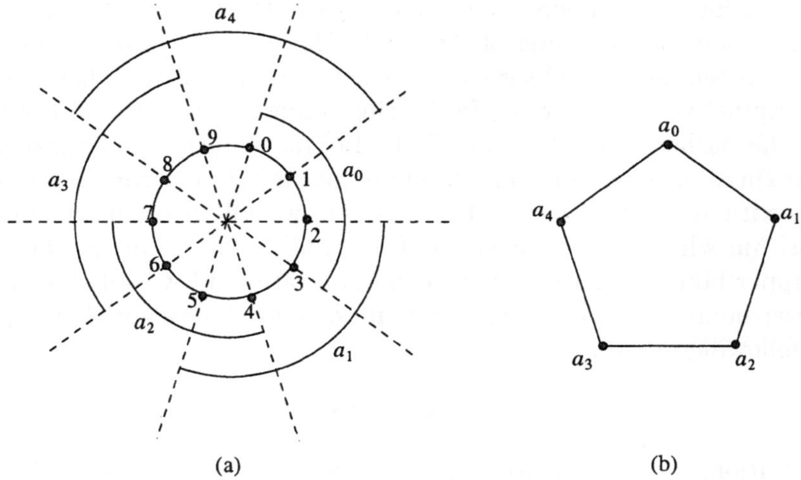

Figure 7.1. The instance with $\chi = 2\ell - 1 = \lfloor \frac{3\omega}{2} \rfloor$: (a) the paths, and (b) their intersection graph.

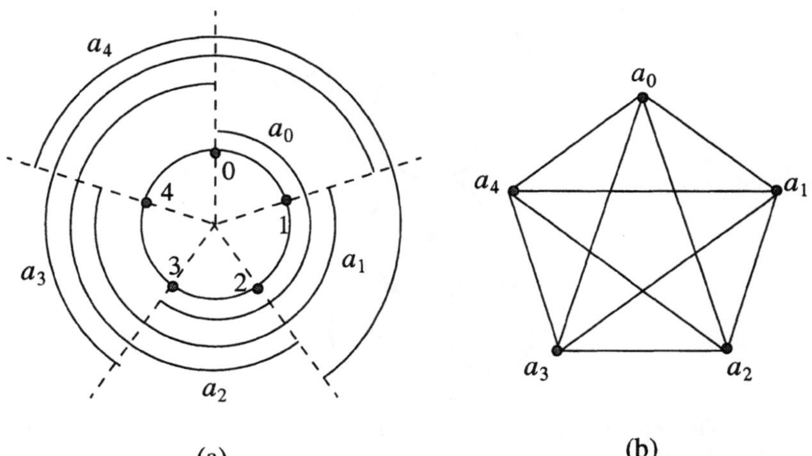

Figure 7.2. The instance with $\chi = \omega = 2\ell - 1$: (a) the paths, and (b) their intersection graph.

vertices have two vertex-disjoint paths if and only if they are within the same ring. From any tree of rings, a tree, referred to its *skeleton*, can be constructed in the following manner. The vertex set of the skeleton tree consists of the original vertices in the tree of rings with degree (even) at least four and a number of new vertices (the black squares in Figure 7.3), each of which corresponds to a unique ring in the tree of rings (imagine the new node as the center of the ring). The edge in the skeleton exists only between an original vertex and a new vertex, and the edge exists if the original vertex is the inside the ring represented by the new vertex (see the dashed lines in Figure 7.3). In [48], a rather complicated 2-approximation algorithm is proposed to give a path coloring in a tree of rings with at most 2ω colors. In Section 3., we will give a much simpler algorithm which uses at most $\min\{\omega + \ell, 3\ell\}$ colors. This implies and an upper bound of 2 on its approximation ratio and we will also prove a lower bound of 1.5 on its approximation ratio. This algorithm implies the following relation,

$$\chi \leq \min\{3\ell, \omega + \ell\}.$$

In addition, we will construct an instance with $\chi = \omega = 3\ell$ in Section 3..

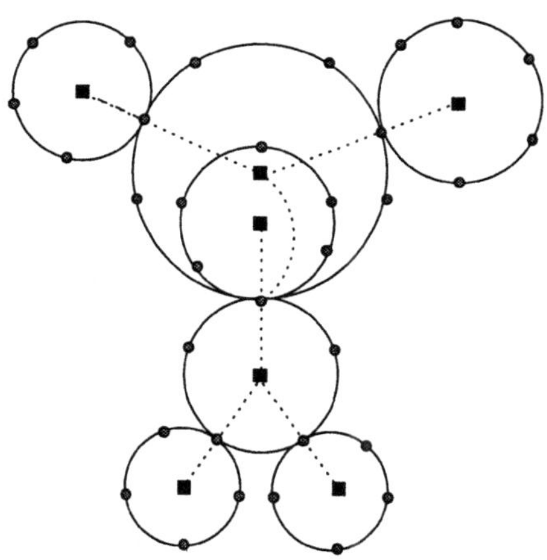

Figure 7.3. The tree of rings depicted in solid arcs, and its skeleton (tree) depicted in dashed lines with black squares representing the centers of the rings.

A remark on this relation is that it can *not* be derived from works in [129] which addressed the routing *and* wavelength assignment in trees of rings. In [129], the algorithm for routing *and* wavelength assignment has the luxury to choose the path for each communication request. It first removes a link from each ring to obtain a tree, and then routes all requests along the tree. The load of such routing is at most twice the minimum load. Finally, it applies the path coloring algorithm for trees presented in [129], which uses colors of at most 1.5 times the load in the tree. Thus the total number of colors is at most three times of the minimum load. On the other hand, our result can lead to an algorithm for routing and routing wavelength assignment which also uses at most three times the minimum load. In fact, we can first apply the algorithm presented in [137] on each ring to get a routing with minimum load, and then apply our algorithm on the obtained paths.

The rest of the chapter proceeds as follows. In Section 2. we present a circular-arc coloring algorithm which uses at most $\min\{2\ell - 1, \omega + \lfloor \frac{\ell}{2} \rfloor\}$ colors. In Section 3. we present a path coloring algorithm in trees of rings which uses at most $\min\{\omega + \ell, 3\ell\}$ colors. In addition, its approximation ratio is shown to be at least 1.5. Instances are also constructed to show that both χ and ω can be as large as 3ℓ. Finally Section 4. summarizes this chapter.

2. CIRCULAR-ARC COLORING

Let $\mathcal{P}$ be a set of circular arcs on a ring oriented *clockwise*. Each arc a is denoted by $(t(a), h(a))$ where $t(a)$ and $h(a)$ are the *tail* and *head* of the arc a respectively. W.l.o.g., assume the link between node 0 and node 1 has the maximum load. We first describe Karapetian's algorithm presented in [90].

- **Initialization:** Calculate the clique number ω of $\mathcal{P}$. Adds $\omega - \ell$ "dummy" arcs from node 0 to node 1. Let $\mathcal{P}'$ denote the obtained paths. Then both the load and the clique number of $\mathcal{P}'$ are equal to ω. Sort the ω arcs containing the link from node 0 to node 1 in the *increasing* order of their heads and denote them by $a_1, a_2, \cdots, a_\omega$.
- **Phase 1**: Sequentially select $\lfloor \frac{\omega}{2} \rfloor$ disjoint subsets $A_1, A_2, \cdots, A_{\lfloor \frac{\omega}{2} \rfloor}$ of $\mathcal{P}' \setminus \{a_1, a_2, \cdots, a_\omega\}$ as follows . Suppose $A_1, A_2, \cdots, A_{i-1}$ have been obtained. The set A_i is selected by repeatedly adding an arc from the remaining arcs *not* in $\{a_1, a_2, \cdots, a_\omega\}$ with the smallest (leftmost) tail which does not intersect any arcs in A_i until no further arcs can be added. Note that the arcs added clockwise.
- **Phase 2**: For any $1 \leq i \leq \omega$, initialize $B_i = \{a_i\}$. Sequentially complete $B_\omega, B_{\omega-1}, \cdots, B_1$ as follows. Suppose $B_\omega, B_{\omega-1}, \cdots, B_{i+1}$ have been obtained. The set B_i is selected by repeatedly adding an arc from the remaining arcs in with the largest (rightmost) head which does not intersect any arcs in B_i until no further arcs can be added. Note that the arcs added counterclockwise.

It was proved in [90] after **Phase 2** all arcs are colored. Thus Karapetian's algorithm uses $\lfloor \frac{3\omega}{2} \rfloor$ colors. Our algorithm, as described below, is simpler and more natural in the sense that we avoid calculating the clique number and adding the dummy arcs. Instead, we add a new phase which is essentially a simple interval coloring.

- **Initialization:** Sort the ℓ arcs containing the link from node 0 to node 1 in the *increasing* order of their heads and denote them by $a_1, a_2, \cdots, a_\ell$.
- **Phase 1**: Sequentially select $\lfloor \frac{\ell}{2} \rfloor$ disjoint subsets $A_1, A_2, \cdots, A_{\lfloor \frac{\ell}{2} \rfloor}$ of $\mathcal{P} \setminus \{a_1, a_2, \cdots, a_\ell\}$ as follows. Suppose $A_1, A_2, \cdots, A_{i-1}$ have been obtained. The set A_i is selected by repeatedly adding an arc (or interval) from the remaining arcs *not* in $\{a_1, a_2, \cdots, a_\ell\}$ with the smallest (leftmost) tail which does not intersect any arcs in A_i until no further arcs can be added. Note that the arcs added clockwise.
- **Phase 2**: For any $1 \leq i \leq \ell$, initialize $B_i = \{a_i\}$. Sequentially complete $B_\ell, B_{\ell-1}, \cdots, B_1$ as follows. Suppose $B_\ell, B_{\ell-1}, \cdots, B_{i+1}$ have been obtained. The set B_i is selected by repeatedly adding an arc from the remaining arcs in with the largest (rightmost) head which does not intersect any arcs in B_i until no further arcs can be added. Note that the arcs added counterclockwise.
- **Phase 3**: Apply any interval coloring algorithm to color the remaining arcs.

Note that the colors used by our algorithm is equal to the $\lfloor \frac{\ell}{2} \rfloor + \ell = \lfloor \frac{3\ell}{2} \rfloor$ plus the load of the arcs after **Phase 2**. So our analysis reduces to bounding the load of the arcs after **Phase 2**. Our approach is based on the techniques developed in [90] with finer analysis. Since the original paper [90] was written in Russian and our proof is more concise, we nevertheless provide the detailed proof here for the completeness.

We begin with the analysis of the properties of the arcs after **Phase 1**.

LEMMA 7.1 *For any $a \notin \left(\{a_1, a_2, \cdots, a_\ell\} \cup \left(\bigcup_{j=1}^{i-1} A_j\right)\right)$ with $1 < i \leq \lfloor \frac{\ell}{2} \rfloor + 1$, $h(a_i) \leq t(a)$ and the link between $t(a)$ and $t(a)+1$ is contained in some arc in A_j for any $1 \leq j < i$.*

Proof. For any $1 \leq j \leq i-1$, A_j must an arc which contains the link between $t(a)$ and $t(a)+1$, for otherwise the link would have been selected earlier. If $t(a) < h(a_i)$, then the link between $t(a)$ and $t(a)+1$ is also contained in the arcs $a_i, a_{i+1}, \cdots, a_\ell$. Thus the load of the link between $t(a)$ and $t(a)+1$ is at least $\ell+1$, which is impossible. ■

The next lemma analyzes the properties of the arcs after **Phase 2**.

LEMMA 7.2 *For any $a \notin (\bigcup_{i=1}^{m} A_i) \cup \left(\bigcup_{i=1}^{\ell} B_i\right)$, there exist some $i \geq \lfloor \frac{\ell}{2} \rfloor + 2$ such that a_i contains the link between $t(a)$ and $t(a)+1$. Furthermore, if a_i contains the link between $t(a)$ and $t(a)+1$, then the tail of each arc in $B_i \setminus \{a_i\}$ is less than $h(a)$.*

Proof. We first observe that a is not contained in $a_{\lfloor \frac{\ell}{2} \rfloor+1}$, and consequently $a_1, a_2, \cdots, a_{\lfloor \frac{\ell}{2} \rfloor}$ from Lemma 7.1. If the link between $t(a)$ and $t(a)+1$ is not contained in any arc a_i for $\lfloor \frac{\ell}{2} \rfloor + 2 \leq i \leq \ell$, then each set B_i must have an arc which contains the link between $h(a)-1$ and $h(a)$, and thus the load of the this link is at least $\ell+1$, which is impossible. Thus the first part of the lemma is true.

Let k be the smallest number such that the link between $t(a)$ and $t(a)+1$ is contained in a_k. Then $k \geq m+2$. We prove by contradiction that the tail of each arc in $\bigcup_{i=k}^{\ell}(B_i \setminus \{a_i\})$ is less than $h(a)$. Assume to the contrary. Let b be the arc in $\bigcup_{i=k}^{\ell}(B_i \setminus \{a_i\})$ such that $h(a) \leq t(b)$ and $t(b)$ is the smallest. Assume that $b \in B_{i_0}$ for some $k \leq i_0 \leq \ell$. For any $1 \leq i \leq k-1$, each B_i has an arc b_i that contains the link between $h(a)-1$ and $h(a)$. Let I_1 be the set of i's with $1 \leq i \leq k-1$ such that b_i contains the link between $t(b)$ and $t(b)+1$. Note that for any $i \in I_1$, the arc b_i could be a_i. Let I_2 be the set of i's with $k \leq i \leq \ell$ and $i \neq i_0$ such that the set B_i has an arc containing the link between $t(b)$ and $t(b)+1$. Since b is not a_{i_0}, the link between $t(b)$ and $t(b)+1$ is also contained in $\lfloor \frac{\ell}{2} \rfloor$ arcs selected in **Phase 1** from Lemma 7.1. In addition, this link is contained in the arc b itself. Thus

$$\ell \geq \left\lfloor \frac{\ell}{2} \right\rfloor + |I_1| + |I_2| + 1,$$

which implies that

$$|I_1| + |I_2| \leq \left\lceil \frac{\ell}{2} \right\rceil - 1.$$

We next show that at least one B_i for some $i \in \{k, k+1, \cdots, \ell\} \setminus I_2$ does not have any arc that contains the link between $h(a)-1$ and $h(a)$.

Assume to the contrary that for any $i \in \{k, k+1, \cdots, \ell\} \setminus I_2$, there is an arc b_i in B_i that contains the link between $h(a) - 1$ and $h(a)$. Then for any $i \notin I_1 \cup I_2$, the arc b_i can't be a_i. Let b' be the arc in the set

$$\{a\} \cup \{b_i \mid i \notin I_1 \cup I_2\}$$

such that $t(b')$ is the largest. Then the link between $t(b')$ and $t(b') + 1$ is contained in m arcs in **Phase 1**, and the above $1 + \ell - (|I_1| + |I_2|)$ arcs. Thus

$$\begin{aligned}
\ell &\geq \left\lfloor \frac{\ell}{2} \right\rfloor + 1 + \ell - (|I_1| + |I_2|) \\
&\geq \left\lfloor \frac{\ell}{2} \right\rfloor + 1 + \ell - \left(\left\lceil \frac{\ell}{2} \right\rceil - 1 \right) \\
&= 2 \left\lfloor \frac{\ell}{2} \right\rfloor + 2,
\end{aligned}$$

which is impossible.

Let j_0 be the smallest number in $\{k, k+1, \cdots, \ell\} \setminus I_2$ such that no arc in B_{j_0} contains the link between $h(a) - 1$ and $h(a)$. For any $i \in \{k, k+1, \cdots, j_0 - 1\} \setminus I_2$, let b_i be the arc in B_i that contains the link between $h(a) - 1$ and $h(a)$. From the definition of b, the portion between $h(a) - 1$ and $t(b) + 1$ if $j_0 \neq i_0$ or the portion between $h(a) - 1$ and $t(b)$ if $j_0 = i_0$ is "blank" in B_{j_0}. Thus for any $i \in \{1, 2, \cdots, j_0 - 1\} \setminus (I_1 \cup I_2)$, the arc b_i must contain the link between $h(a_{j_0}) - 1$ and $h(a_{j_0})$. Let b'' be the arc in the set

$$\{a\} \cup \{b_i \mid i \in \{1, 2, \cdots, j_0 - 1\} \setminus (I_1 \cup I_2)\}$$

such that $t(b'')$ is the largest. Then the link between $t(b'')$ and $t(b'') + 1$ is contained in the above, the arcs $\{a_i \mid j_0 \leq i \leq \ell\}$, and the m arcs selected in **Phase 1**. Therefore,

$$\begin{aligned}
\ell &\geq \left\lfloor \frac{\ell}{2} \right\rfloor + (\ell - j_0 + 1) + 1 + |\{1, 2, \cdots, j_0 - 1\} \setminus (I_1 \cup I_2)| \\
&\geq \left\lfloor \frac{\ell}{2} \right\rfloor + (\ell - j_0 + 1) + 1 + j_0 - 1 - (|I_1| + |I_2|) \\
&= \left\lfloor \frac{\ell}{2} \right\rfloor + 1 + \ell - (|I_1| + |I_2|) \\
&\geq \left\lfloor \frac{\ell}{2} \right\rfloor + 1 + \ell - \left(\left\lceil \frac{\ell}{2} \right\rceil - 1 \right) \\
&= 2 \left\lfloor \frac{\ell}{2} \right\rfloor + 2,
\end{aligned}$$

which is impossible. ■

Now we are ready to bound the load of the arcs after **Phase 2**.

LEMMA 7.3 *The load of the arcs after **Phase 2** is at most*

$$\min\left\{\left\lceil\frac{\ell}{2}\right\rceil - 1, \omega - \ell\right\}.$$

Proof. Let ℓ' denote the load of the arcs after **Phase 2**. Then there is an arc $a \notin \left(\bigcup_{i=1}^{\lfloor \frac{\ell}{2} \rfloor} A_i\right) \cup \left(\bigcup_{i=1}^{\ell} B_i\right)$ such that the link between $h(a)-1$ and $h(a)$ is contained in $\ell'-1$ arcs not in $\left(\bigcup_{i=1}^{\lfloor \frac{\ell}{2} \rfloor} A_i\right) \cup \left(\bigcup_{i=1}^{\ell} B_i\right) \cup \{a\}$. We denote these arcs by $\{c_i \mid 1 \leq i < \ell'\}$. Let k be the smallest number such that the link between $t(a)$ and $t(a)+1$ is contained in a_k. Then $k \geq \lfloor \frac{\ell}{2} \rfloor + 2$ from Lemma 7.2. For any $1 \leq i \leq k-1$, each B_i has an arc b_i that contains the link between $h(a)-1$ and $h(a)$. Note that b_i could be a_i. Then the link between $h(a)-1$ and $h(a)$ is contained in at least $\ell' + k - 1$ links. Thus

$$\ell \geq \ell' + k - 1 \geq \ell' + \left\lfloor\frac{\ell}{2}\right\rfloor + 2 - 1 = \ell' + \left\lfloor\frac{\ell}{2}\right\rfloor + 1,$$

which implies that $\ell' \leq \lceil \frac{\ell}{2} \rceil - 1$.

In the next, we show that $\ell' \leq \omega - \ell$. We construct $H_{k-1}, H_k, \cdots, H_\ell$ sequentially as follows. For any $1 \leq i \leq \omega - \ell$, each C_i has an arc c_i that contains the link between $h(a)-1$ and $h(a)$. Let

$$H_{k-1} = \{a\} \cup \{b_i \mid 1 \leq i \leq k-1\} \cup \{c_i \mid 1 \leq i < \ell'\}.$$

Then all arcs in H_{k-1} intersect pairwise. Suppose that $H_{k-1}, \cdots, H_{i-1}$ for some $k-1 \leq i \leq \ell$ have been constructed and all arcs in each set intersects pairwise. If a_i intersects all arcs in H_{i-1}, set

$$H_i = H_{i-1} \cup \{a_i\}.$$

Otherwise some arc in $H_{i-1} \setminus (\{a_j \mid 1 \leq j \leq \ell\} \cup \{a\})$ does not intersect a_i, and thus $B_i \setminus \{a_i\}$ must be non-empty from the selection of B_i. In this case, let b_i be the arc added to B_i immediately after a_i during phase 2, and set

$$H_i = H_{i-1} \cup \{b_i\}.$$

As $|H_\ell| = \ell + \ell'$, it's sufficient to show that he arcs in H_ℓ intersect pairwise.

We first prove by contradiction that each b_i in H_ℓ contains the link between $h(a) - 1$ and $h(a)$. Assume to the contrary, and let i be the smallest number such that b_i does not contain the link between $h(a) - 1$ and $h(a)$. From Lemma 7.2,

$$t(a) < h(a_i) \leq t(b_i) < h(b_i) < h(a).$$

Furthermore, $h(a) \leq t(a_i)$, for otherwise a_i would intersect all paths in H_{i-1} and thus would be selected. Assume that a_i does not intersect some arc $a' \in H_{i-1}$. Then a' is not any a_j. So a' must contain the link between $h(a) - 1$ and $h(a)$, from the minimality of i. But this implies that a' would be selected to B_i instead of b_i, which is a contradiction.

Now we prove by contradiction that all arcs in H_i intersect pairwise. Assume to the contrary. Let i be the smallest number such that some pair of arcs in H_i do not intersect. Obviously, $i > k - 1$ and $H_i = H_{i-1} \cup \{b_i\}$. This implies that a_i does not intersect some arc $a' \in H_{i-1}$. Then a' is not any a_j and $h(a') \leq h(b_i)$, for otherwise a' would be added to B_i instead of b_i. Thus a' is on the arc from $h(a_i)$ to $h(b_i)$. On the other hand, b_i does not intersect some arc, which must be a_j for some $k - 1 < j < i$. Notice that a_j is on the arc from $h(b_i)$ to $h(a_i)$. Therefore, a' and a_j do not intersect, which contradicts to the fact that all arcs in H_{i-1} are pairwise intersecting. ■

The above lemma leads to the following theorem regarding on the total number of colors used by our algorithm.

THEOREM 7.4 *Our algorithm uses at most* $\min\{2\ell - 1, \omega + \lfloor \frac{\ell}{2} \rfloor\}$ *colors.*

Note that the chromatic number of instance in Figure 7.1 is $2\ell - 1 = \omega + \lfloor \frac{\ell}{2} \rfloor$. Thus the bound in Theorem 7.4 is tight.

3. PATH COLORING IN TREE OF RINGS

Pick any *original* node in the skeleton of the tree of rings as the root. In this rooted skeleton, each original node is called as the *entry* node of the rings represented by its children in the rooted skeleton. Following the breadth-first (or depth-first) search order of the skeleton, we process the rings one bye one by coloring all paths that contains some link in this ring. Fix any ring R in the tree of rings, and orient it clockwise. W.o.l.g., assume that the entry node is 0, and all nodes are numbered clockwise in the increasing order. For any path p that contain at least one link in R, its portion within R is an arc, whose tail and heads are denoted by $t(p)$ and $h(p)$ respectively. An important observation is that all paths that have been colored previously must all go through the entry node. Furthermore, if two paths both contain links in R but do not share any common node in R, then they do not intersect anywhere. The algorithm, referred to as **Iterative Bipartite Matching (IBM)**, for processing the ring R is described below.

- **Initialization**: Keep the old colors for those paths visited previously. Assign a unique (new) color to each path containing the link between node 0 and node 1.

- **Sweeping:** Sweep the nodes in the ring one by one in the clockwise direction from node 1. At each node v, we construct a bipartite graph between the set of current colors and the set of paths p with $t(p) = v$ such that an edge between a color and a path exists if and only if the color can be assigned to this path. We then find a maximal matching in this bipartite graph. Each edge (match) in the obtained matching leads to a color assignment. After that, we assign a new color to any unmatched path, if there is any. This procedure is repeated until all paths are colored.

The above algorithm obviously runs in polynomial time. Before we bound the number of colors used, we recall that an *augmentation* to a bipartite matching M in a graph G is a simple path (of odd length) in G between two unmatched vertices in G which contains edges in M alternatively. It's well known that a bipartite matching is maximal if and only if it has no augmentation. We will heavily use this simple property to prove the next theorem.

THEOREM 7.5 *The algorithm **IBM** uses at most* $\min\{3\ell, \omega + \ell\}$ *colors.*

Proof. Consider the *first* time when the maximum color m is used. Let M denote the obtained maximal matching, and F denote the unmatched paths. Then F is non-empty and $|F|$ new colors are added. Let $f \in F$ be the *longest* one, and I denote the set of colors among the $m - |F|$ old colors such that after the matching no path in this color class contains the link between $h(f)$ and $h(f)+1$. Then among the m colors, $m-|F|-|I|$ (old) colors each have a path that contains the link between $h(f)$ and $h(f)+1$. Thus

$$m-|F|-|I| \leq \ell.$$

So it's sufficient to prove that $|F|+|I| \leq \min\{2\ell, \omega\}$.

We first show that $|F|+|I| \leq 2\ell$. For any $i \in I$, there is some path with color i that intersects f, from the maximality of M. For any $i \in I$, let p_i be the *first* path with color i that intersects f. Note that p_i can't be the first path with color i for any $i \in I$. Let

$$\begin{aligned} I_1 &= \{i \in I \mid t(p_i) = t(f)\}, \\ I_2 &= \{i \in I \mid h(p_i) = t(f)\}, \\ I_0 &= I \setminus (I_1 \cup I_2). \end{aligned}$$

Figure 7.4 (a) illustrates this partition. Then the link between $t(f)$ and $t(f)+1$ is contained in any p_i with $i \in I_0 \cup I_1$, and the link between $h(f)-1$ and $h(f)$ is contained in any p_i with $i \in I_0 \cup I_2$. Thus

$$\begin{aligned} |F|+|I_0|+|I_1| &\leq \ell, \\ |I_0|+|I_2| &\leq \ell. \end{aligned}$$

Sum up these two inequalities, we have

$$|F|+|I| \leq 2\ell-|I_0| \leq 2\ell.$$

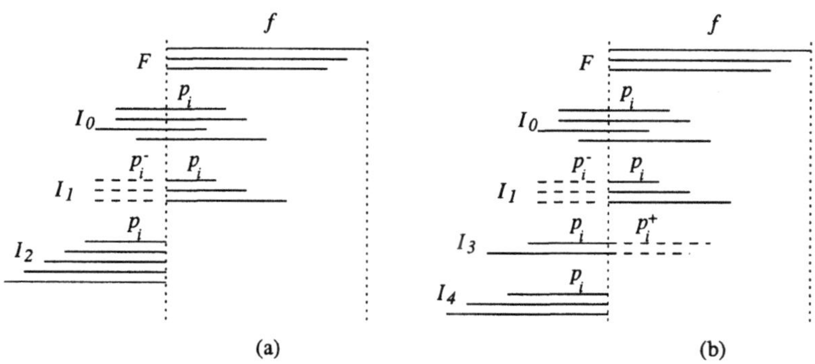

Figure 7.4. The partition of I: (a) $I = I_0 \cup I_1 \cup I_2$, and (b) the further partition of I_2 into I_3 and I_4.

In the next we prove that $|F|+|I| \leq \omega$ by constructing $|I|$ paths with colors in I which together with the $|F|$ paths in F are pairwise intersecting. For any colored path p, we use p^- and p^+ to denote its predecessor and successor, if they exists, in its color class respectively. Then for any $i \in I_1$, p_i^- exists and it does not intersect f .

Consider the following set of $|F| + |I|$ paths

$$F \cup \{p_i \mid i \in I\}.$$

If I_2 is empty or every path in $\{p_i \mid i \in I_2\}$ intersects all paths in $F \cup \{p_i \mid i \in I_1\}$, these paths are pairwise intersecting. Otherwise, we partition I_2 into I_3 and I_4 such that each path in $\{p_i \mid i \in I_3\}$ does not intersect some path in $F \cup \{p_i \mid i \in I_1\}$, and every path in $\{p_i \mid i \in I_4\}$ intersects all paths in $F \cup \{p_i \mid i \in I_1\}$ (see Figure 7.4 (b)). Then I_3 is non-empty. We claim that each color in I_3 is matched in M. Assume to the contrary that a color $i_3 \in I_3$ is unmatched. Then p_{i_3} must intersect every path in F, for otherwise M can be increased by adding the edge between i_3 and this path. Thus there is $i_1 \in I_1$ such that p_{i_1} and p_{i_3} do not intersect. However, in this case, M has an augmentation starting from i_3 followed by p_{i_1}, i_1 and f , since $p_{i_1}^-$ and f do not intersect. This contradicts to the maximality of M, and thus our claim is true.

Consider the following set of $|F| + |I|$ paths

$$F \cup \{p_i \mid i \in I_0 \cup I_1 \cup I_4\} \cup \{p_i^+ \mid i \in I_3\}.$$

If every path in $\{p_i \mid i \in I_4\}$ intersects all paths in $\{p_i^+ \mid i \in I_3\}$, these paths are pairwise intersecting. Otherwise, we will further split I_4 according to the following general procedure. Suppose that we have come up with a partition of I

$$I_0, I_1, I_3, \cdots, I_{2k-3}, I_{2k-1}, I_{2k}$$

which have the following properties.

- Each path in $\{p_i^- \mid i \in I_1\}$ does not intersect f.
- Each path in $\{p_i \mid i \in I_3\}$ does not intersect some path in

$$F \cup \{p_i \mid i \in I_1\}.$$

- For any $1 < j < k$, each path in $\{p_i \mid i \in I_{2j+1}\}$ does not intersect some path in $\{p_i^+ \mid i \in I_{2j-1}\}$.
- Some path in $\{p_i \mid i \in I_{2k}\}$ does not intersect some path in

$$\{p_i^+ \mid i \in I_{2k-1}\}.$$

- Each path in $\{p_i \mid i \in I_{2k}\}$ intersect all paths in

$$F \cup \{p_i \mid i \in I_0 \cup I_1\} \cup \left\{ p_i^+ \mid i \in \bigcup_{j=1}^{k-2} I_{2j+1} \right\}.$$

We split I_{2k} into I_{2k+1} and I_{2k+2} such that each path in

$$\{p_i \mid i \in I_{2k+1}\}$$

does not intersect some path in

$$\{p_i \mid i \in I_{2k-1}\},$$

and every path in

$$\{p_i \mid i \in I_{2k+2}\}$$

intersects all paths in

$$\{p_i \mid i \in I_{2k-1}\}.$$

Then I_{2k+1} is non-empty. We claim that each color in I_{2k+1} is matched in M. Assume to the contrary that a color $i_{2k+1} \in I_{2k+1}$ is unmatched. We construct an augmentation to M starting from color i_{2k+1}. For $j = k, k-1, \cdots, 2$, let $i_{2j-1} \in I_{2j-1}$ be such that $p^+_{i_{2j-1}}$ does not intersect $p_{i_{2j+1}}$, and attach $p^+_{i_{2j-1}}$ and i_{2j-1} to the augmentation. If p_{i_3} does not intersect some path $f' \in F$, then attach f' to the augmentation. Otherwise, let $i_1 \in I_1$ such that p_{i_1} and p_{i_3} do not intersect, and attach p_{i_1}, i_1 and f to the augmentation, since $p^-_{i_1}$ and f do not intersect. In either case, we obtain an the augmentation to M, which is a contradiction. Therefore the claim is true.

Consider the following set of $|F| + |I|$ paths

$$F \cup \{p_i \mid i \in I_0 \cup I_1 \cup I_{2k+2}\} \cup \left\{ p_i^+ \mid i \in \bigcup_{j=1}^{k-1} I_{2j+1} \right\}.$$

If each path in $\{p_i \mid i \in I_{2k+2}\}$ intersects all paths in $\{p_i^+ \mid i \in I_{2k+1}\}$, these paths are pairwise intersecting. Otherwise we repeat the previous procedure. Note that

$$I \supsetneq I_2 \supsetneq I_4 \supsetneq \cdots \supsetneq I_{2k} \supsetneq I_{2k+2} \supsetneq \cdots.$$

The above iterations will eventually terminate, and we then obtain a set of $|F| + |I|$ pairwise intersecting paths. ■

An immediately corollary of Theorem 7.5 is an upper bound of 2 on the approximation ratio of our algorithm. In the next, we construct a "bad" instance as shown in Figure 7.5 to get a lower bound of 1.5. Let k be any *even* number greater than two. Fix a ring R^* of $2k+1$ nodes. Each node i is attached with another ring R_i of three nodes consisting of the nodes i, i' and i''. The node i' and i'' are further attached two rings R_i' and R_i'' of three nodes respectively. We then fix a link e_i' in R_i' which is incident to the node i', and a link e_i'' in R_i'' which is incident to the node i''. Let p_i' be the path that starts from the link e_i', and then visits the nodes $i, i+1, (i+1)'', (i+1)'$ and ends at the link e_{i+1}'. Let p_i'' be the path that starts from the link e_i'', and then visits the nodes $i, i+1, (i+1)', (i+1)''$ and ends at the link e_{i+1}''. Let P_i be the set of paths consisting of $\frac{k}{2}$ copies of p_i' and $\frac{k}{2}$copies of p_i''. Let $\mathcal{P}$ be the union of $P_0, P_1, \cdots, P_{2k}$. Figure 7.5 (a) shows the paths constructed in this way for $k=2$. The the load of $\mathcal{P}$ is $\ell = k$. In the next, we show that its clique number is $\omega = 2k$ and the chromatic number is $\chi = 2k+1$.

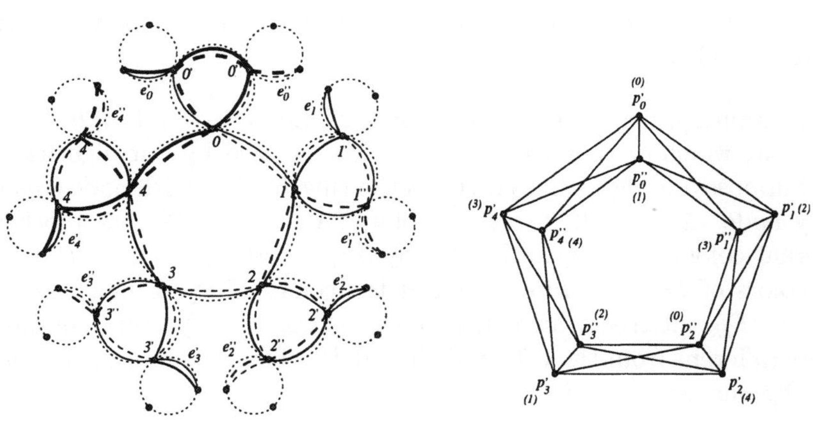

Figure 7.5. The instance with $\chi = \omega + 1 = 2\ell + 1$: (a) the paths, and (b) the intersecting graph and its coloring.

The intersecting graph of $\mathcal{P}$ is illustrated in Figure 7.5 (b). It has the property that for any i and j, if $|i-j| \bmod (2k+1) = 1$, then all paths in P_i and P_j are pairwise intersecting; otherwise no path in P_i intersects any path in P_j. Thus $\omega \geq 2k$. Now consider any clique in the intersecting graph. If all paths are within some P_i, then its size is at most k. So we assume that the clique contains some path in P_i and some path in P_j. Then $|i-j| \bmod (2k+1) = 1$ and w.l.o.g. assume that $j = i+1$. Then all paths must be within P_i and P_{i+1}, as any other path either does not intersect any path in P_i or does not intersect any path in P_{i+1}. Thus the size of the clique is at most $2k$. This implies that the clique number $\omega = 2k$.

Now we show that $\chi = 2k + 1$. First of all, $\chi \geq \omega = 2k$. Secondly χ can't be $2k$. Assume to the contrary. Then all paths in P_0 and P_1 get a unique color. Then P_2 must get the colors of P_0, P_3 must get the colors of P_1, and so on. At last, P_{2k-1} get the colors of all colors in P_1, and so P_{2k} get the color of all colors in P_0. However, all paths in P_{2k} and P_0 are pairwise intersecting. Thus the coloring is improper. Therefore, $\chi > 2k$. Finally, we give a proper coloring of $\mathcal{P}$ with $2k+1$ as follows (see Figure 7.5 (b) in which the colors are given with the parentheses): P_i gets the k colors

$$\{(ik + j) \bmod (2k + 1) \mid 0 \leq j < k\}.$$

Then for any $0 \leq j < 2k$, P_i and P_{i+1} get different colors. In addition, as P_{2k} gets the color $\{k+1, k+2, \cdots, 2k\}$ and P_0 gets the color $\{0, 1, \cdots, k-1\}$, P_{2k} and P_0 also get different colors. Thus the coloring is proper. Therefore, $\chi = 2k + 1$.

In summary, for the instance $\mathcal{P}$, $\omega = 2\ell$ and $\chi = \omega + 1 = 2\ell + 1$. In the next, we show that the number of colors used by our algorithm is $3k$. Suppose our algorithm processes the ring R^* first with node 0 as the entry node. At node 0, P_0 gets k colors. At node 1, P_1 gets another k different colors. Subsequently, for any $0 \leq i < 2k$, if i is even P_i reuses the colors of P_0; otherwise P_i reuses the colors of P_1. Finally, as each path in P_{2k} intersects all paths in P_{2k-1} and P_0, P_{2k} gets another k colors different from the colors of P_0 and P_1. Thus the number of colors used by our algorithm is

$$3k = 3\ell = \omega + \ell = \frac{3}{2}\chi - \frac{3}{2}.$$

From this we can draw conclusion.

LEMMA 7.6 *The bound on the number of colors used by our algorithm given in Theorem 7.5 is tight, and the approximation ratio of our algorithm is at least $\frac{3}{2}$ and at most 2.*

Theorem 7.5 implies that

$$\chi \leq \min\{3\ell, \omega + \ell\}.$$

However it is unknown whether χ could be as large as 3ℓ or $\omega+\ell$. The instance shown in Figure 7.6 indicates that both χ and ω could be as large as 3ℓ. The instance is similar to the one shown in Figure 7.5, except that the size of the ring R^* is also three. These 3ℓ paths are pairwise intersecting, while their load is 3ℓ. Thus for this instance, $\chi=\omega=3\ell$. Unfortunately, we have not been able to find any instance with $\chi=\omega+\ell$ at this moment.

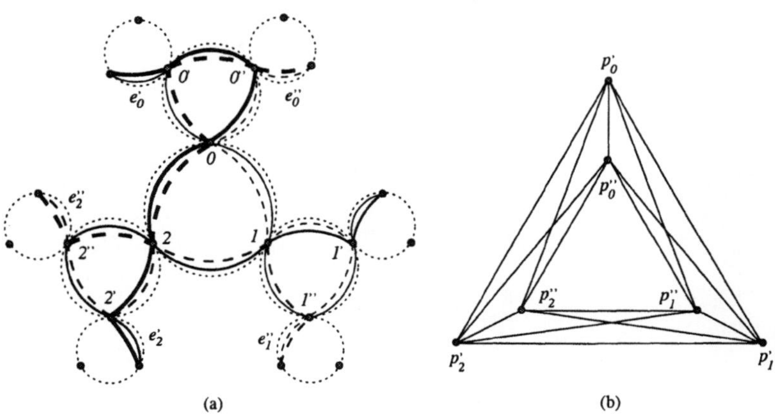

Figure 7.6. An instance with $\chi=\omega=3\ell$: (a) the paths and (b) the intersecting graph which is a clique.

4. SUMMARY

This chapter studies path coloring in the two most practical network topologies, ring and tree of rings. For path coloring in rings, we present a simple algorithm which uses at most $\min\left\{2\ell-1,\omega+\left\lfloor\frac{\ell}{2}\right\rfloor\right\}$ colors. Comparing to Tucker's algorithm [151] which uses at most $2\ell-1$ colors and Karapetian's algorithm [90] which uses at most $\left\lfloor\frac{3\omega}{2}\right\rfloor$ colors, our algorithm leads to the stronger relation

$$\chi\leq\min\left\{2\ell-1,\omega+\left\lfloor\frac{\ell}{2}\right\rfloor\right\}$$

by a single algorithm as both χ and ω can be as large as $2\ell-1$. Furthermore, the equality holds for certain instances. For path coloring in trees of rings, we present an algorithm which uses at most $\min\{3\ell,\omega+\ell\}$ colors. We show that the approximation ratio of this algorithm is between

1.5 and 2. Thus it leads to the relation

$$\chi \leq \min \{3\ell, \omega + \ell\}.$$

We also provide instances to show that both χ and ω can be as large as 3ℓ. However, it remains open whether χ can be as large as $\omega + \ell$.

Chapter 8

MAXIMIZATION OF THROUGHPUT

1. INTRODUCTION

This chapter studies the *maximal throughput problem* in wavelength switched network. An instance of this problem consists of a wavelength switched network, the number of available wavelength channels supported in the optical fibers, and pairs of nodes in this network that wish to communicate. The objective is to find maximal number of connection pairs that can be supported in this network simultaneously. The maximal throughput problem in linear topology can be solved in polynomial time by a simple greedy algorithm [120]. If the input communication pairs have prederternined paths, the maximal throughput problem remains NP-hard in rings and meshes. A polynomial-time $\frac{3}{2}$-approximation algorithm for rings is given [120]and it was shown in [120] that unless P=NP the problem for meshes is inapproxiable within m^δ for some constant $\delta > 0$ where m is the number of communication pairs. The on-line version of the maximal throughput problem has been studied in [14]. The algorithm designed in [14] was transformed from on-line virtual circuit routing algorithms at the expenses of an additive term of one in the competitive ratio. This allows to state the existence of an $O(\log n)$ competitive algorithm for those topologies, including trees and meshes, for which there exists a competitive $O(\log n)$ competitive algorithm for on-line virtual circuit routing.

This chapter studies the complexities and approximation algorithms for the off-line maximal throughput problem in various settings [160]. Our algorithms follow the similar framework in [14] but we come up with a stronger transformation which reduces the approximation ratios for the off-line maximal throughput problem. The remaining of this chapter proceeds as follows. Section 2. gives a graph-theoretic formulation of this problem and characterizes the computational complexities of the maximal throughput problem in various topologies. Section 3. reduces the maximal throughput problem into the well-known maximal disjoint-path problem which provides a general framework for the approximation design. Section 4. studies the approximation algorithms and their performances in the rings, trees, meshes and the general topologies. Finally section 5. concludes this chapter.

2. COMPUTATIONAL COMPLEXITIES

We model the underlying wavelength-switched optical network as an undirected graph $G = (V, E)$ with W wavelength channels over each link in E. A communication *request* is a pair of vertices in V. A set of requests is said to be *routable* in G with W wavelengths if there is a valid lightpath assignment to them with W wavelengths. The maximal throughput problem can then be stated as follows: Given an undirected graph G, a set of communication requests R, the number of available wavelengths over each fiber link W, find a maximal-cardinality subset R' of R that is routable in G with W wavelengths.

When G is a chain, the maximal throughput problem can be solved in polynomial time by using a coloring result given in [171]. However, when G is a ring, tree or mesh, we will show that the maximal throughput problem is NP-complete.

LEMMA 8.1 *Let $\mathcal{G}$ be any infinite set of graphs. If the decision version of the minimal routing and wavelength assignment problem is NP-Complete for the graphs in $\mathcal{G}$, then the decision version of the maximal throughput problem is also NP-Complete for the graphs in $\mathcal{G}$.*

Proof. The decision versions of the maximal throughput problem can be stated as follows:

Given an instance of graph G in $\mathcal{G}$, a set of requests R, the number of wavelengths W and a bound b, whether b requests in R can be routed in G with at most W colors.

The decision versions of the minimal valid routing and wavelength assignment problem can be stated as follows:

Given an instance of graph G in $\mathcal{G}$, a set of requests R and a bound W, whether all requests in R can be routed in G with at most W colors.

It's obvious that by setting b to be the cardinality of R, the decision version of the minimal routing and wavelength assignment problem can be reduced to the decision version of the maximal throughput problem. Therefore, if the decision version of the valid routing and wavelength assignment problem is NP-Complete for the graphs in $\mathcal{G}$, then the decision version of the maximal throughput problem is also NP-Complete for the graphs in $\mathcal{G}$. ■

As the decision version of the minimal routing and wavelength assignment problem is NP-complete in rings, trees and meshes [59, 60], we have the following corollary.

COROLLARY 8.2 *The maximal throughput problem is NP-complete in in rings, trees, and meshes.*

3. REDUCTION TO MAXIMAL DISJOINT-PATH PROBLEM

When the number of available wavelengths W is one, the maximal throughput problem reduces to the *maximal disjoint-path problem*: Given a network and pairs of nodes that wish to communicate, find the maximal number of pairs that can be connected by disjoint paths. This problem is well known to be computationally difficult. Deciding whether all pairs can be so connected is one of the Karp's original NP-complete problems [91]; it remains NP-complete even when the underlying graph is mesh [99]. For the convenience of description, we call a set of requests to be *independent* if they can be connected by disjoint paths. With this notion, the maximal disjoint-path problem can be interpreted as to find a maximal independent subset of the given set of requests.

Given an algorithm $\mathcal{A}$ for the maximal disjoint-path problem, we can build a *greedy* algorithm $\mathcal{A}'$ based on $\mathcal{A}$ as follows. We first run the algorithm $\mathcal{A}$ on the request set R to get an optimal or a near-optimal independent subset of requests in R. Such set of requests are assigned with the first wavelength channel, and then they are removed from the requests set. We repeatedly run the algorithm $\mathcal{A}$ on the remaining requests to find an independent subset of requests and assign each them by a unique wavelength. Such process repeats until we run out of wavelength or no more request is left. This algorithm will be denoted by $\mathcal{A}'$ and is formally stated as follows.

Greedy Algorithm $\mathcal{A}'$ based on $\mathcal{A}$
  for $i = 1$ ***to*** W ***do***
    run $\mathcal{A}$ on R and output an independent subset A_i of R;
    assign wavelength i to the requests in A_i;
    $R = R \setminus A_i$;
    if $R == \emptyset$ ***exit;***
end-of-algorithm

In the next, we will analyze the performance ratio of the algorithm $\mathcal{A}'$ for the maximal throughput problem in term of the performance ratio of the algorithm $\mathcal{A}$ for the maximal disjoint-path problem.

THEOREM 8.3 *If $\mathcal{A}$ has performance ratio of ρ, then $\mathcal{A}'$ has performance ratio of $\frac{1}{1-e^{-\frac{1}{\rho}}}$.*

Proof. We first prove the theorem for the case that the algorithm $\mathcal{A}$ for the maximal disjoint-path problem is deterministic.

Let R be any set of requests. Let

$$O = O_1 \cup O_2 \cup \cdots \cup O_W$$

be the set of the requests in R accepted by an optimal algorithm where O_i is the set of requests in R that are assigned with wavelength i for $1 \leq i \leq W$. Let

$$A = A_1 \cup A_2 \cup \cdots \cup A_W$$

be the set of requests of R accepted by $\mathcal{A}'$ where A_i is the set of requests in R that are assigned with wavelength i by the algorithm $\mathcal{A}'$ for $1 \leq i \leq W$.

For any $1 \leq i \leq W$, consider the set of requests

$$R_i = R \setminus (\cup_{k=1}^{i-1} A_k).$$

As $\mathcal{A}$ has performance ratio of ρ and A_i is the set of requests in R_i generated by the algorithm $\mathcal{A}$, $\rho|A_i|$ is at least the cardinality of the maximal independent subset of requests in R_i. On the other hand, for each $1 \leq j \leq W$, the requests in O_j are independent, so are the set of requests in $O_j \setminus (\cup_{k=1}^{i-1} A_k)$. As $O_j \setminus (\cup_{k=1}^{i-1} A_k)$ is a subset of R_i, the cardinality of the maximal independent subset of requests in R_i is at least $|O_j \setminus (\cup_{k=1}^{i-1} A_k)|$. Thus for any $1 \leq i, j \leq W$,

$$\rho\,|A_i| \geq \left|O_j \setminus (\cup_{k=1}^{i-1} A_k)\right|.$$

Summing up the above inequalities for $1 \leq j \leq W$ and using the fact that the sets O_j are pairwise disjoint, we have that for any $1 \leq i \leq W$,

$$\begin{aligned}
\rho W\,|A_i| &\geq \sum_{j=1}^{W} \left|O_j \setminus (\cup_{k=1}^{i-1} A_k)\right| \\
&= \sum_{j=1}^{W} \left(|O_j| - \left|O_j \cap (\cup_{k=1}^{i-1} A_k)\right|\right) \\
&= \sum_{j=1}^{W} |O_j| - \sum_{j=1}^{W} |O_j \cap (\cup_{k=1}^{i-1} A_k)| \\
&= |O| - |O \cap (\cup_{k=1}^{i-1} A_k)| \\
&\geq |O| - |\cup_{k=1}^{i-1} A_k| \\
&= |O| - \sum_{k=1}^{i-1} |A_k|.
\end{aligned}$$

This implies that for any $1 \leq i \leq W$,

$$|O| \leq \sum_{k=1}^{i-1} |A_k| + \rho W |A_i|.$$

Therefore,

$$|O| \leq \min_{i=1}^{W} \left(\sum_{k=1}^{i-1} |A_k| + \rho W |A_i| \right).$$

As

$$\frac{|O|}{|A|} \leq \min_{i=1}^{W} \left(\sum_{k=1}^{i-1} \frac{|A_k|}{|A|} + \rho W \frac{|A_i|}{|A|} \right),$$

we will bound from above the right-hand side of the above inequality in the next .

Let $x = (x_1, x_2, \cdots, x_W)$ with $x_i = \frac{|A_i|}{|A|}$, for $1 \leq i \leq W$. As the sets A_i are pairwise disjoint, we have

$$\sum_{i=1}^{W} x_i = 1, x_1 \geq x_2 \geq \cdots \geq x_w \geq 0.$$

Let

$$S = \left\{ x \mid \sum_{i=1}^{W} x_i = 1, x_1 \geq x_2 \geq \cdots \geq x_w \geq 0 \right\}.$$

In the following we prove that for any $x \in S$,

$$\min_{i=1}^{W} \left(\sum_{k=1}^{i-1} x_k + \rho W x_i \right) \leq \frac{1}{1 - \left(1 - \frac{1}{\rho W}\right)^W}.$$

For each $0 \leq i \leq W$, let

$$\alpha_i = \frac{\left(1 - \frac{1}{\rho W}\right)^{W-i}}{\rho W \left[1 - \left(1 - \frac{1}{\rho W}\right)^W\right]}.$$

Then

$$\sum_{i=1}^{W} \alpha_i = 1$$

and for each $0 \leq i \leq W$,

$$\rho W \alpha_i + \sum_{j=i+1}^{W} \alpha_j = \frac{1}{1 - e^{-\frac{1}{\rho}}}.$$

Therefore,

$$\begin{aligned}
&\min_{i=1}^{W} \left(\sum_{k=1}^{i-1} x_k + \rho W x_i \right) \\
&\leq \sum_{i=1}^{W} \alpha_i \left(\sum_{k=1}^{i-1} x_k + \rho W x_i \right) \\
&= \sum_{i=1}^{W} \left(\rho W \alpha_i + \sum_{j=i+1}^{W} \alpha_j \right) x_i \\
&= \frac{1}{1 - e^{-\frac{1}{\rho}}} \sum_{i=1}^{W} x_i \\
&= \frac{1}{1 - e^{-\frac{1}{\rho}}}
\end{aligned}$$

And the equation holds if and only if all the W terms $\sum_{k=1}^{i-1} x_k + \rho W x_i$, $1 \le i \le W$ are equal which happens if and only if

$$x_i = \frac{\left(1 - \frac{1}{\rho W}\right)^{i-1}}{\rho W \left[1 - \left(1 - \frac{1}{\rho W}\right)^W\right]}.$$

for $1 \le i \le W$. Therefore,

$$\frac{|O|}{|A|} \le \frac{1}{1 - (1 - \frac{1}{\rho W})^W} \le \frac{1}{1 - e^{-\frac{1}{\rho}}}.$$

The last inequality follows from the fact that the function $(1 - \frac{1}{x})^x$ converges monotonically from below to e^{-1} as $x \to \infty$.

If the algorithm $\mathcal{A}$ is randomized, the same analysis goes as well. The only modification that needs to be made in the above analysis is to replace the cardinality of a set by the expectation of the cardinality of that set. We omit the details over here. ■

Similar technique used in the above proof has been used in [44]. One remark made here is that $\frac{1}{1-e^{-\frac{1}{\rho}}} < \rho + 1$ for all $\rho \ge 1$. In fact it's well know that for all $x > 0$

$$e^x > 1 + x.$$

Substituting x with $\frac{1}{\rho}$ gives

$$e^{\frac{1}{\rho}} > 1 + \frac{1}{\rho}$$

which implies

$$e^{-\frac{1}{\rho}} < \frac{\rho}{\rho + 1}.$$

Thus

$$1 - e^{-\frac{1}{\rho}} > \frac{1}{\rho + 1}$$

or equivalently

$$\frac{1}{1 - e^{-\frac{1}{\rho}}} < \rho + 1.$$

4. APPROXIMATION ALGORITHMS AND ANALYSES

In the previous section, we gave a general framework for the design of approximation algorithms by using algorithms for maximal disjoint-path problem. In this section, we will give approximation algorithms for maximal throughput problem on rings, trees, meshes and general topologies.

4.1 RINGS

We first consider the maximal disjoint-path problem over rings. We will establish the connections between the maximal disjoint-path problem over rings and the maximal disjoint-path problem over chains. The maximal disjoint-path problem over chains is equivalent to the well-studied maximum independent set problem in interval graphs, which can be solved in polynomial time.

Let R be a set of requests and G be a ring. Consider any fixed link e in a ring G. Then for any independent subset S of requests is R, at most one request can be routed through the link e. If there is no request in S that is routed through the link e, then all requests in S can be routed in the chain derived from G by the removal of e. Now we consider the case that there is one request, say r, in S that is routed through e. We denote the unique arc of G containing the link e that connects the request r by $arc(r, e)$. Then the remaining requests in $S \setminus \{r\}$ must be able connected by disjoint paths in the chain $G \setminus arc(r, e)$, derived from the ring G by removing all links in $arc(r, e)$.

Let $\mathcal{A}_c$ be any polynomial algorithm for the maximal disjoint-path problem in chains. For a chain G and a set of requests R, we denote by $\mathcal{A}_c(G, R)$ the output generated by the algorithm $\mathcal{A}_c$ over the instance of G and R. Then based on the above discussion, we can get the following polynomial algorithm $\mathcal{A}_r$ for the maximal disjoint-path problem in rings.

```
Algorithm A_r
    let e be the first link of G;
    MIS = A_c(G \ {e}, R);
    for each r ∈ R do
        IS = {r} ∪ A_c(G \ arc(r, e), R \ {r});
        if |MIS| < |IS|, then MIS = IS;
    output MIS;
end-of-algorithm
```

Let $\mathcal{A}'_r$ be the greedy algorithm based on $\mathcal{A}_r$ following the general framework in Section 3.. As the algorithm $\mathcal{A}_r$ can find a set of maximal disjoint-paths in rings, we have the following theorem.

THEOREM 8.4 *The greedy algorithm $\mathcal{A}'_r$ is a $\frac{e}{e-1}$-approximation algorithm for the maximal throughput problem in wavelength-switched optical ring networks.*

4.2 TREES

One of the unique properties of the tree is the unique path between any pair of nodes. Such property significantly reduces the complexity of the maximal disjoint-path problem. In fact, if the tree is a star, then the finding the maximal set of disjoint paths is equivalent to finding a maximum matching in the demand graph, which can be solved in polynomial time. Based on this observation, [64] gave a polynomial algorithm for the maximal disjoint-path problem by solving a sequence of maximum matching problems. Here we briefly describes the idea of this algorithm.

The algorithm starts by rooting the tree at an arbitrary vertex. It then does two passes over the tree level-by-level, an *upward pass* followed by a *downward pass.* In the upward pass, at each vertex v a number of maximal matching problems are solved. We first solve a maximum matching problem to route requests whose routes go through v and are within the subtree rooted at v. However, it may also be advantageous to route a request along the edge $(v, parent(v))$. This will be strictly advantageous only if we can still route the previous requests in the subtree rooted at v. We determine the requests giving strict advantage by solving a maximum matching problem for each choice of request routed on the edge

$(v, parent(v))$. The requests giving strict advantage are thought of as originating at v itself.

In the downward pass, we start at the root, fixing requests. The vertex $parent(v)$ decides which request gets routed on $(v, parent(v))$. Once this is done vertex v fixes the requests routed on the edges to its children.

We use $\mathcal{A}_t$ to denote the previous algorithm for the maximal disjoint-path problem in trees, and let $\mathcal{A}'_t$ be the greedy algorithm based on $\mathcal{A}_t$ following the general framework in section 4. As the algorithm $\mathcal{A}_t$ can find a set of maximal disjoint-paths in trees, we have the following theorem.

THEOREM 8.5 *The greedy algorithm* $\mathcal{A}'_t$ *is a* $\frac{e}{e-1}$*-approximation algorithm for the maximal throughput problem in wavelength-switched optical tree networks.*

4.3 MESHES

As mentioned at the beginning of Section 3., the maximal disjoint-path problem is NP-complete when the underlying graphs are meshes. [96] give a constant-factor randomized approximation algorithm for the maximal disjoint path problem in a broader class of graphs called *uniformly densely embedded Euclidean graphs*, which includes two-dimensional mesh and many other planar and locally planar interconnection networks. Their algorithm involves complex and delicate techniques. All requests are classified into four categories: long and heavy, long and light, short and heavy, and short and light. By approximating the set of maximal disjoint paths in each category and taking the one with the most disjoint paths, they obtained a constant-factor approximation for the overall problem. The interested readers can refer [96] for the details.

We use $\mathcal{A}_m$ to denote the randomized approximation algorithm for the maximal disjoint-path problem in meshes given in [96], and let $\mathcal{A}'_m$ be the greedy algorithm based on $\mathcal{A}_m$ following the general framework in section 4. As the algorithm $\mathcal{A}_m$ is a constant-factor randomized approximation for the maximal disjoint-path problem in meshes, we have the following theorem.

THEOREM 8.6 *The greedy algorithm $\mathcal{A}'_m$ is a constant-factor approximation algorithm for the maximal throughput problem in wavelength-switched optical mesh networks.*

A remark is that the hidden constant of the performance guarantee of the algorithm $\mathcal{A}_m$ could be very large, so is the hidden constant of the performance guarantee of the algorithm $\mathcal{A}'_m$.

4.4 GENERAL TOPOLOGIES

In the previous section, the topologies of the underlying network studied are some special ones. In this section, we will study the approximation algorithm for the maximal throughput for the general topologies. As before, we begin with the algorithm for finding a set of maximal disjoint paths.

A straightforward and intuitive algorithm for the maximal disjoint-path problem is the shortest-path-first algorithm. We call a request a *shortest* request if it has the shortest distance in term of the number of links. At each step, the algorithm selects one shortest request and then remove the shortest path connecting the selected request from the graph. Such process repeats until no request can be connected. Such algorithm is greedy in the sense that by selecting the request with shorter path, the more edges remains and thus we can potentially further support more requests. We use $\mathcal{A}_g$ to denote such greedy algorithm. The algorithm $\mathcal{A}_g$ is formally stated as follows.

Algorithm $\mathcal{A}_g$
    $P = \emptyset$;
    $i = 0$;
    while R is not separated **do**
        $i = i + 1$;
        Let r_i be a shortest request in R;
        Let p_i be a shortest path connecting r_i;
        $R = R \setminus \{r_i\}$;
        $P = P \cup \{p_i\}$;
        $E(G) = E(G) \setminus \{p_i\}$;
    output R and P;
end-of-algorithm

For any instance (G, R) of maximal disjoint-path problem, let $\bar{d}(G, R)$ be the minimal average length of any set of maximal disjoint paths connecting requests in R. Then we have the following lemma on the performance of the algorithm $\mathcal{A}_g$.

LEMMA 8.7 *For any instance (G, R) of the maximal disjoint path problem, the algorithm $\mathcal{A}_g$ is a $\bar{d}(G, R)$-approximation.*

Proof. Let O be a set of maximal disjoint paths connecting requests in R whose average length is $\bar{d}(G, R)$, and let

$$P = \{p_1, p_2, \cdots, p_{|P|}\}$$

be the set of disjoint paths output by the algorithm $\mathcal{A}_g$, where p_i is the i-th path selected. Then any path in O must share edges with some path in P because the removal of edges of the paths in P separate all other requests. Let O_1 be the set of paths in O that share edges with p_1, and for each $2 \leq i \leq |P|$, let O_i be the set of paths in $O \setminus (O_1 \cup \cdots \cup O_{i-1})$ that share edges with p_i. Then the sets

$$O_1, O_2, \cdots, O_{|P|}$$

form a partition of O, i.e.,

$$O = O_1 \cup O_2 \cup \cdots \cup O_{|P|},$$

and

$$O_i \cap O_j = \emptyset \text{ if } i \neq j.$$

Furthermore, each path in O_i has length of at least $|O_i|$. This is because p_i has length of at least $|O_i|$ and in turn each path in O_i must have length of at least $|O_i|$ for otherwise it would have been selected by the greedy algorithm $\mathcal{A}_g$ instead of p_i.

Since $\bar{d}(G, R)$ be the average length of paths in O, the total number of edges consumed by the paths in O is $\bar{d}(G, R)|O|$. On the other hand, as all paths in O are edge-disjoint, the total number of edges consumed by the paths in O is at least

$$\sum_{i=1}^{|P|} |O_i| \cdot |O_i| = \sum_{i=1}^{|P|} |O_i|^2 \geq \frac{\left(\sum_{i=1}^{|P|} |O_i|\right)^2}{|P|} = \frac{|O|^2}{|P|},$$

where the inequality follows from the well-known Cauchy-Schwartz inequality. Therefore,

$$\bar{d}(G,R)|O| \geq \frac{|O|^2}{|P|}$$

which implies

$$\frac{|O|}{|P|} \leq \bar{d}(G,R).$$

■

We denote by $\mathcal{A}'_g$ the greedy algorithm based on $\mathcal{A}_g$ following the general framework in section 4. Consider an instance of the maximal throughput problem (G, R, W). Let

$$A = A_1 \cup A_2 \cup \cdots \cup A_w$$

be the set of requests output by the algorithm $\mathcal{A}'_g$, where the set of requests in A_i are assigned with wavelength i. Let

$$R_1 = A_1$$

and

$$R_i = R \setminus (A_1 \cup \cdots \cup A_{i-1}$$

for $2 \leq i \leq W$. We define

$$\bar{d}_A = \max_{i=1}^{W} \bar{d}(G, R_i).$$

We have the following theorem on the performance guarantee of $\mathcal{A}'_g$.

THEOREM 8.8 *Let (G, R, W) be an instance of the maximal throughput problem and A be the set of requests output by the algorithm $\mathcal{A}'_g$. Then for any optimal solution O, we have*

$$\frac{|O|}{|A|} \leq \frac{1}{1 - e^{-\frac{1}{\bar{d}_A}}}.$$

5. CONCLUSION

In this chapter, we have studied the maximal throughput problem on wavelength-switched optical networks. Such problem is proved to be NP-complete in general topologies and remains NP-complete even when the underlying topologies are rings, trees or meshes. We have given constant-factor approximation algorithms for the maximal throughput problem in rings, trees and meshes, and an approximation algorithm for the same problem in the general topologies whose performance guarantee is a function of the average length of the set of disjoint paths. The key technique in design our approximation algorithms is the reduction of the maximal throughput problem in wavelength-switched optical network to the maximal disjoint-path problem.

Chapter 9

OPTIMAL PLACEMENT OF WAVELENGTH CONVERTERS

1. INTRODUCTION

In wavelength switched optical networks without any wavelength conversion [119], the wavelength assignment must meet the wavelength continuity constraint, i.e., the same wavelength is allocated on all of the links in the path established for a connection [4, 13, 100, 114, 128, 129]. Such constraint can be relaxed when wavelength converters are placed at certain nodes. If a node of the network contains a converter, any path that passes through this node may change its wavelength. In a network with converters, the wavelengths are assigned to individual links of all paths, with the restriction that the same wavelength is allocated on all of the links in any sub-path that does not pass through a converter. Clearly wavelength assignments in networks with converters can sometimes be more efficient (i.e. use fewer wavelengths) than optimal wavelength assignments for the same set of paths when no converters are available. One extreme example is that if each node contains a wavelength converter, the number of wavelengths required by any routing is reduced down to the natural *congestion* or *load* bound, defined to be the maximum number of paths passing through any one link in the network. Another extreme example is that the placing a converter at any node of a WDM ring has the same effect that the number of wavelengths required for any routing is equal to its load [166].

However, it is not always necessary to place a wavelength converter at each node such that the number of wavelengths required by any set of paths is equal to its link load. As observed in [166], it is sufficient to place a converter at a single arbitrary node in a WDM ring to achieve this objective [166]. Motivated by this observation, the following question was raised in [166]: in a general WDM network, how to choose a minimum number of nodes to place wavelength converters so that the number of wavelengths required by any set of paths is equal to its link load? A set S of nodes in a network is said to be *sufficient* [166] if placing converters at the nodes in S can ensure that the number of wavelengths required by any set of paths equals to its link load. The *minimum sufficient set problem* is NP-complete in general WDM networks [166]. Furthermore, it is as hard as the *minimum vertex cover (MVC) problem* in undirected graphs [97]. Since providing an approximation ratio better than 2 for the MVC problem is a long-standing open problem, this indicates that improving on the performance guarantee of two for minimum sufficient set problem will be difficult as well.

While the work in [97] provided approximation solutions for general WDM networks, we notice that the topologies of most practical WDM networks are not general. In particular, trees, tree-connected rings and tree of rings are of more practical concrete relevance to the telecommunications industry. For practical reasons, backbone telecommunication networks need to reflect irregularity of geography, non-uniform clustering of users and traffic, hierarchy of services, dynamic growth, etc. In addition, wide-area multiwavelength technology is evolving around current signal wavelength networking architectures and existing fiber networks. These are mainly SONET rings and tree-like interconnection of such rings [16, 15, 132]. In this chapter, we will show that the minimum sufficient set problem in these special topologies can be solved in polynomial time. Our algorithms are based on the reduction of the minimum sufficient set problem to the MVC problem established in [97]. They are very efficient and easy to implement.

The remaining of this paper is arranged as follows. In Section 2., we first introduce some basic terminologies and the reduction from the minimum sufficient set problem to the MVC problem. In Section 3., we present a polynomial-time algorithm which finds a minimum sufficient set problem in tree networks. In Section 4., we present a polynomial-time algorithm which finds a minimum sufficient set problem in tree-connected rings. In Section 5., we present a polynomial-time algorithm which finds a minimum sufficient set problem in tree of rings. Finally Section 6. summarizes this chapter.

2. PRELIMINARIES

A WDM network in this paper is a bi-directed graph $G = (V, E)$: one for which $(u, v) \in E$ if and only if $(v, u) \in E$. Let $G_s = (V, E')$ denote the *skeleton* of the network G, the undirected graph obtained from G by replacing each bi-directed pair of edges with a single undirected edge. By partial abuse of terminology, we will say a set is sufficient in G_s if and only if it is sufficient in G. A vertex v is referred to as a *branching node* if its degree in G_s is greater than 2, a *relay node* if its degree in G_s is equal to 2, or a *leaf node* if its degree in G_s is equal to 1. We will assume that G_s is connected and contains at least one branching node, since otherwise G_s is either a path or a cycle, and the minimum sufficient set can be solved trivially. We say that a node of a path P is an *internal node* in this path if it is not one of the two endpoints.

From the graph G_s, we construct another undirected graph $G_C = (V_C, E_C)$, referred to as the *contraction* of the graph G_s, as follows: V_C consists of all branching nodes in G_s. For any two branching nodes u and v, (u, v) is an edge in E_C if and only if there exists a path in G_s between u and v such that all internal nodes in this path are relay nodes. Note that G_C may have self-loops, which we retain as part of the graph. The following lemma establishes the connection between the minimum sufficient set in G and the minimum vertex cover in G_C.

LEMMA 9.1 *[97] Any minimum vertex cover of G_C is also a minimum sufficient set of G.*

A consequence of Lemma 9.1 is that in order to find a minimum sufficient set in a graph G, one only has to find a minimum vertex cover of its contraction, a potentially simpler undirected graph. In general, the minimum vertex cover problem is NP-complete and it has a 2-approximation algorithm. But it is fixed-parameter tractable: whether a graph has a vertex cover of size at most k can be decided with time $O(f(k) \cdot p(n))$ (see e.g. [51]) where p is a polynomial function. If the graphs are planar, a polynomial-time approximation scheme exists [17, 18]. However, as will show in this paper, when the graph is a tree or a tree of rings, a minimum vertex cover can be found in polynomial time.

3. MINIMUM SUFFICIENT SET IN TREES

In this section, we consider the WDM networks whose underlying topologies are trees. It's obvious that the contraction of any tree is also a tree with the additional property that all internal nodes have degrees of at least three. Thus according to Lemma 9.1 the minimum sufficient set can be obtained by finding a MVC in the contraction tree. In the next we will present a general polynomial-time algorithm to find MVCs in trees.

We call an *internal* node of a tree to be a *leaf-root* if one of its neighbors is a leaf node. If a tree has no leaf-root then it must be a single edge and its MVC consists of an arbitrary node of this edge. If a tree contains a leaf-root, then the next lemma indicates that there is a MVC which contains this leaf-root.

LEMMA 9.2 *Let G be a tree with a leaf root u. Then G has a MVC which contains u.*

Proof. We prove the lemma by contradiction. Assume the lemma were not true. Let $\mathcal{C}$ be a MVC of G which contains the *most* number of leaf-roots of G, but does not contain u. Let v be any leaf node that is a neighbor of u. Clearly, v must be in $\mathcal{C}$, for otherwise the edge (u, v) would not be covered by $\mathcal{C}$. Consider

$$\mathcal{C}' = (\mathcal{C} - \{v\}) \cup \{u\}.$$

Then $\mathcal{C}'$ is also a MVC of G, and contains one more leaf-roots of G than $\mathcal{C}$. This contradicts the selection of $\mathcal{C}$. Thus the lemma is true. ■

Let G be any tree and $\mathcal{C}$ be any MVC of G which contains a leaf-root u of G. Let G' be the graph obtained from G by removing u and all its incident edges. Then G' is a forest and it is referred to as the *residue* of G. Let $\mathcal{C}' = \mathcal{C} \setminus \{u\}$. Then for each component tree T' of G', $\mathcal{C}' \cap T'$ is also a MVC of T'. Thus we can apply Lemma 9.2 to all component trees of G' recursively. Based on this observation, we have the following recursive algorithm to find a MVC of a tree described in Table 9.1.

The algorithm **MVC_Tree** can be implemented efficiently. The details are omitted over here.

Algorithm: MVC_Tree
Input: *a tree G;*
Output: *a MVC of G;*
begin
  if *G is a single edge*
    return an arbitrary node of the edge;
  else
    add a leaf-root of G to the VC of G;
    obtain the residue graph G′ of G;
    find the MVC of all componets of G′ recursively;
    add them to the VC of G;
end

Table 9.1. Recursive algorithm to find a MVC of a tree.

4. MINIMUM SUFFICIENT SET IN TREE-CONNECTED RINGS

The tree-connected rings, illustrated in Figure 9.1, is an interconnection topology widely used in the telecommunications industry. In this topology, each node is within a ring and these rings are interconnected via a tree-like topology. It's easy to see that the contraction of any tree-connected rings is also a tree-connected rings. As the MVC of the contraction graph provides an optimal sufficient set of the original graph, we will provide a polynomial-time algorithm that finds a MVC in an arbitrary tree-connected rings.

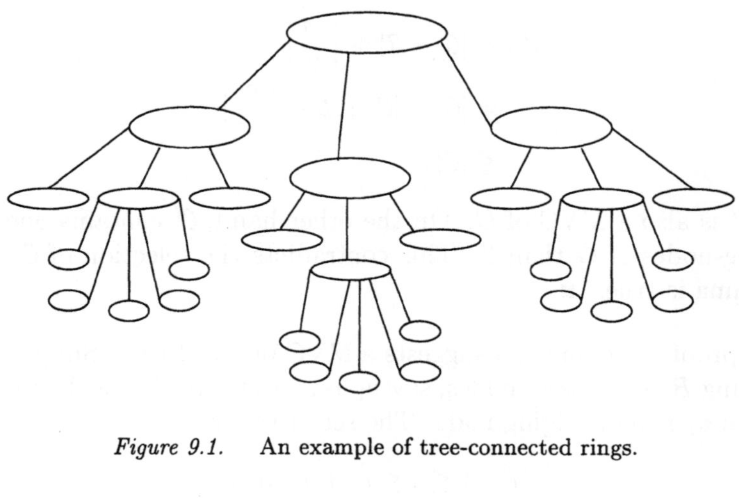

Figure 9.1. An example of tree-connected rings.

It's well-known that there are at least two leaf-nodes in any tree. Similarly, one can show that in any tree-connected rings, there exist at least two rings in which all nodes have degree of two except one whose degree is three. Such rings are referred to as *leaf-rings.* The only node in a leaf-ring whose degree is three is called a ***bridging-node.*** Suppose that a leaf-ring contains m nodes. Then any MVC contains at least $\lceil \frac{m}{2} \rceil$ nodes in this leaf-ring. In the next we show that there is always a MVC which contains the bridging node of any leaf ring.

LEMMA 9.3 *Let G be a tree-connected rings. Then G has a MVC which contains all bridging-nodes of all leaf-rings in G.*

Proof. We prove the lemma by contradiction. Assume the lemma were not true. Let C be a MVC of G which contains the *most* number of bridging-nodes of G, and let u be any bridging-nodes of G that is not in C. Let R be the leaf-ring containing the bridging-node u. Then R can't be a self-loop, for otherwise u must be in any VC of G and thus in C too. So R contains $m \geq 2$ nodes, say $u = v_1, v_2, \cdots, v_m$ in the clockwise order. Clearly C must contain at least $\lceil \frac{m}{2} \rceil$ nodes in R, i.e.,

$$|C \cap R| \geq \left\lceil \frac{m}{2} \right\rceil .$$

Consider

$$C' = (C - R) \cup \{v_i : 1 \leq i \leq m, i \text{ is odd}\} .$$

Then C' is also a VC of G, and

$$\begin{aligned} |C'| &= |C - R| + \left\lceil \frac{m}{2} \right\rceil \\ &= |C| - |C \cap R| + \left\lceil \frac{m}{2} \right\rceil \\ &\leq |C| . \end{aligned}$$

Thus C' is also a MVC of G. On the other hand, C' contains one more bridging-nodes of G than C. This contradicts the selection of C. Thus the lemma is true. ■

The proof of Lemma 9.3 suggests a MVC of a leaf-ring. Suppose that a leaf-ring R contains m nodes, say $v_1, v_2, \cdots, v_m$ in the clockwise order in which v_1 is its bridging node. The set of vertices

$$\{v_i : 1 \leq i \leq m, i \text{ is odd}\}$$

is called a *canonical vertex cover* of the leaf-ring R. Then Lemma 9.3 indicates that for any leaf-ring R, there is a MVC $\mathcal{C}$ of G which contains its canonical VC. Let $G-R$ denote the graph obtained from G by removing all node in R and their incident edges. Let $\mathcal{C}-R$ denote the vertex set obtained from $\mathcal{C}$ by removing the canonical VC of R. Then $G-R$ is also a tree-connected rings and $\mathcal{C}-R$ is a MVC of G'. On the other hand, the union of the canonical VC of R and any MVC of $G-R$ is also a MVC of G. Based on this observation, we have the following incremental algorithm to find the MVC of a tree-connected rings listed in Table 9.2.

```
Algorithm: MVC_Tree_Connected_Rings
Input: a tree-connected rings G;
Output: a MVC of G;
begin
  C = ∅; // the VC of G
  while G is not empty
    find a leaf-ring R in G;
    add the canonical VC of R to C;
    G = G − R;
  output C;
end
```

Table 9.2. Algorithm to find a MVC of a tree-connected rings.

The algorithm **MVC_Tree_Rings** removes one leaf-ring at each incremental step and add its canonical VC. With carefully selected data structures and implementation, its run-time can be linear in the network size. The details are omit over here.

From the algorithm **MVC_Tree_Rings**, we can explicitly count the cardinality of any MVC of a tree-connected rings.

THEOREM 9.4 *Let $G=(V,E)$ be a tree-connected rings, and k be the number of odd-sized rings in G. Then the cardinality of any MVC of G is $\frac{|V|+k}{2}$.*

Proof. Let $R_1, R_2, \cdots, R_\ell$ be the component rings in G. According to our above algorithm, each ring will eventually become a leaf-ring and its canonical VC will be added to the MVC of G. Note that the canonical

VCs of different component rings are disjoint. Thereby, the cardinality of any MVC of G is

$$\sum_{i=1}^{\ell} \left\lceil \frac{|R_i|}{2} \right\rceil = \sum_{i=1}^{\ell} \frac{|R_i|}{2} + \frac{k}{2} = \frac{1}{2} \sum_{i=1}^{\ell} |R_i| + \frac{k}{2}$$
$$= \frac{|V|}{2} + \frac{k}{2} = \frac{|V| + k}{2}.$$

■

As a corollary of Theorem 9.4, the cardinality of any minimum sufficient set is at least half of the number of branching-nodes. In order to reduce the number of converters needed, the topology should be carefully designed. For an example, we can choose the topology such that the number of branching nodes in each component ring to be even.

5. MINIMUM SUFFICIENT SET IN TREES OF RINGS

The contraction of any tree of rings is also a tree of rings with the property that the nodal degrees are all even numbers greater than two. In addition, the contraction has at least two rings which are self-loops. These rings are also referred to as *leaf-rings.* Note that any VC of the contraction must contain the unique node in each loop. However, after removing all edges incident to the vertices selected as part of the VC, the residue graph may no longer be a tree of rings, and indeed it may even become disconnected. However each connected component can be treated as subgraph of a tree of rings, which can be obtained by choosing a subset of rings and remove a link from each of these rings. A connected component obtained in this way is referred to as a *sub-tree of rings.* In a sub-tree of rings, we can similarly define leaf-roots as in trees and leaf-rings as in tree-connected rings. A *leaf-root* is an internal node with at least one neighbor as a leaf. A *leaf-ring* is a ring which contains at most one node with nodal degree more than two. For each leaf-ring, one can define its *canonical VC* as for tree-connected rings. By the similar argument, we can show that if a sub-tree of rings has non-empty leaf-roots, then it has a MVC that contains all leaf-roots. If a sub-tree of rings has no leaf-roots, then it must contains at least two leaf-rings. Pick any leaf-ring. Then the sub-tree of rings has a MVC that contains the canonical VC of the picked leaf-ring. Notice that a tree of rings can also be treated as a sub-tree of rings. So we propose the following recursive algorithm for MVC of sub-tree of rings as describe in Table 9.3.

Algorithm: MVC_Sub-tree_of_Rings
Input: *a subtree of rings G;*
Output: *a MVC of G;*
begin
  if *G is a ring,*
    return an arbitrary MVC of the ring;
  if *G is a single edge,*
    return an arbitrary node of the edge;
  if *G has a leaf-root,*
    add the leaf-root of G to the VC of G;
  else
    find a leaf-ring of G;
    add the canonical VC of the leaf ring to the VC of G;
  obtain the residue graph G';
  find the MVC of all components of G' recursively;;
  add them to the VC of G;
end

Table 9.3. Recursive algorithm to find a MVC of a tree of rings.

6. SUMMARY

The minimum sufficient set problem is in general NP-complete and is as hard as the MVC problem. However, the underlying topologies of most WDM networks in the telecommunications industry are built around trees, rings, tree-connected rings, and tree of rings. For these topologies, this paper showed that both minimum sufficient problem and the MVC problem can be solved in polynomial time. Efficient algorithms have been provided for the MVC problem in these topologies which in turn are used to solve the minimum sufficient set problem in these topologies.

IV
SONET/WDM OPTICAL NETWORKS

Chapter 10

OPTIMAL GROOMING OF ARBITRARY TRAFFIC IN WDM BLSRS

1. INTRODUCTION

WDM self-healing rings inherit both network architecture and the fault restoration mechanism from the SONET self-healing rings. The bandwidth of these networks are managed at two complementary levels, the fine level through SONET ADMs and the coarse level through OADMs. Typically, the traffic demand between two nodes has a lower rate (e.g., OC-3) than the transmission capacity (e.g., OC-48) of individual wavelength channels. The SONET ADMs are used to groom a number of low-rated traffic streams to form a logical SONET ring which is carried in a unique wavelength. The OADMs are used to add, drop or bypass individual wavelengths, and to provide fault protection and restoration as described in Chapter 1. Since there are no duplicated electronics, exactly one SONET ADM is required for a node to participate in logical SONET ring. Thus the total number of SONET ADMs by a node is equal to the number of logical SONET rings that contain this node. Alternatively, the number of ADMs required by a logical SONET ring is equal to the size of this SONET ring, i.e., the number of nodes it contains. As SONET ADMs are the dominant cost factors in a WDM self-healing ring, a fundamental bandwidth management problem is how to route and groom the traffic demands to minimize the total SONET ADM cost. This problem is referred to as the *minimum ADM cost problem.*

In this chapter, we assume that the WDM self-healing ring is a bidirectional line-switched ring (BLSR) with either two fiber rings (i.e.,

BLSR/2) or four fiber rings (BLSR/4). All wavelength channels are assumed to have the same line-speed, denoted by g. By allowing parallel traffic streams between the same pair of nodes, each traffic steam is assumed to have a *unit* traffic demand. However, each traffic stream can occur between any *arbitrary* pair of nodes, as opposite to the all-to-all traffic and one-to-all traffic. Two versions of the minimum ADM cost problem will be addressed. In the first version, each traffic stream has its predetermined routing, such as short-path routing. Thus we can deal with the two (working) fiber rings separately. In each fiber ring, a traffic stream can be represented as a (directed) circular arc over the fiber ring. Therefore an instance of the minimum ADM cost problem can be represented by a set A of circular arcs in a clockwise ring. Depending on the implementation, the circular arcs may or may not be allowed to be split. If the splitting is not allowed, a valid traffic grooming corresponds to a partition of A into groups of g-colorable circular arcs. If the splitting is allowed, then a valid traffic grooming consists of a choice of splitting each arc of A, thus obtaining A', and then a partition of A' into groups of g-colorable circular arcs. In either case, each group of g-colorable circular arcs can be carried in a wavelength and thus form a logical SONET ring. This version is referred to as the *arc-version* the minimum ADM cost problem.

In the second version, the routing of each traffic stream is not given in advance. If each traffic stream is symmetrically duplex and its two portions in opposite directions must be routed along the same path (in opposite directions), we can treat the two working fiber rings as one (clockwise) ring, and each traffic stream as a (undirected) chord. Otherwise, each traffic stream is a simplex traffic demand from its origin node to its destination node. These simplex traffic demands can still be represented by undirected chords in a clockwise oriented ring, and their routings are then given as clockwise arcs over this ring. Indeed, for a simplex traffic demand from node i to node j, a routing of the chord (i, j) with the arc from i to j indicates choosing the clockwise physical ring for the traffic demand, and a routing of the chord (i, j) with the arc from j to i indicates choosing the counterclockwise physical ring for the traffic demand. Thus in either case, an instance of the minimum ADM cost problem can be represented by a set C of chords in a clockwise ring. Similarly, depending on the implementation, the chords may or may not be allowed to be split. If the splitting is not allowed, a valid traffic routing and grooming corresponds to a partition of C into groups of g-colorable chords (a set of chords is said to be g-colorable if they can be routed as g-colorable circular arcs). If the splitting is allowed, then a valid traffic routing and grooming consists of a choice of splitting each chord of C, thus obtaining C', and then a partition of C' into groups of g-colorable chords. In either case, each group of g-colorable chords can be carried in a wavelength and thus form a logical SONET ring. This version is referred to as the *chord-version* minimum ADM cost problem.

Both versions of the minimum ADM cost problem are NP-hard even for any fixed constant $g \geq 1$ as will be shown in Section 2.. Due to hardness of this problem, we adopt the following two-phased approach.

Generation of Primitive Rings: Split the arcs or chords as necessary if the splitting is allowed, and partition the resulting arcs or chords into groups of non-overlapping arcs or chords (a set of arcs are said to be *non-overlapping* if their interiors have empty intersection; a set of chords are said to be *non-overlapping* if they can be routed as non-overlapping arcs). Each group of non-overlapping arcs or chords in the partition can be arranged in a single ring, referred to as a *primitive ring*. The cost of each primitive ring is defined as the number of nodes appearing in this primitive ring, and the cost of a partition is defined as the sum of the costs of all primitive rings within this partition. The objective in this phase is to find a partition with minimum cost.

Grooming of Primitive Rings: Partition those primitive rings that are constructed in the first phase into groups of at most g primitive rings. Each group then forms a logical SONET ring. Once again, the ADM cost of each group or SONET ring is the number of nodes contained in this group, and the total ADM cost of a partition is the sum of ADM costs of all groups in this partition. The objective of this phase is thus to find a partition with the minimum total ADM cost.

This two-phased approach tries to attack the original minimum ADM cost problem by solving two sub-problems. However, the combination of an optimal solution in each phase, even if it could be found, does not necessarily lead to an overall optimal solution.

This chapter is organized as follows. In Section 2., we prove the NP-hardness of the minimum ADM cost problem for any fixed g. In Section 3., we present two lower bounds on the minimum ADM cost. In Section 4., we reduce the optimum primitive ring generation problem into the optimal valid chain generation problem. In Sections 5. through 8., we provide a number of heuristics for various optimal valid chain generations and analyze their approximation ratios. In Section 9., we study the complexity and approximation algorithms for optimal grooming of primitive rings. Finally in Section 10., we summarize this chapter.

2. COMPUTATIONAL COMPLEXITIES

Let A be any set of circular arcs over a ring. Without loss of generality, we assume that each node in the ring is one endpoint of some arc in A, for otherwise we can remove this node from the ring without affecting the ADM cost. Thus the ring size can be implied from A. A is said to be *uniformly loaded* if the number of circular arcs passing through each link is the same. In particular, if there are exactly L circular arcs in A passing through each link, A is said to be *L-uniformly loaded.* The following NP-hardness result was proved in [63].

LEMMA 10.1 *It is NP-hard to decide whether a set of L-uniformly loaded arcs can be partitioned into L groups such that the arcs in each group form a whole circle.*

Using this NP-hardness result, we can prove the NP-hardness of the arc-version minimum ADM cost problem.

THEOREM 10.2 *For any constant integer $g \geq 1$, the arc-version minimum ADM cost problem is NP-hard.*

Proof. Let A be any set of L-uniformly loaded circular arcs. Let A' be the set of arcs obtained by making g copies of A, and $opt_g(A')$ be its minimal ADM cost. We show that $opt_g(A') = |A|$ if and only if A can be partitioned into L groups such that the arcs in each group form a whole circle.

Consider any optimal grooming of A'. Each ADM can terminate at most $2g$ arcs. As each arc is terminated by two ADMs, cumulatively there are at least $2|A'| = 2g|A|$ arcs terminated by all ADMs. Hence

$$opt_g(A') \geq \frac{2g|A|}{2g} = |A|,$$

and the equality holds if and only if no arc is split and each ADM terminates exactly $2g$ arcs. As in each logical SONET ring at most

g arcs can cross over any link, each ADM terminates exactly $2g$ arcs if and only if there are exactly g copies of one arc between any neighboring ADMs in each logical SONET ring. Thus $opt_g(A') = |A|$ if and only if each logical SONET ring is exactly g copies of a subset of A which forms a whole circle. This implies that $opt_g(A') = |A|$ if and only if A can be partitioned into L subsets each of which forms a whole circle.

From Lemma 10.1, the NP-hardness of the arc-version minimum ADM cost problem follows. ■

In the following, we show the NP-hardness of chord-version minimum ADM problem for any fixed constant g.

THEOREM 10.3 *For any constant integer $g \geq 1$, the chord-version minimum ADM cost problem is NP-hard.*

Proof. Let A be any set of arcs over a ring R. We orient the ring and each circular arcs in the clockwise direction. For each arc $a \in A$, we add a unique node v_a in the ring R within the arc a and create two chords, $(t(a), v_a)$ and $(v_a, h(a))$, as illustrated in Figure 10.1. We use R' and C to denote the obtained ring and set of chords respectively. From the construction,

$$\begin{aligned} |R'| &= |R| + |A|, \\ |C| &= 2|A|. \end{aligned}$$

Let C' be the set of chords obtained by making g copies of C, and let $opt_g(C')$ be its minimal ADM cost. We show that $opt_g(C') = 2|A|$ if and only if A can be partitioned into L groups such that the arcs in each group form a whole circle.

Consider any optimal routing and grooming of C'. Each ADM can terminate at most $2g$ chords. As each chord is terminated by two ADMs, cumulatively there are at least $2|C'| = 2g|C|$ chords terminated by all ADMs. Hence

$$opt_g(B') \geq \frac{2g|C|}{2g} = |C| = 2|A|,$$

and the equality holds if and only if no chord is split and each ADM terminates exactly $2g$ chords. As in each logical SONET ring at most g arcs can cross over any link, each ADM terminates exactly $2g$ chords if and only if there are exactly g copies of one chord between any neighboring ADMs in each logical SONET ring. Thus $opt_g(C') = 2|A|$ if and only if each logical SONET ring is g copies of a subset of C which forms a whole circle. Therefore $opt_g(C') = 2|A|$ if and only if C can be partitioned into subsets each of which forms a whole circle. Notice that in any such partition, the two chords created from each arc must be in the same subset. Thus $opt_g(C') = 2|A|$ if and only if the arcs in A can be partitioned into subsets each of which forms a whole circle.

From Lemma 10.1, the NP-hardness of the arc-version minimum ADM cost problem follows. ■

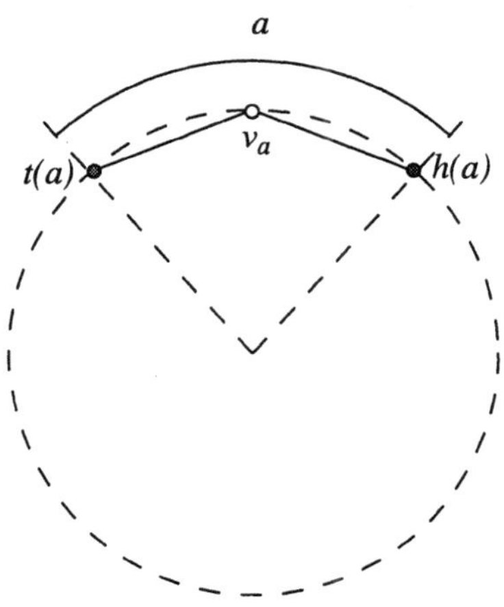

Figure 10.1. Replace each arc by two chords: the filled circles are the endpoints of the arc, and the empty cycle is the new node inserted.

By setting $g = 1$ in Theorem 10.2 and Theorem 10.3, we obtain that both versions of optimal primitive ring generation are NP-hard.

3. GENERAL LOWER BOUNDS

In this section, we present some general lower bounds on the minimum ADM cost. One straightforward but rather loose lower bound can be obtained in the following way. We begin with the arc version. Let A be the set of input circular arcs. For any node i, let $\sigma_A(i)$ denote the total number of circular arcs in A that originate from node i, and $\tau_A(i)$ denote the total number of circular arcs in A that terminate at node i. Then the node i must use at least

$$\left\lceil \frac{\max(\sigma_A(i), \tau_A(i))}{g} \right\rceil$$

ADMs. Hence the total ADM cost is at least

$$\sum_{i=0}^{n-1} \left\lceil \frac{\max(\sigma_A(i), \tau_A(i))}{g} \right\rceil.$$

This lower bound is a generalization of the one in given in [67] for $g = 1$.

Now we consider the chord version. Let C be the set of input chords. For any node i, let $\deg_C(i)$ denote the total number of chords in A that contain node i as one endpoint. As each ADM can terminate at most $2g$ chords, node i must use at least $\left\lceil \frac{\deg_C(i)}{2g} \right\rceil$ ADMs. Hence the total ADM cost is at least

$$\sum_{i=0}^{n-1} \left\lceil \frac{\deg_C(i)}{2g} \right\rceil.$$

In the remaining of this section, we develop another lower bound by calculating the *maximum ADM efficiency*, as defined later. We notice that similar lower bounding technique was used in [115] for only $g = 4, 16$, but no general lower bound was established in [115]. The lower bound derived below is general and applicable to any values of n and g. Given any two positive integers n and g, let $A(n, g)$ denote the maximal number of arcs with *different* endpoint pairs over a ring of n nodes whose load is g, and let

$$E(n, g) = \frac{A(n, g)}{n}.$$

Then the *maximal ADM efficiency* with line-speed g is defined as

$$E(g) = \max_{n \geq 2} E(n, g).$$

Intuitively, $E(g)$ puts an upper bound on the average number of arcs that can be carried by one ADM. Thus the total number of ADMs required by a set of arcs (or chords) with different endpoint pairs is at least the total number of arcs (or chords respectively) divided by the $E(g)$.

When g is small, $E(g)$ can be calculated easily. For example, when $g = 1$ or 2,

$$\begin{aligned} A(n,1) &= n, \\ E(n,1) &= E(1) = 1; \\ A(n,2) &= \begin{cases} \left\lfloor \frac{3n}{2} \right\rfloor & \text{if } n \geq 5, \\ \left\lfloor \frac{3n}{2} \right\rfloor - 1 & \text{if } n \leq 4, \end{cases} \\ E(2) &= \frac{3}{2}. \end{aligned}$$

In general, the calculation of $E(g)$ relies on the concept of *canonical* set of arcs. A set of arcs with different endpoint pairs is said to be *canonical* if it satisfies the following property: if an arc of length ℓ is in this set, then all arcs of length less than ℓ are also in this set. It's easy to show that any set of arcs over a ring can be transformed to a canonical set of arcs of the *same* cardinality over the same ring with the same or less load. Thus there is always a canonical set of arcs over a ring of n nodes whose load is g. A canonical set of arcs can be generated in the following greedy manner: we first add all arcs of length one, then we add all arcs of length two and so on until we can not add all arcs of some length; in this case we add as many arcs of such length as possible. The following lemma gives the load of the set of arcs of length no more than k.

LEMMA 10.4 *For any $\ell \leq n-1$, the load of all circular arcs of length no more than ℓ over an n-node ring is $\frac{\ell(\ell+1)}{2}$.*

Proof. For any $k \leq n-1$, there are n arcs of length k. These n arcs of length k contribute a load of k to each link. Thus the load contributed to each link by those $n\ell$ arcs of length at most ℓ is

$$1 + 2 + \cdots + \ell = \frac{\ell(\ell+1)}{2}.$$

■

Let ℓ be the largest integer satisfying that $\frac{\ell(\ell+1)}{2} \leq g$. If $n \leq \ell + 1$, then the load of all $n(n-1)$ circular arcs over an n-node ring is no more than g, and thereby

$$\begin{aligned} A(n,g) &= n(n-1) \\ E(n,g) &= n - 1 \leq \ell - 1 \end{aligned}$$

Now we assume that $n \geq \ell + 2$. If $g = \frac{\ell(\ell+1)}{2}$, then the load of all $n\ell$ circular arcs of length at ℓ over an n-node ring is exactly g, and thereby

$$A(n,g) = n\ell,$$
$$E(n,g) = \ell.$$

If $g > \frac{\ell(\ell+1)}{2}$, then $A(n,g) - n\ell$ is equal to the maximal number of circular arcs of length $\ell+1$ which contribute a load of no more than $g - \frac{\ell(\ell+1)}{2}$ to each link. The cumulative link load contributed by these circular arcs of length $\ell+1$ is at most $n\left(g - \frac{\ell(\ell+1)}{2}\right)$. As each circular arc of length $\ell+1$ contributes a unit load to $\ell+1$ links, the total number of such circular arcs is

$$A(n,g) - n\ell \leq \frac{n\left(g - \frac{\ell(\ell+1)}{2}\right)}{\ell+1} = n\left(\frac{g}{\ell+1} - \frac{\ell}{2}\right).$$

On the other hand, it's obvious that

$$A(n,g) - n\ell \geq \left\lfloor \frac{n}{\ell+1} \right\rfloor \left(g - \frac{\ell(\ell+1)}{2}\right).$$

Hence

$$n\ell + \left\lfloor \frac{n}{\ell+1} \right\rfloor \left(g - \frac{\ell(\ell+1)}{2}\right) \leq A(n,g) \leq n\left(\frac{g}{\ell+1} + \frac{\ell}{2}\right),$$
$$\ell + \frac{1}{n}\left\lfloor \frac{n}{\ell+1} \right\rfloor \left(g - \frac{\ell(\ell+1)}{2}\right) \leq E(n,g) \leq \frac{g}{\ell+1} + \frac{\ell}{2}.$$

In particular, when n is a multiple of $\ell+1$,

$$A(n,g) = n\left(\frac{g}{\ell+1} + \frac{\ell}{2}\right),$$
$$E(n,g) = \frac{g}{\ell+1} + \frac{\ell}{2}.$$

Therefore

$$E(g) = \frac{g}{\ell+1} + \frac{\ell}{2}.$$

The next lemma summarizes the above discussion.

g	ℓ	$E(g)$
2	1	$\frac{3}{2}$
4	2	$\frac{7}{3}$
8	3	$\frac{7}{2}$
16	5	$\frac{31}{6}$

Table 10.1. Maximal ADM efficiency with various link capacities.

LEMMA 10.5 *For any positive integer g, the maximal ADM efficiency is $\frac{g}{\ell+1} + \frac{\ell}{2}$ where ℓ is the largest integer satisfying that $\frac{\ell(\ell+1)}{2} \leq g$.*

Table 10.1 lists the values of $E(g)$ when $g = 2, 4, 8$ and 16.

The maximal node efficiency provides an upper bound on the maximal ADM efficiency. As the total number of ADMs required is at least the total traffic divided by the maximal ADM efficiency, we have the following lower bound on the minimal ADMs required.

LEMMA 10.6 *Let g be the line-speed of each wavelength. Let ℓ is the largest integer satisfying that $\frac{\ell(\ell+1)}{2} \leq g$. If the splitting is not allowed, then the minimum ADM cost of any set A of different arcs is at least*

$$\left\lceil \frac{|A|}{\frac{g}{\ell+1} + \frac{\ell}{2}} \right\rceil,$$

and the minimum ADM cost of any set C of different chords is at least

$$\left\lceil \frac{|C|}{\frac{g}{\ell+1} + \frac{\ell}{2}} \right\rceil.$$

4. REDUCTION FROM PRIMITIVE RINGS TO VALID CHAINS

4.1 WALK, TRAIL, PATH, CIRCUIT

To avoid confusion, we first present the standard definitions of some basic concepts in graph theory. A *walk* in a graph G is a sequence

$$W := (v_0, e_1, v_1, e_2, v_2, \cdots, e_\ell, v_\ell),$$

where $v_0, v_1, \cdots, v_\ell$ are vertices of G, $e_1, e_2, , \cdots, e_\ell$ are edges G, and e_i is the edge joining v_{i-1} and v_i for $i = 1, \cdots, \ell$. In a simple graph,, the edges in a walk is completely specified by its sequence of vertices $(v_0, v_1, \cdots, v_\ell)$. The walk W is *closed* if $n > 0$ and $v_0 = v_n$, and *open* otherwise. The *length* of W is the number of edges, namely ℓ, and is denoted by $|W|$; the *parity* of W is the parity of its length. A walk of length ℓ is called as an ℓ-walk. If W is open, the vertex v_0 is the tail of W, the vertices $v_1, \cdots, v_{\ell-1}$ its *internal* vertices and the vertex v_ℓ its head. A *trail* is a walk with all its edges distinct; a *path* is a open walk with all its vertices distinct; a *circuit* is a closed trail of positive length whose vertices are all distinct. It is obvious that a trail can be decomposed into paths and circuits. The definitions of these concepts in a directed graph are the same as in a graph with the exception that the edge must be traversed in the correct direction. In an open directed walk W, v_0 is referred to as the tail of W and is denoted by $t(W)$; v_n is referred to as the head of W and is denoted by $h(W)$.

An *Eulerian trail* in a graph is a trail which includes every edge. A closed Eulerian trail is also called an *Eulerian tour.* A (directed) graph is said to be *Eulerian* if it has an Eulerian tour. It is well known that a graph is Eulerian if and only if it is connected and every vertex has even degree, and a directed graph is Eulerian if and only if it is connected and the in-degree and out-degree of any vertex are equal. An Eulerian tour can be generated in polynomial time.

A *circuit cover* of a (directed) graph G is a collection of vertex-disjoint circuits which together cover all vertices of G. If G is a weighted (directed) graph, then the weight of a circuit cover of G is the sum of the weights of all edges in this circuit cover. It is well-known that a minimum-weighted circuit cover of G can be found in polynomial time by a reduction to minimum-weight perfect matching in a bipartite graph.

4.2 RING, ARC, CHORD

We assume that a ring network consists of n nodes numbered clockwise by $0, 1, \cdots, n-1$. It is oriented clockwise and is treated as a directed graph. All arithmetic involving nodes is performed implicitly using modulo n operations. The link from the node i to node $i+1$ in the ring is referred to as link i.

An (clockwise) arc a over a ring is represented by an *ordered* pair $(t(a), h(a))$, where $t(a)$ is the tail of a and $h(a)$ is the head of a. Two arcs are said to be *intersecting* (or *overlapping*) if they contain a common link of the ring, *disjoint* if they do not contain any common node of the ring, *adjacent* if they are not intersecting but share at least one endpoint, and *complementary* if they are not intersecting and share two endpoints. An arc is *red* if it does not contain the link $n - 1$, and *blue* otherwise. Given a set of arcs S, the *blue number* of S, denoted by $b(S)$, is the number of blue arcs in S. Sometimes we write $b(a)$ instead of $b(\{a\})$.

A chord in a ring is specified by an *unordered* pair (i, j) where i and j are the two endpoints of the chord. *Routing* a chord means selecting from its two endpoints a tail and a head, thus replacing the chord by an arc. Two chords are said to be *crossing* if they have different endpoints and a walk along the ring visits their endpoints alternatively. Two chords are said to be *parallel* if they have different endpoints and are not crossing. Two chords are said to be *adjacent* if they have at least one common endpoint, and be *identical* if they have the same endpoints. Note that two adjacent but not identical pair of chords have a unique routing without intersection.

4.3 CHAINS OF ARCS OR CHORDS

Let A be any set of arcs over a ring. It induces naturally a directed graph $G(A)$ with the nodes of the ring as its vertex set and A as its edge set. A trail in $G(A)$ is also called a *trail of arcs* in A. A trail of arcs induces a (clockwise) walk over the ring starting from its origin to its termination. A trail of arcs is said to be a *chain of arcs* if any pair of consecutive arcs are non-overlapping. A chain of arcs is said to be *valid* if all arcs in this chain are non-overlapping. It is obvious that a trail of arcs can be decomposed into chains and a chain of arcs can be decomposed into valid chains of arcs. A chain is said to be *blue* if it contains a blue arc. Note that any valid closed chain has blue number one, while a valid open chain has blue number at most one.

Let C be any set of chords. It induces naturally an undirected graph $G(C)$ with the nodes of the ring as its vertex set and C as its edge set. A trail in $G(C)$ is also called as a *trail of chords* in C. A trail of chord is said to be a *chain of chords* if either itself or its reverse induces a (clockwise) walk over the ring in which the two arcs corresponding to two consecutive chords are non-overlapping. A *routing* of a chain of chords is the (clockwise) walk induced by either itself or its reverse in which two arcs corresponding to consecutive chords are non-overlapping. Note that if a chain of chords is a path of at least two chords or a circuit of at least three chords, it has a unique routing. A chain of a single chord or two identical chords can have two different routings. A chain of chords is said to be *valid* if it has a routing in which all arcs corresponding to the chords are non-overlapping. It is obvious that a trail of chords can be decomposed into chains of chords, and a chain of chords can be decomposed into valid chains.

For simplicity, we also refer to the length of a chain (of arcs or chords) as its *size*, and the number of vertices in a chain as its *cost*. Thus the cost of a closed chain is exactly its size, and the cost of an open chain is one plus its size. Let $\mathcal{P}$ be a collection of chains. The *cost* of $\mathcal{P}$ is defined as the sum of the costs of all chains in $\mathcal{P}$, which equals to the total number of arcs in the chains in $\mathcal{P}$ plus the total number of open chains in $\mathcal{P}$. The *fit graph* of $\mathcal{P}$, denoted by $F(\mathcal{P})$, is a weighted undirected graph in which the vertex set is $\mathcal{P}$, there is an edge between two chains P_1 and P_2 if and only if P_1 and P_2 can be merged into a larger valid chain, and the weight of the edge between P_1 and P_2 is equal to the number of endpoints shared by P_1 and P_2.

For both arc-version and chord version of the optimum primitive ring generation problem, we can restrict the solutions to partitions of the input arcs or chords into valid chains, referred to as *valid chain generations*. This restriction does not change the optimum value, but may requires larger number of primitive rings. The number of primitive rings can be further reduced by treating each chain as an arc and applying Tucker's for circular-arc coloring [151]. For the arc-version, this processing can reduce the number of required wavelengths within twice the minimum. Indeed, suppose that the load of the original arcs is l and the number of (valid) closed chains is k. Then the load of those open chains is $l-k$. The Tucker's algorithm for circular-arc coloring guarantees that the number of primitive rings is at most $2(l-k)-1$. Thus the total number of primitive rings is at most $k+2(l-k)-1=2l-k-1$, which is within twice the minimum.

Notice that the cost of chains produced a valid chain generation is the number of input arcs or chords plus the number of open chains. Therefore, an optimal solution corresponds to the a valid chain generation with minimal number of open valid chains.

5. VALID CHAIN GENERATION FROM UNSPLITABLE ARCS

Given a set of arcs A, as in Section 3. we use $\sigma_A(i)$ to denote the total number of arcs in A that originate from node i, and $\tau_A(i)$ to denote the total number of arcs in A that terminate at node i. The *surplus* of a node i with respect to S is defined as $\delta_A(i) = \sigma_A(i) - \tau_A(i)$. Note that the surplus might be negative and in fact $\sum_{i=0}^{n-1} \delta_A(i) = 0$. An open chain in A with negative surplus at its origin and positive surplus at its termination is said to be *tight with respect to A*. The *deficiency* of a node i is $d_A(i) = \frac{1}{2}|\delta_A(i)|$. The deficiency of a set of arcs A is $d(A) = \sum_{i=0}^{n-1} d_A(i)$. In all these notations, the subscripts may be omitted if understood from the context.

Notice that in any primitive ring generation from A, each node i requires at least $\max\{\sigma_A(i), \tau_A(i)\}$ ADMs. Thus the total number of ADMs required by any primitive ring generation from A is at least

$$\begin{aligned}
&\sum_{i=0}^{n-1} \max\{\sigma_A(i), \tau_A(i)\} \\
&= \sum_{i=0}^{n-1} \frac{\sigma_A(i) + \tau_A(i) + |\sigma_A(i) - \tau_A(i)|}{2} \\
&= \frac{\sum_{i=0}^{n-1} (\sigma_A(i) + \tau_A(i))}{2} + \frac{\sum_{i=0}^{n-1} |\sigma_A(i) - \tau_A(i)|}{2} \\
&= |A| + d(A).
\end{aligned}$$

So $|A| + d(A)$ is a lower bound on the minimum ADM cost required by any primitive ring generation from A. This fact was first noticed by [67].

Fix a set of arcs A. Let OPT denote the optimal valid chain generation from A and let *opt* denote the cost of OPT. Consider any valid chain generation $\mathcal{P}$ of A. As discussed in Section 4., the cost of $\mathcal{P}$ is $|A|$ plus the open chains in $\mathcal{P}$. Since the number of open chains in $\mathcal{P}$ can never exceed $|A|$, the cost of $\mathcal{P}$ is at most $2|A|$. Note that $|A|$is a lower bound on *opt*. Thus the cost of $\mathcal{P}$ is at most $2 \cdot opt$. This implies that two is a trivial upper bound on the approximation ratio of *any* valid chain generation. In the next, we provide a less trivial upper bound on the cost of those valid chain generations in which any pair of (valid) chains can not be merged into a larger valid chain.

LEMMA 10.7 *If any pair of chains of arcs in $\mathcal{P}$ can not be merged into a larger valid chain, the cost of $\mathcal{P}$ is at most $\frac{7}{4} \cdot opt$.*

Proof. We call an arc *unmatched* in $\mathcal{P}$ if it forms a 1-chain in $\mathcal{P}$. Then out of any two consecutive arcs in any (valid) chain of OPT at most one is unmatched in $\mathcal{P}$. Let P be any (valid) chain in OPT. If P is closed, then at most $\left\lfloor \frac{|P|}{2} \right\rfloor$ arcs in P are unmatched in $\mathcal{P}$. If P is open, then at most $\left\lceil \frac{|P|}{2} \right\rceil$ arcs in P are unmatched in $\mathcal{P}$. Let i denote the unmatched arcs in $\mathcal{P}$, and k denote the number of odd open chains in OPT. Then $i \leq \frac{|A|+k}{2}$. Thus the total number of (valid) chains in $\mathcal{P}$ is at most

$$i + \left\lfloor \frac{|A| - i}{2} \right\rfloor \leq \frac{|A| + i}{2} \leq \frac{|A| + \frac{|A|+k}{2}}{2} = \frac{3}{4}|A| + \frac{k}{4}.$$

Thus the cost of $\mathcal{P}$ is at most

$$|A| + \left(\frac{3}{4}|A| + \frac{k}{4}\right) = \frac{7}{4}|A| + \frac{k}{4} \leq \frac{7}{4}(|A| + k) \leq \frac{7}{4} \cdot opt.$$

■

By repeatedly merging two non-overlapping chains which share at least one endpoint into a larger (valid) chain, one can convert any valid chain generation into one in which any pair of (valid) chains can not be merged into a larger valid chain. Lemma 10.7 implies that any algorithm appended with this post-processing will have an approximation ratio of at most $\frac{7}{4}$. In the next, we will present several algorithms whose approximation ratios beat $\frac{7}{4}$.

5.1 CUT AND MERGE

If all circular arcs in A do not cross over some link of the ring, then an optimal solution can be obtained in polynomial time. In fact by removing this link from the ring, we can treat all arcs as intervals along a line. A simple **Greedy Sweeping** algorithm for an optimal partition of these intervals into valid chains works as follows [67]. At the left-most node, each interval starting from this node forms a 1-chain. Subsequently at each node, if an interval starting from this node can be merged with an existing chain, then merge them to form a larger chain; otherwise create a new 1-chain from this interval. This procedure is repeated until no interval is left. The optimality of this algorithm follows from the fact that each node i uses $\max\{\sigma_A(i), \tau_A(i)\}$ ADMs.

The **Assign First** heuristic presented in [67] initially puts each circular arc that passes through a carefully selected link into a unique 1-chain. The remaining circular arcs, which form an interval graph, are then greedily grouped into valid chains as above. It was claimed in [67] that cost of **Assign First** is at most

$$\sum_{i=0}^{n-1} \max\{\sigma_A(i), \tau_A(i)\} + \min_{i=0}^{n-1}\{x_A(i) + \min\{\sigma_A(i), \tau_A(i)\}\}$$

where $x_A(i)$ is the number of arcs in A that pass though node i. Before we explain why the analysis and the algorithm itself are incorrect, we first show by a counter-example that above expression is incorrect even as an upper bound. Let $n = 4k$ and

$$A = \{(i, (i + 2k + 1) \bmod n) \mid 0 \leq i < n\}$$

Each circular arc in A traverses $2k + 1$ links, which is more than half way around the ring. Thus, each circular arc from a 1-chain. Thus the cost of the unique valid chain generation is $2n = 8k$. On the other hand, $\sigma_A(i) = \tau_A(i) = 1$ and $x_A(i) = 2k$. According to the above upper bound, the cost of Assign-First is at most $n + 2k + 1 = 6k + 1$, which is impossible. Thus the performance analysis of the **Assign-First** given in [67] must be incorrect.

The **Assign First** heuristic is problematic itself. Indeed, after all circular arcs passing through a carefully selected link are put into unique 1-chains, the remaining circular arcs can not be assigned greedily because the circular arc concerned at any step of the algorithm may possibly overlap with some circular arc(s) passing through the selected common link. However, if we take into this into consideration, then the number of primitive rings where a node i appears might be more than $\max\{\sigma_A(i), \tau_A(i)\}$ and thus the original analysis given in [67] does not hold any more.

In the following, we present a modified **Assign-First**, referred to as **Cut And Merge (CM)**. It is a slight revision from the well-known Tucker's algorithm [151] for circular-arc coloring. It consists of the following two steps:

- Step 1: Choose a link i with minimum load, and let B_i denote the set of circular arcs in A that pass through link i. Partition $A \setminus B_i$ into an optimal set of chains using the Greedy Sweeping algorithm. Let $\mathcal{P}'$ denote the obtained chains.

- Step 2: Construct a weighted bipartite graph between B_i and $\mathcal{P}'$ as follows: if an arc a in B_i can be merged with a chain P in $\mathcal{P}'$, add an edge between a and P with weight equal to their common endpoints. Find a maximum-weighted bipartite matching in the resulting bipartite graph. Merge each matched pair of arc and chain into a larger chain. This process is repeated until no further matching (or merging) can be obtained.

We first use the following example to illustrate the algorithm and to derive a lower bound on its approximation ratio.

EXAMPLE 10.8 *Let $n = 4$, and $A = A_1 \cup A_2 \cup A_3 \cup A_4$ where*

$$A_1 = \{(0,2), (2,0)\}, A_2 = \{(1,2), (2,1)\},$$
$$A_3 = \{(1,3), (3,1)\}, A_4 = \{(3,0), (0,3)\},$$

as illustrated in Figure 10.2 (a). Since the two arcs in each A_i form a closed valid chain for $1 \leq i \leq 4$, $opt = |A| = 8$. Since each link has the same load equal to four, during Step 1 link 0 may be chosen as the link to be removed, and

$$B_0 = \{(0,2), (2,1), (3,1), (0,3)\}.$$

The algorithm, if unlucky, would partition $A \setminus B_0$ into two valid chains

$$\{(1,2), (2,0)\}, \{(1,3), (3,0)\}.$$

During Step 2, the four arcs in B_0 can not be merged with either of the above two chains. So they will form four 1-chains alone. Thus totally six open chains are generated (see Figure 10.2 (b)). Their total ADM cost is

$$8 + 6 = 14 = \frac{7}{4} \cdot opt.$$

*This implies that the approximation ratio of **CM** is at least $\frac{7}{4}$.*

From Lemma 10.7 and Example 10.8, we have the following theorem.

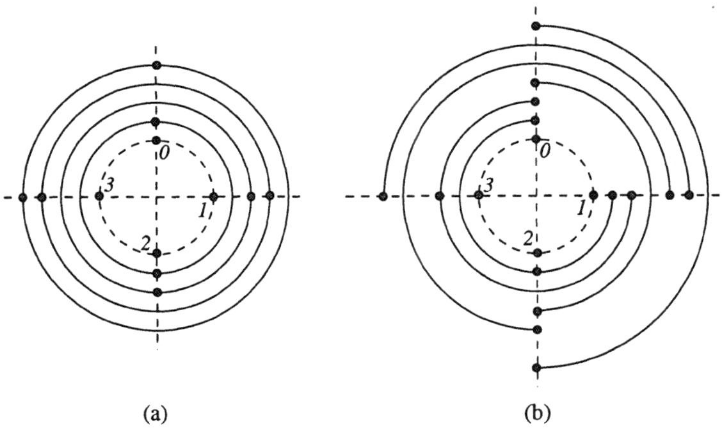

Figure 10.2. The instance in Example 10.8: (a) the optimal solution and (b) the solution produced by the algorithm **Cut and Merge**.

THEOREM 10.9 *The approximation ratio of **CM** is exactly 1.75.*

Example 10.8 suggests that it might improve the performance by taking out all complementary pairs of arcs before applying the algorithm **CM**. Indeed, it is easy to verify that there is an optimal solution in which these complementary arcs each form a closed chain. We refer the algorithm which takes out all complementary pairs of arcs and then applies **CM** to the remaining arcs by **Preprocessed Cut-And-Merge (PCM)**.

The next lemma provides an upper bound on the approximation ratio of **PCM**.

LEMMA 10.10 *The approximation ratio of **PCM** is at most $\frac{5}{3}$.*

Proof. Let A' denote the set of arcs in A that survive after the preprocessing. Then all arcs in $A \setminus A'$ form closed valid 2-chains and thus has total cost of $|A \setminus A'|$. Let OPT be any optimal solution in which all arcs in $A \setminus A'$ form closed valid 2-chains.

Without loss of generality, assume link 0 has the least load. Let B_0 be the set of arcs in A' that pass through link 0. Let B_0' denote the set of arcs in B_0 that form 1-chains in OPT, and $B_0'' = B_0 \setminus B_0'$. Thus each arc in B_0'' is within some chain of length at least two, and no two arcs in B_0'' can be in the same chain in OPT. Let P be any chain in OPT that contains an arc in B_0''. The cost of P is at least three, since if it is closed its length must be at least three, and if it is open, its length must be at least two. Since the total cost of these chains which contain arcs in B_0'' is at most $opt - |A \setminus A'| - 2|B_0'|$, we have:

$$|B_0''| \leq \frac{1}{3}\left(opt - |A \setminus A'| - 2|B_0'|\right).$$

Note that OPT induces a partition of $A' \setminus B_0$ into chains of cost at most $opt - |A \setminus A'| - 2|B_0'|$. On the other hand, Step 1 obtains an optimal partition of $A' \setminus B_0$ into valid chains. Therefore, the total cost of our solution is at most

$$\begin{aligned}
&|A \setminus A'| + \left(opt - |A \setminus A'| - 2|B_0'|\right) + 2|B_0| \\
&= opt + 2|B_0''| \\
&\leq opt + \frac{2}{3}\left(opt - 2|B_0'|\right) \\
&\leq \frac{5}{3}opt - \frac{4}{3}|B_0'| \leq \frac{5}{3}opt.
\end{aligned}$$

■

The next examples indicates that even after the preprocessing, the approximation ratio of **PCM** is still at least $\frac{3}{2}$.

EXAMPLE 10.11 *Let $n = 8$, and $A = A_1 \cup A_2 \cup A_3 \cup A_4$ where*

$$\begin{aligned}
A_1 &= \{(0,3), (3,4), (4,0)\}, \\
A_2 &= \{(0,6), (6,7), (7,0)\}, \\
A_3 &= \{(1,2), (2,3), (3,1)\}, \\
A_4 &= \{(1,5), (5,6), (6,1)\},
\end{aligned}$$

as illustrated in Figure 10.3 (a). Since the three arcs in each A_i form a closed valid chain for $1 \leq i \leq 4$, $opt = |A| = 12$. Note that all links have load equal to four. Thus during Step 1, link 0 may be chosen as the cutting node, and

$$B_0 = \{(0,3), (0,6), (3,1), (6,1)\}.$$

The algorithm, if unlucky, would partition $A \setminus B_0$ into two valid chains

$$\begin{aligned}
&\{(1,2), (2,3), (3,4), (4,0)\}, \\
&\{(1,5), (5,6), (6,7), (7,0)\}.
\end{aligned}$$

During Step 2, the four arcs in B_0 can not be merged with either of the above two chains. So they will form four 1-chains alone. Thus totally six open chains are generated (see Figure 10.3 (b)). Their total ADM cost is

$$12 + 6 = 18 = \frac{3}{2} \cdot opt.$$

This implies that the approximation ratio of ***PCM*** *is at least $\frac{3}{2}$.*

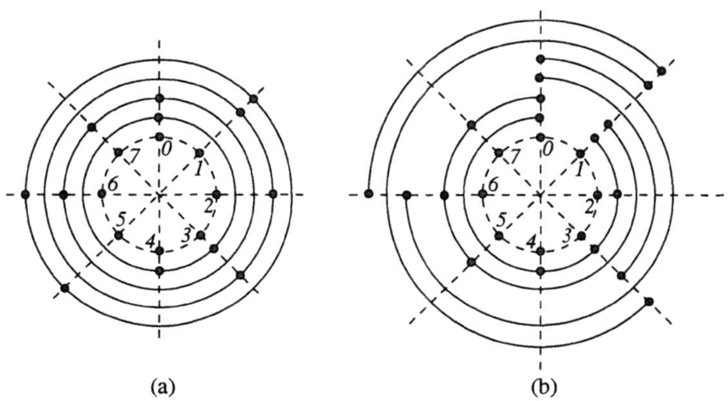

Figure 10.3. An instance used in Example 10.11: (a) the optimal solution and (b) the solution produced by the algorithm **PCM**.

From Lemma 10.23 and Example 10.11, we have the following theorem.

THEOREM 10.12 *The approximation ratio of* ***PCM*** *is between $\frac{3}{2}$ and $\frac{5}{3}$.*

5.2 TRAIL SPLITTING

We start by describing the a trail splitting algorithm based on Eulerian tour, referred to as **ET-TS**.

1. Eulerian Tour Phase: Add a set of fake arcs A' such that $|A'| = d(A)$ and $d(A \cup A') = 0$. This can be easily done by adding one by one fake arc from a node i with $\sigma_A(i) > \tau_A(i)$ to a node j with $\sigma_A(j) < \tau_A(j)$, thus each fake arc decreasing the deficiency by one. Now the directed

graph $G(A \cup A')$ is Eulerian. Find an Eulerian tour of $G(A \cup A')$. Remove all fake arcs from the Eulerian tour to obtain $d(A)$ trails of $G(A)$.

2. Trail Decomposing Phase: Decompose each trail into simple paths and circuits of $G(A)$. Each simple path induces an open chain, and each cycle induces a closed chain.

3. Chain Split Phase: Split each invalid open chain into valid chains by walking along the chain from the first arc and generating a valid chain whenever overlap occurs; split each invalid closed chain into valid chains by walking along the chain from *each* arc and generating a valid chain whenever overlap occurs, and then choose the one with the smallest number of open chains.

4. Chain Merging Phase: Repeatedly merge any pair of open valid chains into a larger valid chain until no more merging can occur.

From Lemma 10.7, the approximation ratio of **ET-TS** is at most 1.75. The next example shows it is also at least 1.75.

EXAMPLE 10.13 *Let $n = 2k + 1$ for some $k > 1$, and A consist of the following $2n$ arcs*

$$\{a_i = (i, i + k + 1) \mid 0 \leq i < n\} \cup$$
$$\{a_i' = (i + k + 1, i) \mid 0 \leq i < n\},$$

*as illustrated in Figure 10.4 (a). Note that for any $0 \leq i < n$, the arcs a_i and a_i' are complementary and thus form a closed valid chain. So $opt = |A| = 2n$. On the other hand, **ET-TS** may produce at Eulerian Tour Phase an Eulerian tour*

$$a_0, a_{k+1}, a_1, a_{k+2}, a_2, \cdots, a_k,$$
$$a_k', a_{2k}', a_{k-1}', a_{2k-1}', a_{k-2}', \cdots, a_0'.$$

It is then broken into two circuits at Trail Decomposing Phase:

$$a_0, a_{k+1}, a_1, a_{k+2}, a_2, \cdots, a_k;$$

and

$$a_k', a_{2k}', a_{k-1}', a_{2k-1}', a_{k-2}', \cdots, a_0'.$$

Both circuits leads to an invalid closed chain. At Chain Split Phase, the first invalid closed chain is split into $n = 2k + 1$ open valid 1-chains

$$\{\{a_i\} \mid 0 \leq i < n\};$$

the second invalid closed chain is split into k open valid 2-chains

$$\{\{a'_i, a'_{i+k}\} \mid 1 \leq i \leq k\}$$

and one open valid 1-chain $\{a'_0\}$. At Chain Merging Phase, the two open chains $\{a_0\}$ and $\{a'_0\}$ are merged into a closed valid chain $\{a_0, a'_0\}$. So totally $n + k = 3k + 1$ chains are obtained, among which $3k$ are open (see Figure 10.4 (b)). The total ADM cost of all these chains is

$$\begin{aligned} |A| + 3k &= 2(2k+1) + 3k = 7k + 2 \\ &= \frac{7}{4} \cdot opt - \frac{3}{2}. \end{aligned}$$

*Thus the approximation ratio of **ET-TS** is at least $\frac{7}{4}$.*

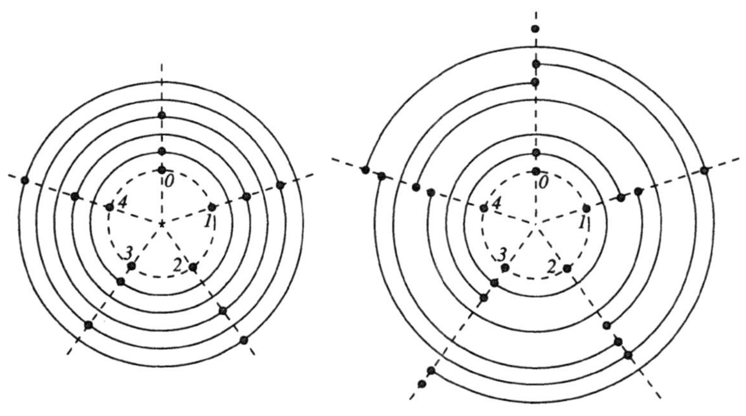

Figure 10.4. An instance used in Example 10.13: (a) the optimal solution and (b) the solution produced by the algorithm **ET-TS**.

From Lemma 10.7 and Example 10.13, we have the following theorem.

THEOREM 10.14 *The approximation ratio of **ET-TS** is exactly 1.75.*

Example 10.13 suggests that it might improve the performance by taking out all complementary pairs of arcs before applying the algorithm

ET-TS, as done in the algorithm **PCM**. However, the next example shows that after this preprocessing the approximation ratio is still at least 1.5.

EXAMPLE 10.15 *Let $n = 2(2k+1)$ for some $k > 1$, and A consist of the following $3(2k+1)$ arcs*

$$\left\{a_i = (2i, 2(i+k+1)) \mid 0 \le i < \frac{n}{2}\right\} \cup$$
$$\left\{a_i' = (2(i+k+1), 2i-1) \mid 0 \le i < \frac{n}{2}\right\} \cup$$
$$\left\{a_i'' = (2i-1, 2i) \mid 0 \le i < \frac{n}{2}\right\},$$

*as illustrated in Figure 10.5 (a). Note that for any $0 \le i < n$, the three arcs a_i, a_i', a_i'' form a closed valid chain. So opt $= |A| = 3(2k+1)$. On the other hand, **ET-TS** may produce at Eulerian Tour Phase an Eulerian tour*

$$a_0, a_{k+1}, a_1, a_{k+2}, a_2, \cdots, a_k, a_{2k},$$
$$a_k', a_k'', a_{2k}', a_{2k}'', a_{k-1}', a_{k-1}'',$$
$$a_{2k-1}', a_{2k-1}'', a_{k-2}', a_{k-2}'', \cdots, a_0', a_0''.$$

This trail is then broken into two circuits at Trail Decomposing Phase:

$$a_0, a_{k+1}, a_1, a_{k+2}, a_2, \cdots, a_k, a_{2k};$$

and

$$a_k', a_k'', a_{2k}', a_{2k}'', a_{k-1}', a_{k-1}'',$$
$$a_{2k-1}', a_{2k-1}'', a_{k-2}', a_{k-2}'', \cdots, a_0', a_0''.$$

Both circuits leads to an invalid closed chain. At Chain Split Phase, the first invalid closed chain is split into $\frac{n}{2} = 2k+1$ open valid 1-chains

$$\left\{\{a_i\} \mid 0 \le i < \frac{n}{2}\right\};$$

the second invalid closed chain is split into k open valid 4-chains

$$\left\{\{a_i', a_i'', a_{i+k}', a_{i+k}''\} \mid 1 \le i \le k\right\}$$

and one open valid 2-chain $\{a_0', a_0''\}$. For an example, see Figure 10.5 (b). At Chain Merging Phase, the two open chains $\{a_0\}$ and $\{a_0', a_0''\}$ are merged into a closed valid chain $\{a_0, a_0', a_0''\}$. So totally $3k+1$ chains *are obtained, among which* $3k$ *are open. The total ADM cost of all these chains is*

$$|A| + 3k = 3(2k+1) + 3k = 9k + 3$$
$$= \frac{3}{2} \cdot opt - \frac{3}{2}.$$

Thus the approximation ratio of ***ET-TS*** *is at least* $\frac{3}{2}$.

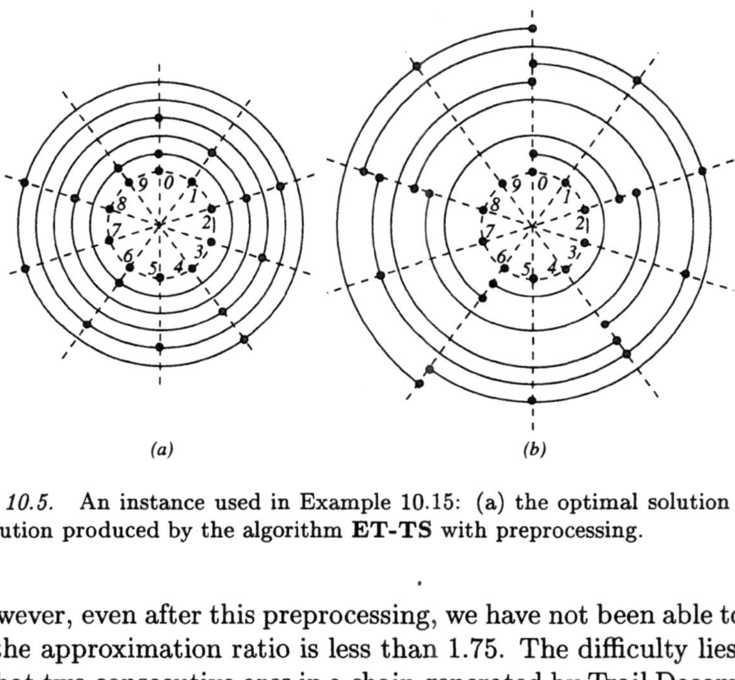

Figure 10.5. An instance used in Example 10.15: (a) the optimal solution and (b) the solution produced by the algorithm **ET-TS** with preprocessing.

However, even after this preprocessing, we have not been able to prove that the approximation ratio is less than 1.75. The difficulty lies in the fact that two consecutive arcs in a chain generated by Trail Decomposing Phase in **ET-TS** may intersect with each other. This implies that the size of an invalid open chain can be as low as two and an invalid closed chain can be as low as three. To avoid this situation, we propose below a new phase which would replace the Eulerian Tour Phase in **ET-TS** and can guarantee that any two consecutive arcs in a chain generated by Trail Decomposing Phase in **ET-TS** do not intersect with each other.

Since we can take all complementary pairs of arcs out of A, we assume that no pairs of arcs in A are complementary. We define a weighted directed graph $H(A)$ over A as follows. The vertex set of $H(A)$ is A. For any pair of non-overlapping arcs a_1 and a_2 in A, add one link from a_1 to a_2 and one link from a_2 to a_1. If a_1 and a_2 do not share any endpoints, the weights of both links are set to two. If $h(a_1) = t(a_2)$, the weight of the link from a_1 to a_2 is set to one and the weight of the link from a_2 to a_1 is set to two. If $h(a_2) = t(a_1)$, the weight of the link from a_2 to a_1 is set to one and the weight of the link from a_1 to a_2 is set to two. In addition, there is one loop link with weight two at each arc. Note that $H(A)$ has a circuit cover. Furthermore, any valid chain generation induces naturally a circuit cover of $H(A)$ whose weight is equal to the cost of the valid chain generation. Thus the weight of a minimum-weighted circuit cover of $H(A)$ is a lower bound of the minimum ADM cost.

Using these definitions, we propose the following algorithm, referred to as **MCC-TS**, for valid chain generation. This algorithm replaces Eulerian Tour Phase in **ET-TS** by the following MCC Phase:

- Construct the graph $H(A)$, and find a minimum-weighted circuit cover of $H(A)$.
- Remove all links of weight two from the minimum-weighted circuit cover and obtain a collection of paths and circuits. Each path (circuit respectively) induces an trail (closed trail respectively) along the ring of $G(A)$.

One key property of is that in any chain generated by the Trail Decomposition Phase, any two consecutive arcs do not intersect. This property implies that the size of any invalid open chain is at least three, and the size of any invalid closed chain is at least five. This observation leads to the following lemma.

LEMMA 10.16 *The approximation ratio of **MCC-TS** is at most* 1.6.

Proof. Let ω denote the weight of a minimum-weighted circuit cover of $H(A)$. Then the total cost of the chains generated by Trail Decomposition Phase is exactly ω. The splitting of invalid chains into valid chains at Chain Split Phase will increase the cost. Let P be any invalid chain, and i be the number of valid chains split from P. Then the splitting of P into i valid chains increases the cost by $i-1$ if P is open, or i if P is closed. From the construction of the graph $H(A)$, the sizes of all valid chains obtained from P, except the last one, are at least two. Thus

$$i \leq \left\lceil \frac{|P|}{2} \right\rceil \leq \frac{|P|+1}{2}.$$

So P contributes to an additional cost of at most $\frac{|P|-1}{2}$ if P is open, or at most $\frac{|P|+1}{2}$ if P is closed. Let j be the number of invalid closed chains generated by Trail Decomposition Phase. Then the cost of the valid chains generated by Chain Split Phase is at most $\omega + \frac{|A|+j}{2}$. As the size of any invalid closed chain is at least five, $j \leq \frac{|A|}{5}$. So the cost of the valid chains produced by our algorithm is at most

$$\begin{aligned}\omega + \frac{|A|+j}{2} &\leq \omega + \frac{|A| + \frac{|A|}{5}}{2} = \omega + \frac{3}{5}|A| \\ &\leq opt + \frac{3}{5} \cdot opt = \frac{8}{5} \cdot opt,\end{aligned}$$

as both ω and $|A|$ are lower bounds of opt. Thus the lemma is true. ∎

In the next, we will derive a lower bound on the approximation ratio of **MCC-TS**. We notice that the execution of **MCC-TS** on the instance given in Example 10.15 will always lead to an optimal solution. Thus Example 10.15 can not provide us a good lower bound. So instead we developed the following example.

EXAMPLE 10.17 *Let $n = 2(2k+1)$ for some $k > 1$, and A consist of the following $\frac{3n}{2}$ arcs*

$$\begin{aligned}&\left\{a_i = \left(2i, 2i + \frac{n}{2}\right) \mid 0 \leq i < \frac{n}{2}\right\} \cup \\ &\left\{a_i' = \left(2i + \frac{n}{2}, 2i + \frac{n}{2} + 1\right) \mid 0 \leq i < \frac{n}{2}\right\} \cup \\ &\left\{a_i'' = \left(2i + \frac{n}{2} + 1, 2i\right) \mid 0 \leq i < \frac{n}{2}\right\},\end{aligned}$$

as illustrated in Figure 10.6 (a). Note that for any $0 \leq i < n$, the three arcs a_i, a_i', a_i'' form a closed valid chain. So $opt = |A| = \frac{3n}{2}$. On the other hand, ***MCC-TS*** *may produce at MCC Phase two circuits*

$$a_k'', a_{2k}'', a_{k-1}'', a_{2k-1}'', \cdots, a_1'', a_{k+1}'', a_0'';$$

and

$$\begin{aligned}&a_0, a_0', a_{k+1}, a_{k+1}', a_1, a_1', a_{k+2}, a_{k+2}', \cdots, \\ &a_{k-1}, a_{k-1}', a_{2k}, a_{2k}', a_k, a_k'.\end{aligned}$$

Each circuit induces an invalid closed chain. At Chain Split Phase, the first invalid closed chain is split into k open valid 2-chains

$$\left\{\left\{a_i'', a_{i+k}''\right\} \mid 1 \leq i \leq k\right\}$$

and one open valid 1-chain $\{a''_0\}$. *Since the* $2k+1$ *arcs*

$$\{a_i \mid 0 \leq i \leq 2k\}$$

are pairwise intersecting, at least $2k+1$ *open valid chains must be used by the second closed invalid chain. One solution is the following:*

$$\left\{\{a_i, a'_i\} \mid 0 \leq i \leq 2k\right\}.$$

At Chain Merging Phase, the two open chains $\{a_0\}$ *and* $\{a'_0, a''_0\}$ *are merged into a closed valid chain* $\{a_0, a'_0, a''_0\}$. *So totally* $3k+1$ *chains are obtained, among which* $3k$ *are open (see Figure 10.6 (b)). The total ADM cost of all these chains is*

$$\begin{aligned} |A| + 3k &= 3(2k+1) + 3k = 9k+3 \\ &= \frac{3}{2} \cdot opt - \frac{3}{2}. \end{aligned}$$

Thus the approximation ratio of ***MCC-TS*** *is at least* $\frac{3}{2}$.

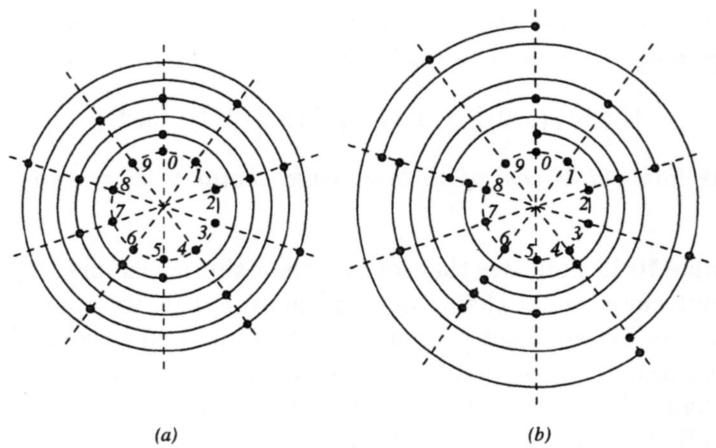

Figure 10.6. An instance used in Example 10.17: (a) the optimal solution and (b) the solution produced by the algorithm **ET-TS** with preprocessing.

From Lemma 10.16 and Example 10.17, we have the following theorem.

THEOREM 10.18 *The approximation ratio of* ***MCC-TS*** *is between* 1.5 *and* 1.6.

Both Example 10.15 and Example 10.17 suggest the preprocessing by repeatedly taking out valid closed chains of sizes at most three. However, it is no longer guaranteed that there is an optimal solution that contains a specified valid closed chain. This can be illustrated by the following Example.

EXAMPLE 10.19 *Let* $n = 6$, *and* $A = A_1 \cup A_2 \cup A_3$ *where*

$$A_1 = \{(0,2), (2,5), (5,0)\},$$
$$A_2 = \{(0,3), (3,4), (4,0)\},$$
$$A_3 = \{(1,2), (2,4), (4,1)\},$$

as illustrated in Figure 10.7 (a). Since the three arcs in A_i *form a closed valid chain for any* $1 \leq i \leq 3$, $opt = |A| = 9$. *If the preprocessing takes out the following closed valid chain*

$$\{(0,2), (2,4), (4,0)\},$$

then the remaining 6 *arcs do not contain any closed valid chain. Since the three arcs* $(0,3)$ *and* $(2,5)$ *and* $(4,1)$ *are pairwise intersecting, at least three open valid chains must be used by the remaining six arcs. One (optimal) solution is the following (see Figure 10.7 (b)):*

$$\{(2,5), (5,0)\}, \{(4,1), (1,2)\}, \{(0,3), (3,4)\}.$$

Thus the total ADM cost of all these valid chains is $12 = \frac{4}{3} \cdot opt$.

Example 10.19 indicates that for any algorithm which first repeatedly taking out closed valid chains until no more is left, its approximation ratio can not be less than $\frac{4}{3}$. Note that Examples 10.15 and 10.17 can be generalized to show that even if an instance does not contain any closed chain of sizes at most l, the approximation ratios of **ET-TS** and **MCC-TS** is at least $1 + \frac{3}{2(l+1)}$. In particular, if $l = 3$, the lower bound on the approximation ratio is 1.375. It would be interesting to explore the effect of preprocessing by repeatedly taking out closed valid chains on the approximation ratios of both **ET-TS** and **MCC-TS**.

5.3 ITERATIVE MATCHING

We start by analyzing a simple algorithm referred to as **Iterative Matching (IM)** proposed in [106]. This algorithm maintains a set of

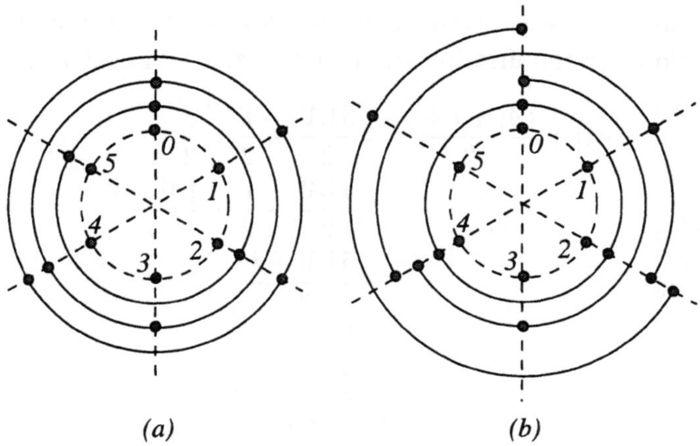

Figure 10.7. The instance in Example 10.19: (a) the optimal solution and (b) a solution produced after the preprocessing.

valid chains of arcs $\mathcal{P}$ throughout its execution. Initially $\mathcal{P}$ consists of 1-chains each of which is an arc in A. While the fit graph $F(\mathcal{P})$ of $\mathcal{P}$ has non-empty edge set, we find a maximum matching M in $F(\mathcal{P})$ and then merge each matched pair of chains of arcs in M into a larger chain. When $F(\mathcal{P})$ has empty edge set, $\mathcal{P}$ is output as the valid chain generation.

It is obvious that the algorithm **IM** has polynomial run-time. In the next we show that its approximation ratio is at most $\frac{5}{3}$.

LEMMA 10.20 *The approximation ratio of* **IM** *is at most* $\frac{5}{3}$.

Proof. From any valid chain P in OPT, a matching of cardinality $\left\lfloor \frac{|P|}{2} \right\rfloor$ can be obtained. Let i be the number of odd open chains in OPT, and j be the number of odd closed chains in OPT. Then from the chains in OPT, we can obtain a matching of cardinality $\frac{|A|-i-j}{2}$. Thus the cardinality of any maximum matching obtained in the first iteration is at least $\frac{|A|-i-j}{2}$, and consequently after the first iteration, the total number of valid chains is at most

$$|A| - \frac{|A|-i-j}{2} = \frac{|A|+i+j}{2}.$$

Note that any odd closed chain must contain at least three arcs. Thus $j \leq \frac{|A|-i}{3}$. So after the first iteration, the total ADM cost is at most

$$\begin{aligned} |A| + \frac{|A|+i+j}{2} &= \frac{3|A|}{2} + \frac{i}{2} + \frac{j}{2} \\ &\leq \frac{3|A|}{2} + \frac{i}{2} + \frac{|A|-i}{6} \\ &= \frac{5|A|+i}{3} \\ &\leq \frac{5}{3}(|A|+i) \\ &\leq \frac{5}{3} \cdot opt. \end{aligned}$$

Therefore, the approximation ratio of **IM** is at most $\frac{5}{3}$. ■

In the next, we will use a bad example to show that the approximation ratio of **IM** is at least $\frac{3}{2}$.

EXAMPLE 10.21 *Let $n = 5$ and $A = A_1 \cup A_2$ where*

$$\begin{aligned} A_1 &= \{(0,1), (1,3), (3,0)\}, \\ A_2 &= \{(0,2), (2,4), (4,0)\}, \end{aligned}$$

*as illustrated in Figure 10.8 (a). Note that the three arcs in A_1 form a closed valid chain, so do the three arcs in A_2. Thus $opt = |A| = 6$. On the other hand, the algorithm **IM** may output the following open valid chains (see Figure 10.7 (b)):*

$$\{(4,0), (0,1)\}, \{(0,2), (2,4)\}, \{(1,3), (3,0)\}.$$

*The cost of these three chains is $9 = \frac{3}{2} \cdot opt$. This example can be scaled to the rings with size multiple of five. Thus the approximation ratio of **IM** is at least $\frac{3}{2}$.*

From Lemma 10.20 and Example 10.21, we have the following theorem.

THEOREM 10.22 *The approximation ratio of the algorithm **IM** is between $\frac{3}{2}$ and $\frac{5}{3}$.*

From the proof of Lemma 10.20, the closed chains play a critical role in bounding the total ADM cost. In particular, if *OPT* does not

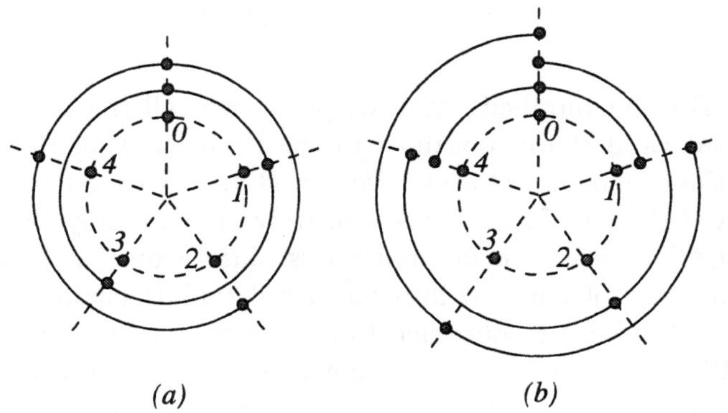

Figure 10.8. The instance in Example 10.21: (a) the optimal solution and (b) a solution produced produced by the algorithm **IM**.

contain any closed chains, the total ADM cost after the first iteration is at most $\frac{3}{2}\cdot opt$. Based on this observation, we propose the **Preprocessed Iterative Matching (PIM)** algorithm, which runs in two phases:

1. Preprocessing Phase: repeatedly take valid closed chains out of the remaining arcs until no more closed chain can be obtained from the remaining arcs.

2. Matching Phase: apply the algorithm **IM** to the remaining arcs.

The following procedure is used by the Preprocessing Phase to obtain a closed valid chain containing a specified arc a, if there is any, from a set of arcs S. We build a directed acyclic graph (DAG) that consists of only those arcs in S that do not overlap with a. Obviously there is a path from $h(a)$ to $t(a)$ in the DAG if and only if there is a closed valid chain in S that contains a. By using *breadth-first search* in the DAG, we can obtain a path, if there is any, from the from $h(a)$ to $t(a)$. Once this path is obtained, we merge it with a to obtain a closed valid chain. The Preprocessing Phase runs as follows:

- Find the link e which has the minimum link load, and let A_e denote the number of arcs that contain the link e. Set $i \leftarrow 1$.

- While A_e is non-empty: pick any $a \in A_e$, and set $A_e \leftarrow A_e \setminus \{a\}$; if A has a valid closed chain P containing a, set $P_i \leftarrow P$, $A \leftarrow A \setminus P$, and $i \leftarrow i + 1$.

It is obvious that the algorithm has polynomial run-time. In the next, we show that its approximation ratio of is at most 1.5.

LEMMA 10.23 *The approximation ratio of **PIM** is at most* $\frac{3}{2}$.

Proof. For the simplicity of description, we call the arcs appearing in the closed valid chains obtained in the Preprocessing Phase as *blue* arcs, and the others as *red* arcs. We use B and R to denote the set of blue arcs and the set of red arcs respectively. Then in any closed (valid) chain in OPT, at least one circular arc is blue. From OPT, we remove all blue arcs and obtain a collection of red (valid) chains. Note that the number of red chains obtained from a closed chain P is at most the number of blue arcs in P; the number of red chains obtained from an open chain P is at most the number of blue arcs in P plus one. Thus the total number of red chains is at most $|B|$ plus the total number of open chains, and consequently is at most

$$|B| + opt - |A| = opt - |R|\,.$$

Let k be the number of *odd* red chains. Following the same argument in the proof of Lemma 10.20, from the red chains we can generate a matching of cardinality $\frac{|R|-k}{2}$. Thus the cardinality of any maximum matching obtained in the first iteration in the Matching Phase is at least $\frac{|R|-k}{2}$, and consequently after this first iteration, the total number of valid red chains is at most

$$|R| - \frac{|R|-k}{2} = \frac{|R|+k}{2}.$$

So after the first maximum matching in the Matching Phase, the total ADM cost is at most

$$\begin{aligned} |A| + \frac{|R|+k}{2} &\le |A| + \frac{|R| + opt - |R|}{2} \\ &= |A| + \frac{opt}{2} \le opt + \frac{opt}{2} \\ &\le \frac{3}{2} \cdot opt. \end{aligned}$$

Therefore the approximation ratio of **PIM** is at most $\frac{3}{2}$. ■

The instance in Example 10.19 can be used to derive a lower bound of $\frac{4}{3}$ on the approximation ratio of **PIM**. Thus we have the following theorem.

THEOREM 10.24 *The approximation ratio of **PIM** is between* $\frac{4}{3}$ *and* $\frac{3}{2}$.

6. VALID CHAIN GENERATION FROM SPLITABLE ARCS

In this section, we assume that each arc is allowed to be split. A splitting of an arc a is a replacement of the arc a by several arcs $a_1, a_2, \cdots, a_k$ such that $t(a) = t(a_1)$, $h(a) = h(a_k)$, and $h(a_i) = t(a_{i+1})$ for any $1 \leq i < k$. The arcs split from an original arc are called *fragments.* A chain of arcs is said to be *unsplit* if none of its arcs is a fragment and to be *split* otherwise.

Obviously, the optimum with splits is at most the optimum without splits. The benefit of splitting is illustrated by an example presented in [67] for the arc version with splits. In this example, the optimum with splits is $\frac{3}{4}$ the optimum without splits. In fact, optimum with splits could be $\frac{2}{3}$ the optimum without splits, as shown by the following example: $A = \{a_1, a_2, a_3\}$ over a 3-node ring, with $a_1 = (0, 2)$ (here (i, j) represents the arc from node i to node j), $a_2 = (2, 1)$, and $a_3 = (1, 0)$. The optimum without splits is 6, while with splits a solution of 4 can be obtained by splitting a_2 into $a_2' = (2, 0)$ and $a_2'' = (0, 1)$. After the splitting, an optimal partition consists of $\{a_1, a_2'\}$ and $\{a_2'', a_3\}$. See Figure 10.9.

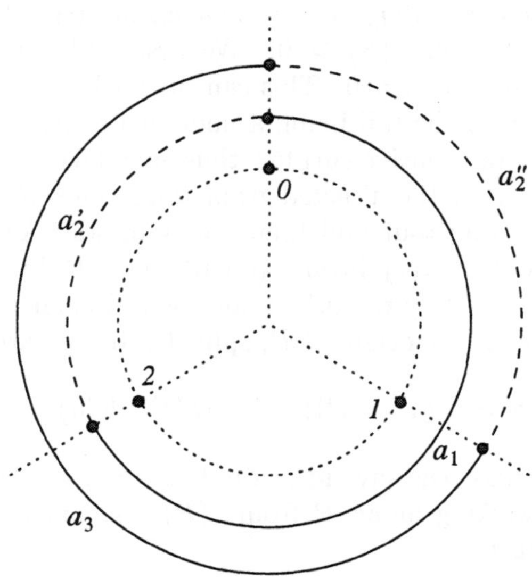

Figure 10.9. Arc a_2 is split into a_2' and a_2'' to obtain a solution with splits of cost 4.

Actually, the example above is extreme, in the sense that any solution with splits can be converted, by simply putting alone in a wavelength any arc which is split, into a solution without splits of cost at most 50% bigger. But how the two costs of optimum relate is not as interesting as how the best solutions we can find relate.

A heuristic called **Cut-First** was proposed in [67] for the arc version with splits. However, little is known about its approximation ratio. In this section we introduce a new polynomial-time approximation algorithm. The algorithm combines greedy ideas with Eulerian rounding, a technique similar to the Cut-First heuristic of [67]. We prove our algorithm has approximation ratio at most $\frac{5}{4}$. An example shows that the new algorithm has approximation ratio at least $\frac{10}{9}$. By comparison, when arcs are not splitable, the best known algorithm has approximation ratio in between $\frac{4}{3}$ and $\frac{3}{2}$.

For the purpose of the proof, we use arcs which are not in A. Such an auxiliary arc is called a *fake*, and has an origin and a termination like any other arc. For a set of arcs A, we use $opt(A)$ to denote the value of the optimum solution with A as the input. If A is understood, we use only *opt*.

We start by describing the last and most interesting phase of the algorithm, *Eulerian rounding*. Let S be a set of arcs. We consider two cases. In the first case, $d(S) > 0$. We first add a set of $d(S)$ fake arcs F such that $d(S \cup F) = 0$. This can be easily done by adding one by one fake arcs with the tail being a node of positive surplus and the head being a node of negative surplus, thus each fake arc decreasing the deficiency by one. Now the directed graph with edges $S \cup F$ is Eulerian. Choosing any Eulerian tour and then removing all fake arcs results in $d(S)$ open chains. For every invalid (open) chain P, break it into valid chains as follows (see Figure 10.10): for each circular arc a in P that passes through $t(P)$, the origin of P, split it into two arcs

$$a' = (t(a), t(P)), a'' = (t(P), h(a)).$$

After these splittings, the invalid chain P is then be decomposed into valid chains by walking along P from $t(P)$ and output a valid chain whenever reaching $t(P)$.

In the second case, $d(S) = 0$, and thus the directed graph with edges S is Eulerian. Choosing any Eulerian tour. Let i be any node which is

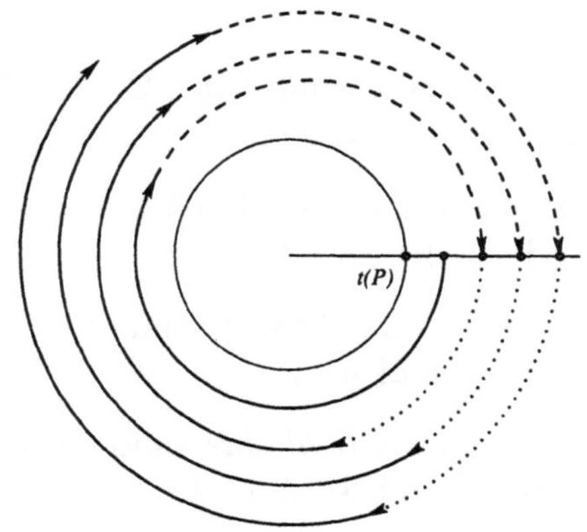

Figure 10.10. Eulerian rounding. Arcs are split into two fragments when they pass through node $t(P)$: the dashed fragment closes a chain, while the dotted fragment starts another chain.

the origin of some arc. For any circular arc a in the oriented Eulerian tour that passes through i, split it into two arcs

$$a' = (t(a), i), a'' = (i, h(a)).$$

After these splittings, the oriented Eulerian tour is then be decomposed into valid (closed) chains by walking along the oriented Eulerian tour from node i and output a valid (closed) chain whenever reaching node i.

LEMMA 10.25 *The solution produced from Eulerian rounding for S has cost at most $|S| + b(S) + d(S)$.*

Proof. We first consider the case that $d(S) > 0$. Let P be any invalid chain. Note that among all valid chains generated by Eulerian rounding from P, exactly one is open and all others are closed. Since the chain P must pass through the link $(n-1, 0)$ each time before passing through $t(P)$, the number of splits is at most $b(P)$. So the total cost of these valid chains generated from P is at most $|P| + b(P) + 1$. Since there are total $d(S)$ open chains, the total cost of all valid chains is at most $|S| + b(S) + d(S)$.

Now we consider the case that $d(S) = 0$. In this case, all valid chains are closed, and the number of splits is at most $b(P)$. So the total cost of these valid chains is at most $|S| + b(S)$. ∎

Using the machinery developed before, we propose the following approximation algorithm:

- The input is a set of arcs A.
- Phase 1: While A contains a valid closed 2-chain P, unsplit the arcs in P, output P and set $A \leftarrow A \setminus P$.
- Phase 2: While A contains a valid closed 3-chain P, unsplit the arcs in P, output P and set $A \leftarrow A \setminus P$.
- Phase 3: While A contains a *tight blue* valid (open) 1-chain P, unsplit the arcs in P, output P and set $A \leftarrow A \setminus P$.
- Phase 4: While A contains a *tight blue* valid open 2-chain P, unsplit the arcs in P, output P and set $A \leftarrow A \setminus P$.
- Phase 5: Do the Eulerian rounding of A.

The last phase is the most interesting one. If optimum is "perfect", (has only valid closed chains and no splits), Eulerian rounding cannot guarantee a good ratio. This is the reason why we use the first two phases, which achieve good results when there are many short valid closed chains. The third and fourth phase produce open chains, but decrease the deficiency. Since in any solution, the number of open chains is at least the deficiency, it makes sense to reduce deficiency. We also insist on reducing the number of blue arcs while reducing deficiency in phases three and four, to obtain a good Eulerian rounding in the fifth phase according to Lemma 10.25.

The first phase is very intuitive. Indeed, let B be the set of arcs unsplit and then assigned to valid chains during the first phase. Then there is always an optimum solution which assigns to valid chains the arcs of B exactly the way our algorithm does. This fact follows from the following lemma, which was previously known, but whose proof we include for completeness.

LEMMA 10.26 *Let A be a set of arcs and a_1, a_2 be two complementary arcs of A. Then $opt(A) = opt(A \setminus \{a_1, a_2\}) + 2$.*

Proof. Assume we have a solution which does not put a_1 and a_2 in the same valid closed chain. Let P_1 be the chain containing a_1 and P_2 be the chain containing a_2. Removing a_1 from P_1 leaves us with one of the following two cases:

1. If P_1 is closed: one valid chain, which we call P_1^*, from $h(a_1)$ to $t(a_1)$.

2. If P_1 is open: two non-overlapping valid chains, which we call P_1' and P_1'', and which both could be empty. P_1' starts at $h(a_1) = t(a_2)$ and ends at some node in the interior of a_2, and P_1'' starts at some node in the interior of a_2 and ends at $t(a_1) = t(a_2)$.

Similarly, removing a_2 from P_2 leaves us with either P_2^* or two chains P_2' and P_2''. Then, after putting a_1 and a_2 in the same closed valid chains, depending on which case we are in, we produce one of the following four combinations:

1. P_1^* and P_2^* in one valid closed chain, at a cost of $|P_1^*| + |P_2^*| = |P_1| + |P_2| - 2$. The total cost of P_1 and P_2 is $|P_1| + |P_2|$.

2. P_2'', P_1^*, P_2' in valid open chain, at a cost of $|P_2''| + |P_1^*| + |P_2'| + 1 = |P_1| - 1 + |P_2|$. The total cost of P_1 and P_2 is $|P_1| + |P_2| + 1$.

3. P_1'', P_2^*, P_1' in one valid open chain, at a cost of $|P_1''|+|P_2^*|+|P_1'|+1 = |P_1| - 1 + |P_2|$. The total cost of P_1 and P_2 is $|P_1| + |P_2| + 1$.

4. P_1'' and P_2' in one valid open chain, and P_2'' and P_1' in another valid open chain, at a total cost of at most $|P_1''|+|P_2'|+1+|P_2''|+|P_1'|+1 = |P_1| + |P_2|$. The total cost of P_1 and P_2 is $|P_1| + |P_2| + 2$.

In all the cases, the cost of the solution which puts a_1 and a_2 in the same valid closed chain is at most the cost of the solution which does not put a_1 and a_2 in the same valid closed chain. ■

Based on the lemma above, in the following we assume that no two arcs of the input form a valid closed chain.

THEOREM 10.27 *The algorithm above has performance ratio at most $\frac{5}{4}$.*

First we present a general overview of the proof. We fix OPT, an optimum solution, and based on OPT we give credit to arcs of A, and $d_A(i)$ credit to each node i. The exact initial credit allocation scheme is described later. Before the algorithm starts, as shown in Lemma 10.28, the total credit is at most $\frac{5}{4}opt$. During the execution of phases two, three and four of the algorithm, as the algorithm selects arcs, puts them in valid chains and outputs the chains, (during these phases we only unsplit), we take the credit from these arcs, and in phases three and four, also from the nodes whose deficiency decreases. We use this credit taken to cover the cost of the valid chains the algorithm produces and to give some extra credit to some other arcs, according to a credit transfer scheme to be described later. Finally, by the time we get to the fifth phase, the total credit remaining covers the cost of the output of Eulerian rounding.

We start the proof somehow backwards, discussing this last phase. This would give the motivation for all the previous steps. Let S be the set of remaining arcs after phase four. Assume each node i has $d_S(i)$ and each arc $a \in S$ has $1 + b(a)$ credit. In total, we have $d(S) + |S| + b(S)$ credit and therefore, by Lemma 10.25, we can cover the cost of the Eulerian rounding. All the remaining discussion is about how to make sure that at the end of phase four, each remaining arc a has $1 + b(a)$ credit and each node i has $d_S(i)$ credit.

We now describe the initial credit allocation. Each node i gets $d_A(i)$ credit. The credit allocation for arcs is given by the following scheme. For a set of arcs S, we define OPT_S to be the restriction of OPT to the set of arcs S.

- Initialize S to be the set of original arcs, A.
- Phase 1: As long as possible, do the following: if in OPT_S there is (valid) tight open unsplit 1-chain C, then the arc in C gets 1.5 credit. Remove from S the arc of C.
- Phase 2: As long as possible, do the following: if in OPT_S there is (valid) tight open unsplit 2-chain C, then each arc a in C gets $1 + \frac{3}{4}l(a)$ credit. Remove from S the arcs of C.
- Phase 3: As long as possible, do the following: if in OPT_S there is (valid) closed unsplit 3-chain C, then each arc a in C gets $1 + \frac{3}{4}l(a)$ credit. Remove from S the arcs of C.
- Every arc a of S gets $1 + l(a)$ credit.

We now describe the initial credit allocation. For a set of arcs S, we define OPT_S to be the restriction of OPT to the set of arcs S. Each node i gets $d_A(i)$ credit. Every red arc receives 1 credit. Blue arcs receive credit according to their position in OPT, as given by the following scheme.

- Initialize S to be the initial set of arcs, A.
- Phase 1: As long as possible, do the following: if in OPT_S there is a (valid) tight (with respect to S), blue open unsplit 1-chain P, then the blue arc in P receives 1.5 credit. Remove from S the arc of P.

- Phase 2: As long as possible, do the following: if in OPT_S there is a (valid) tight (with respect to S), blue open unsplit 2-chain P, then the blue arc in P receives 1.75 credit. Remove from S the arcs of P.
- Phase 3: As long as possible, do the following: if in OPT_S there is a (valid) closed unsplit 3-chain P, then the blue arc in P receives 1.75 credit. Remove from S the arcs of P.
- Every remaining blue arc receives 2 credit.

LEMMA 10.28 *The total initial credit is at most* $\frac{5}{4}opt$.

Proof. We consider the chains of OPT one by one. We start with the chains used in the first three phases of the credit allocation scheme above, in the order they are used. Then we continue with the remaining chains of optimum, in some order to be described later.

Take an open chain of OPT as described in the first phase of the scheme. For the arc a, OPT incurs a cost of 2. We give 1.5 credit to a, 0.5 credit to $t(a)$, and 0.5 credit to $h(a)$. Altogether, we give 5/4 times the cost of this chain in OPT. Note that after removing each such arc a from S, both $d_S(h(a))$ and $d_S(t(a))$ drop by 0.5.

Take an open chain of OPT as described in the second phase of the scheme. For the chain $P = \langle a_1, a_2 \rangle$, OPT incurs a cost of 3. We give 1 credit to the red arc, 1.75 credit to the blue arc, 0.5 credit to $h(a_2)$ and 0.5 credit to $t(a_1)$. Altogether, we give 3.75 credit, which is 5/4 times the cost of this chain in OPT. Note that after removing the arcs from P from S, both $d_S(h(a_2))$ and $d_S(t(a_1))$ drop by 0.5.

Take a closed chain of OPT as described in the third phase of the scheme. For the chain $P = \langle a_1, a_2, a_3 \rangle$, OPT incurs a cost of 3. We give 3.75 credit, which is 5/4 times the cost of this chain in OPT. Note that no deficiency changes after removing the arcs from P from S.

Now we start giving credit to the arcs in the set S of remaining arcs, and we also give more credit to the nodes. As described in the scheme, each blue arcs receives 2 credit and each red arcs receives 1 credit. In addition we need to give to each node i exactly $d_S(i)$ credit, as i already received $d_A(i) - d_S(i)$ credit.

Take a closed chain P of OPT_S, consisting only of unsplit arcs. Then the cost of P in OPT is $|P|$. P contains exactly one blue arc, and therefore we give $|P|+1$ credit to the arcs of P. As P was not processed in phase 3 of the scheme, and since we assumed P cannot have only two arcs, P has at least four arcs. Therefore we have $|P|+1 \leq \frac{5}{4}|P|$. Remove the arcs of P from S and note that no deficiency changes as a result of this removal. Repeat the step above as long as possible.

Consider an arc $a \in A$ which is split into $d_1, d_2, \cdots, d_k$. Then a will get enough (at least $1 + b(a)$) credit if each d_i gets at least $0.5 + b(d_i)$ credit. Let S' be the set obtained from S by the splitting optimum does. Note that for any node i, $d_{S'}(i) = d_S(i)$. We continue a credit-giving scheme, using the arcs of S'. It is enough if we give at least $d_{S'}(i)$ credit to each node i, at least $1 + b(a)$ to each original arc a, and at least $0.5 + b(d)$ to each fragment arc d. That is exactly what we are going to do.

Take a closed chain P of $OPT_{S'}$. By this stage, P has some fragment arc. The cost in OPT of P is $|P|$, while we give credit at most $|P| + 0.5$, as exactly one arc in P is blue, but at least one of the arcs of P, being a fragment, saves 0.5 credit. Since $|P| \geq 2$, we have $|P| + 0.5 \leq \frac{5}{4}|P|$, and therefore the credit we give is at most 5/4 times the cost of this chain in OPT. Remove the arcs of P from S' and note that no deficiency changes as a result of this removal. Repeat the step above as long as possible.

At this moment, $OPT_{S'}$ has only open chains.

Take a tight (with respect to S') open chain $P = \langle a_1, a_2, \cdots, a_k \rangle$ of $OPT_{S'}$, consisting only of original (unsplit) arcs. The cost in OPT of P is $|P| + 1$. If P has no blue arc, we give $|P|$ credits to the arcs of P and 0.5 credit to $t(a_1)$ and 0.5 credit to $h(a_k)$, for a total of $|P| + 1$ credit. In the following we assume P has blue arcs. Since P is valid, it has exactly one blue arc. Since P was not considered in phases one and two of the scheme, $|P| \geq 3$. Therefore, we give $|P| + 1$ credit to the arcs of P. We also give 0.5 credit to $t(a_1)$ and 0.5 credit to $h(a_k)$. In total, we give $|P| + 2$ credit, which is at most $\frac{5}{4}(|P| + 1)$. Remove the arcs of P from S' and note that after the removal, both $d_{S'}(t(a_1))$ and $d_{S'}(h(a_k))$ drop by 0.5. Repeat the step above as long as possible.

Take a tight (with respect to S') open chain $P = \langle a_1, a_2, \cdots, a_k \rangle$ of $OPT_{S'}$, where at least one of the $a_i's$ is a fragment arc. The cost in OPT of P is $|P| + 1$. We give credit at most $|P| + 0.5$ to the arcs of P, 0.5 to $t(a_1)$, and 0.5 to $h(a_k)$. In total, we give $|P| + 1.5$ credit, which is at most $\frac{5}{4}(|P| + 1)$, as $|P| \geq 1$. Remove the arcs of P from S' and note that after the removal, both $d_{S'}(t(a_1))$ and $d_{S'}(h(a_k))$ drop by 0.5. Repeat the step above as long as possible.

Take a sequence of open chains $P_1, P_2, \cdots, P_k$ of $OPT_{S'}$, which concatenated form a (non-valid) closed chain. The cost in OPT of P_i is $|P_i| + 1$, and the total cost of the chains $P_1, P_2, \cdots, P_k$, is $k + \sum_{i=1}^{k} |P_i|$. We give $1 + b(a)$ to any arc (original or fragment) in any of these k open chains, at a total cost of at most $k + \sum_{i=1}^{k} |P_i|$ (we used the fact that each P_i has at most one blue arc). Remove the arcs of $P_1, P_2, \cdots, P_k$

from S' and note that after the removal, no deficiency changes. Repeat the step above as long as possible.

Take a sequence of open chains $P_1, P_2, \cdots, P_j$ of $OPT_{S'}$, which concatenated form a (non-valid) tight open chain $P' = \langle a_1, a_2, \cdots, a_k \rangle$. Note that $j > 1$, since if such a chain P_1 existed, would have been considered previously. This implies $|P'| > 1$ Also note that $b(P') \leq j$. The cost in OPT of P' is $|P'| + j$. We give credit $1 + b(a_i)$ to each a_i, 0.5 to $t(a_1)$, and 0.5 to $h(a_k)$. In total, we give at most $|P'| + j + 1$ credit, which is at most $\frac{5}{4}(|P'| + j)$, since $|P'| + j \geq 4$. Remove the arcs of P' from S' and note that after the removal, both $d_{S'}(t(a_1))$ and $d_{S'}(h(a_k))$ drop by 0.5. Repeat the step above as long as possible.

It is easy to see that $S' = \emptyset$ at this moment. This completes the proof of Lemma 10.28. ■

We continue with the proof of Theorem 10.27. But first some definitions. If a blue arc has 2 credit, it is *happy*, otherwise it is *unhappy*. As our credit transfer scheme only takes credit from an arc when the arc is put (in phases 2, 3, and 4 of the algorithm) in a valid chain (which is then output), happy arcs stay happy as long as they are in the current set of arcs. By the initial credit allocation scheme, we have three types of unhappy arcs:

1. Blue arcs given 1.5 credit during the first phase of the initial credit allocation scheme. These arcs are called *unhappy singles.*
2. Blue arcs given 1.75 credit in the second phase of the initial credit allocation scheme. The two arcs from the same open chain of OPT are called *partners*, and they form an *unhappy pair.*
3. Blue arcs given 1.75 credit in the third phase of the initial credit allocation scheme. The three arcs from the same closed chain of OPT are called *partners*, and they form an *unhappy triple.*

For a node i, define $dep(i)$ to be the number of unhappy singles a with $t(a) = i$ plus the number of unhappy pairs $\langle a_1, a_2 \rangle$ with $t(a_1) = i$. As arcs are made happy or assigned to valid chains (which are then output), $dep(i)$ is decreasing during the execution of the algorithm. Similarly, for a node i, define $arr(i)$ to be the number of unhappy singles a with $h(a) = i$ plus the number of unhappy pairs $\langle a_1, a_2 \rangle$ with $h(a_2) = i$. As arcs are made happy or assigned to valid chains (which are then output), $arr(i)$ is decreasing during the execution of the algorithm.

Since only tight (with the respect to the current set of arcs during the credit allocation scheme) chains contribute to arr, initially $arr(i) \leq \delta(i)$.

Similarly, only nodes with initial negative surplus can have nonzero *dep*, and in fact initially $dep(i) \leq -\delta(i)$. Our credit transfer scheme maintains the following *deparr* invariant during the execution of the algorithm (that is, with respect to the current set of arcs): for any node i, if $arr(i) > 0$, then $arr(i) \leq \delta(i)$, and if $dep(i) > 0$, then $dep(i) \leq -\delta(i)$. To maintain the deparr invariant, our credit transfer scheme ensures that whenever (during phases three or four) we choose an open chain, and we increase the surplus of a node i with $dep(i) > 0$, we also decrease $dep(i)$ by either making a unhappy single happy or by making the blue arc of an unhappy pair happy. Similarly, whenever (during phases three or four) we choose an open chain, and we decrease the surplus of a node i with $arr(i) > 0$, we also decrease $arr(i)$ by either making a unhappy single happy or by making the blue arc of an unhappy pair happy. Our credit transfer scheme, as we will see, also makes sure that whenever an arc from an unhappy pair or from an unhappy triple is selected and used by the algorithm, any remaining blue arc from the pair or triple receives enough credit to become happy.

Phase two of the algorithm is designed to eliminate the unhappy triples. Please note that no surplus is changed, and therefore there is no need to worry about *arr* or *dep*. Our credit transfer scheme is as follows: Let $P = \langle a_1, a_2, a_3 \rangle$ be a closed chain selected in phase two of the algorithm, and assume that a_1 is the blue arc. The cost of P in the output of our algorithm is 3. The blue arc a_1 has at least 1.5 credit, and therefore the total credit on these three arcs is at least 3.5. We use 3 credit to cover the output, give to the unhappy partner (if it exists) of a_2 0.25 credit, and give to the unhappy partner (if it exists) of a_3 0.25 credit, thus making happy any remaining unhappy arc which had one of its partners selected. No unhappy triple remains at the end of the second phase.

We proceed to the third phase of the algorithm. A blue arc a is selected such that $\delta(t(a)) < 0$, $\delta(h(a)) > 0$. The algorithm puts a into a valid chain by itself, and outputs this chain of cost 2. We get whatever credit a has, and another 1 credit since we decrease $d(t(a))$ and $d(h(a))$. We need to pay for the output, for unhappy partners separated, and for maintaining the deparr invariant. We have three cases: a is an unhappy single, a is a happy blue arc, or a is a partner in an unhappy pair.

1. If a is an unhappy single, as both $dep(t(a))$ and $arr(h(a))$ are decreasing by 1, the deparr invariant is maintained, and no one is separated, and so we are done.

2. If a is a happy arc, it has 2 credit. After covering the output, 1 credit is left. No partners are separated. However we might need to pay to maintain the deparr invariant, which could be become violated at both $t(a)$ and $h(a)$. We allocate 0.5 to fixing each $t(a)$ and $h(a)$. 0.5 credit is enough to make any unhappy arc happy, and therefore it is enough to make an unhappy single or pair happy, thus decreasing $dep(t(a))$ or $arr(h(a))$.

3. Now assume a is in an unhappy pair. By symmetry, say $\langle a, d\rangle$ is the unhappy pair (the other case being $\langle d, a\rangle$ as the unhappy pair). After covering the output, 0.75 credit is left. Now since the unhappy pair $\langle a, d\rangle$ disappears, $dep(t(a))$ decreases and therefore the deparr invariant is maintained at $t(a)$. We allocate 0.5 credit to fixing the deparr invariant at $h(a)$, and this credit is enough by the same argument used in the case when a is happy.

We claim that no unhappy single arc a can survive the third phase of the algorithm, as shown below. Since $dep(t(a))$ is positive (because of a itself), by the deparr invariant, $\delta(t(a)) < 0$, and similarly $\delta(h(a)) > 0$. Therefore a is eligible to be selected by the algorithm. Thus at the end of the third phase, only unhappy pairs could exist.

We now analyze phase four. Due to the deparr invariant, each unhappy pair is eligible to be chosen by the algorithm in the fourth phase, as long as it is unhappy. A valid open chain $\langle a_1, a_2\rangle$ is selected such that $\delta(t(a_1)) < 0$ and $\delta(h(a_2)) > 0$, and either a_1 or a_2 is blue. This open chain is output, and has a cost of 3. We get whatever credit a_1 and a_2 have, and another 1 credit since we decrease $d(t(a))$ and $d(h(a))$. We need to pay for the output, for unhappy partners separated, and for maintaining the deparr invariant. After paying for the output, we have at least 0.75 credit in hand. We allocate 0.25 credit to fix the deparr invariant at each of $t(a_1)$ and $h(a_2)$, and 0.25 to make happy the blue partner (if any) of the red arc in $\{a_1.a_2\}$. As at this stage, any unhappy blue arc already has 1.75 credit, 0.25 credit is enough to make it happy.

Finally, note that as long as an unhappy pair exists, the algorithm is still in phase four as an eligible pair exists. This fact implies that at the end of phase four, no unhappy arc exists. Thus we have enough credit for the fifth and last phase. This completes the proof of Theorem 10.27.

Obvious practical improvements are: after Phase 4, but before the Eulerian rounding, insert three other phases:

- Phase 4.1. While A contains a valid closed chain P, find one with minimum number of arcs. unsplit the arcs in P, output P, and set $A \leftarrow A \setminus P$.
- Phase 4.2 While A contains a tight valid chain $P = \langle a_1, \cdots, a_k \rangle$, unsplit the arcs in P, output P, and set $A \leftarrow A \setminus P$.
- Phase 4.3 While A contains a closed chain P, find one with $b(P)$ minimum. The Floyd-Warshall algorithm on the graph with vertex set N and arc set A finds this minimum. Do an Eulerian rounding on P (producing $b(P)$ closed chains). Then set $A \leftarrow A \setminus P$.

There is nothing special about the link $n-1$ deciding which arc is blue and which one is not. If enough computing time is available, one should try running the algorithm using each of the remaining $n-1$ links of the ring to decide which arc is blue and which one is not.

We are not able to prove a better approximation ratio based on these improvements. The worst example (for which the practical improvements also do not help) we have is the following: $n = 6$ and A consists of the following nine arcs:

$$a_1 = (0,2), a_2 = (2,5), a_3 = (5,0),$$
$$b_1 = (2,4), b_2 = (4,1), b_3 = (1,2),$$
$$c_1 = (4,0), c_2 = (0,3), c_3 = (3,4).$$

The optimum produces the closed chains

$$\langle a_1, a_2, a_3 \rangle, \langle b_1, b_2, b_3 \rangle, \langle c_1, c_2, c_3 \rangle,$$

of total cost 9. Indeed, the optimum has cost 9, while if our algorithm is unlucky and chooses the closed chain $\langle a_1, b_1, c_1 \rangle$ in phase two, it produces a solution of cost 10. See Figure 10.11.

7. VALID CHAIN GENERATION FROM UNSPLITABLE CHORDS

Given a set of chords C, we use $\deg_C(i)$ to denote the total number of chords in C that contain node i as one endpoint as in Section 3.. It is well known the number of nodes with odd degree is even. The *deficiency* of C, denoted by $d(C)$, is defined as the half of the number of nodes with odd degree. Notice that in any primitive ring generation from C,

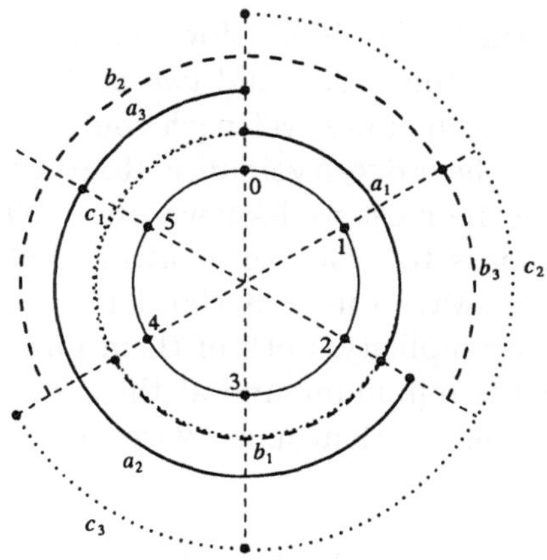

Figure 10.11. The algorithm selects the closed chain $\langle a_1, b_1, c_1\rangle$. Eulerian rounding then splits one arc in two (a_2, for example) and produces a solution of cost 10. Optimum uses the closed chains $\langle a_1, a_2, a_3\rangle$, $\langle b_1, b_2, b_3\rangle$, and $\langle c_1, c_2, c_3\rangle$, of total cost 9.

each node i requires at least $\left\lceil \frac{\deg_C(i)}{2} \right\rceil$ ADMs, the total number of ADMs required by any primitive ring generation from C is at least

$$\frac{1}{2}\sum_{i=0}^{n-1} \deg_C(i) + d(C) = \frac{1}{2} \cdot 2|C| + d(C) = |C| + d(C).$$

So $|C| + d(C)$ is a lower bound on the minimum ADM cost required by any primitive ring generation from C.

Fix a set of arcs C. Let OPT denote the optimal valid chain generation from C and let opt denote the cost of OPT. The same argument to Lemma 10.7 applies to the chord version.

LEMMA 10.29 *If any pair of chains of chords in $\mathcal{P}$ can not be merged into a larger valid chain, the cost of $\mathcal{P}$ is at most $\frac{7}{4} \cdot opt$.*

Thus it is interesting to find heuristics which can beat the 1.75 ratio. A straightforward approach to valid chain generation from chords consists of two separate stages. In the first stage, find a routing of the input chords with the low load or wavelength requirement. In the second stage, apply heuristics developed in the previous section to the obtained arcs. Given a set of chords, its minimum-load routing can be found in polynomial time [137, 166]. However, it is NP-hard to find a minimum-wavelength routing [129], and two is the best approximation ratio of heuristics for minimum-wavelength routing known in the literature. Among them, the Edge-Avoidance Routing and the Shortest-Path Routing [35] are the most well-known ones. The Edge-Avoidance Routing routes all chords to avoid an arbitrary link of the ring. The shortest-path routing always routes a chord that traverses the fewest links. Despite of their simplicity, both of them guarantee that the link load is at most twice the minimum, and at the same time they can lead to a wavelength assignment with at most twice the minimum the number of wavelengths.

We first show that minimum-load routing and minimum-wavelength routing may have the worst performance in terms of the ADM cost.

EXAMPLE 10.30 *Let $n = 2k$ for some $k \geq 2$, and C consists of two copies of $(2i, 2i+1)$ for all $0 \leq i < k$. In order to achieve minimum load (or wavelength), each chord must be routed in the short-way, as shown in Figure 10.12 (a). However, with this routing each chord must form a 1-chain, which would lead the total ADM cost of $2|C|$. On the other hand, one can form a valid closed 2-chain from each pair of identical chords, as shown in Figure 10.12 (b). So the minimum ADM cost is $|C|$.*

Now we examine the performance of the Edge-Avoidance Routing and the Shortest-Path Routing. However, Example 10.30 shows that both of them can cost as high as twice the minimum. Even when there are no identical chords, they can still cost as high as $\frac{5}{3}$ times the minimum as illustrated by the following example.

EXAMPLE 10.31 *Let $n = 3k$ for some $k \geq 2$, and $C = \bigcup_{i=0}^{k-1} C_i$ where*

$$C_i = \{(3i, 3i+1), (3i+1, 3i+2), (3i, 3i+2)\},$$

as illustrated in Figure 10.13. Note that the three chords in each C_i form a closed valid chain. Thus $opt = |C| = 3k$. On the other hand, it is easy to verify that both edge-avoidance routing and shortest-path routing would lead to at least $2k$ open chains. The cost of these three chains is $3k + 2k = 5k = \frac{5}{3} \cdot opt$.

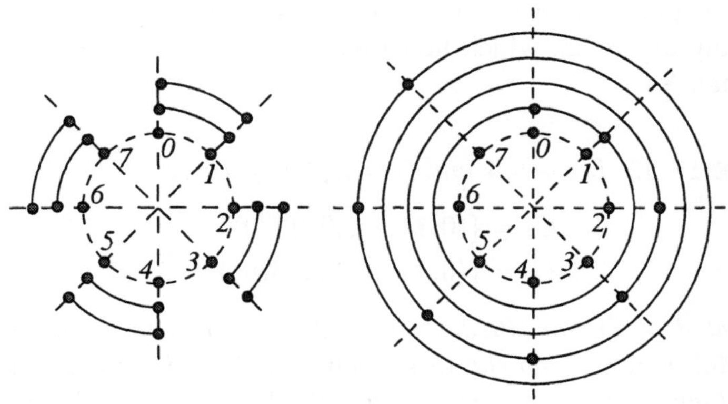

Figure 10.12. An instance used in Example 10.30: (a) the minimum load (or wavelength routing), and (b) the routing with minimum ADM cost.

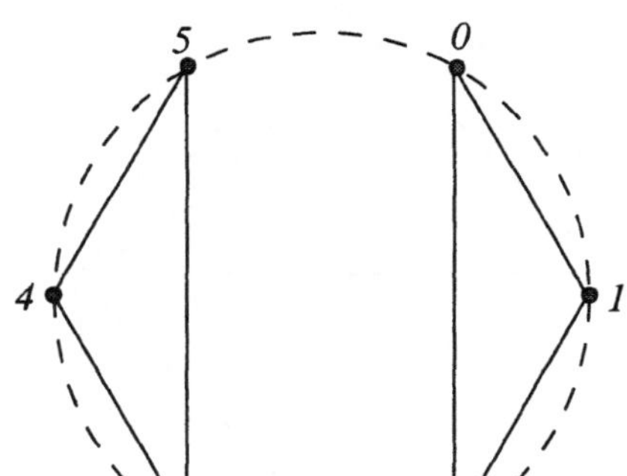

Figure 10.13. An instance used in Example 10.31 with $n = 6$.

The previous two examples imply that heuristics based on separated routing and valid chain generation from arcs can not provide good performance guarantees. In the next, we propose heuristics with integrated routing and chain generation. Before we describe the heuristics, we would like to point out a subtle difference between valid chain generation from arcs and valid chain generation from chords. While there is an optimal valid chain generation from arcs which pair up a given pair of complementary arcs, there may be no optimal valid chain generation from chords which pair up a given pair of identical chords. In fact, the following example indicates that one can not expect the approximation ratio of any algorithm which first take out pairs of identical chords to be less than $\frac{4}{3}$.

EXAMPLE 10.32 *Let* $n = 4$ *and* $C = C_1 \cup C_2$ *where*

$$C_1 = \{(0,1), (1,2), (2,0)\},$$
$$C_2 = \{(0,1), (1,3), (3,0)\},$$

as illustrated in Figure 10.13. Note that the three chords in C_1 *form a closed valid chain, so do the three chords in* C_2. *Thus opt* $= |C| = 6$. *On the other hand, if we take the two identical chords between 0 and 1 to form a closed 2-chain, then the remaining four chords require at least two open chains. The cost of these three chains is* $6 + 2 = 8 = \frac{4}{3} \cdot opt$.

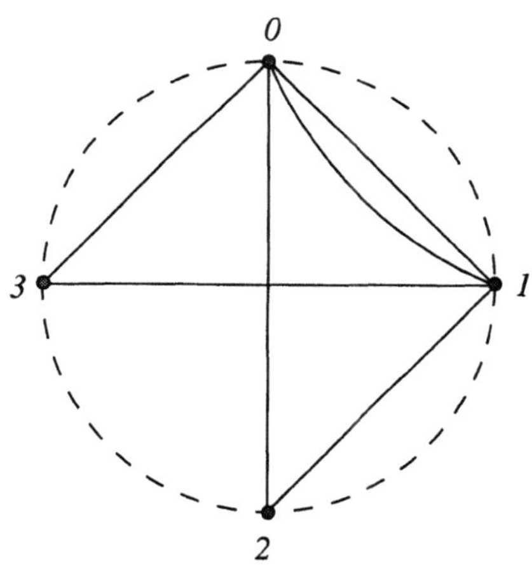

Figure 10.14. Pairing up two identical chords into a closed chain may not give rise to optimal solution.

In summary, both minimum-load routing [137, 166] and minimum-wavelength routing [129] fail to beat the 1.75 ratio. The two well-known routing heuristics, Edge-Avoidance routing and Shortest-Path routing [35], fail to beat the 1.75 ratio as well. Even after the preprocessing by pairing up identical chords, their approximation ratios is still at least $\frac{5}{3} \approx 1.67$. In the remaining of this section, we propose two heuristic approaches based on Eulerian tour decomposition and matching respectively. While its counterpart for arcs performs poorly and requires improvement to get an approximation ratio between 1.5 and 1.6, the Eulerian Tour Decomposition for chords has an approximation ratio of exactly 1.5. The Iterative Matching and its variants have the similar performance as their counterpart for arcs. We prove that the approximation ratio of the basic Iterative Matching is between 1.5 and $\frac{5}{3}$, and the approximation ratio of its improvement by a preprocessing with repeatedly taking out chords which can form a ring is between $\frac{4}{3}$ and 1.5.

7.1 EULERIAN TOUR DECOMPOSITION

In this section we propose the following **Eulerian Tour Decomposition (ETD)** heuristic, which is a counter-part of the **ET-TS** heuristic described in Subsection 5.2. Without loss of generality, we assume that $G(C)$ is connected, for otherwise we can apply the algorithm to each connected component separately. We consider two cases.

In the first case, some nodes in the ring have odd degree. The algorithm ETD runs in the following four steps:

1. Step 1: Divide the set of nodes with odd degree into disjoint pairs. Add one *fake* chord between the two nodes in each pair. Let C' be the set of fake chords. Then $|C'| = d(C)$ and the graph $G(C \cup C')$ is Eulerian. Find an Eulerian tour of $G(C \cup C')$.

2. Step 2: Remove all fake chords from the Eulerian tour to obtain $d(C)$ trails of $G(C)$.

3. Step 3: Split each trail into valid chains by walking along the trail from the first chord and generating a valid chain whenever overlap occurs;

4. Step 4: Repeatedly merge any pair of open valid chains into a larger valid chain until no more merging can occur.

In the second case, all nodes in the ring have even degree. Thus the graph $G(C)$ is Eulerian and has an Eulerian tour. If $|C|$ is even, we apply Step 3 and Step 4 in the first case to any Eulerian tour of $G(C)$. If $|C|$ is odd, we obtain an Eulerian tour of $G(C)$ whose first three chords form a valid chain, if there is any, as follows. Fix a pair of adjacent but non-identical chords c_1 and c_2. Let (u, v) denote the (clockwise) arc induced by the unique routing of the 2-chain consisting of c_1 and c_2. Without of loss of generality, we assume that u is the an endpoint of c_1 and v is the an endpoint of c_2. If there is an chord c with one endpoint being u and the other endpoint being v or inside the arc (v, u), then we can obtain a desired Eulerian tour whose first three chords are c, c_1, and c_2 sequentially. Otherwise if there is an chord c with one endpoint being v and the other endpoint being inside the arc (v, u), then we can obtain a desired Eulerian tour whose first three chords are c_1, c_2, and c. sequentially. Otherwise, we repeat the previous procedure to other unselected pair of adjacent but non-identical chords. Eventually, we either obtain a desired Eulerian tour, or conclude that any three chords can not form a valid chain. In the former situation, we apply Step 3 and Step 4 in the first case to the obtained Eulerian tour. In the latter situation, we apply Step 3 and Step 4 in the first case to any Eulerian tour.

While the approximation ratio of **ET-TS** is 1.75 according to Theorem 10.14, the next lemma shows that the approximation ratio of **ETD** is at most 1.5.

LEMMA 10.33 *The approximation ratio of* ***ETD*** *is at most* 1.5.

Proof. We first consider the first case, i.e. $d(C) \geq 1$. Step 2 generates $d(C)$ trails. Let P be any such trail. Since each pair of consecutive chords in P can form a valid chain, all valid chains except the last one split from P at Step 3 contains at least two chords. Thus at most $\left\lceil \frac{|P|}{2} \right\rceil$ chains are split from P. So totally at most $\frac{|C|+d(C)}{2}$ chains are generated at Step 3,and costs at most

$$|C| + \frac{|C| + d(C)}{2} = \frac{3|C| + d(C)}{2} \leq \frac{3}{2}(|C| + d(C)) \leq \frac{3}{2} \cdot opt.$$

This implies that the approximation ratio of **ETD** is at most 1.5.

In the next we consider the second case, i.e. $d(C) = 0$. If $|C|$ is even, then following the same argument above, at most $\frac{|C|}{2}$ chains are produced. So the total cost is at most

$$|C| + \frac{|C|}{2} = \frac{3}{2}|C| \leq \frac{3}{2} \cdot opt.$$

So we now assume that $|C|$ is odd. If some three chords form a valid chain, let $k \geq 3$ be the number of chords in the first valid chain split from the obtained Eulerian tour. Then again from the previous argument, at most $1 + \left\lceil \frac{|C|-k}{2} \right\rceil$ chains are produced. So the total cost is at most

$$|C| + 1 + \left\lceil \frac{|C| - k}{2} \right\rceil \leq |C| + 1 + \frac{|C| - k + 1}{2}$$
$$= \frac{3\,|C| - k + 3}{2} \leq \frac{3}{2}\,|C| \leq \frac{3}{2} \cdot opt.$$

If any three chords in C can not form a valid chain, we prove by contradiction that $opt > |C|$. Assume to the contrary. Then in any optimal solution, each valid chain is closed, and therefore must be a closed 2-chain. This implies that $|C|$ is even, which is a contradiction. Thus $opt > |C|$. Since exactly $\frac{|C|+1}{2}$ chains are generated, the total cost is at most

$$|C| + \frac{|C| + 1}{2} = \frac{3\,|C| + 1}{2} < \frac{3}{2}\left(|C| + 1\right) \leq \frac{3}{2} \cdot opt.$$

Therefore, in either case the lemma is true. ■

The next example shows that the approximation ratio of **ETD** is at least 1.5.

EXAMPLE 10.34 *Let $n = 2\,(2k+1)$ for some $k > 1$, and C consist of the following $\frac{3n}{2}$ chords*

$$\left\{c_i = \left(2i, 2i + \frac{n}{2}\right) \mid 0 \leq i < \frac{n}{2}\right\} \cup$$
$$\left\{c'_i = \left(2i + \frac{n}{2}, 2i + \frac{n}{2} + 1\right) \mid 0 \leq i < \frac{n}{2}\right\} \cup$$
$$\left\{c''_i = \left(2i + \frac{n}{2} + 1, 2i\right) \mid 0 \leq i < \frac{n}{2}\right\},$$

*as illustrated in Figure 10.15. This instance can be obtained from the instance in Example 10.17 by replacing each arc with a chord between its two endpoints. Note that for any $0 \leq i < n$, the three chords c_i, c'_i, c''_i form a closed valid chain. So $opt = |C| = \frac{3n}{2} = 3\,(2k+1)$. As $d\,(C) = 0$ and $|C|$ is odd, the algorithm **ETD** tries to find an Eulerian tour in which the first three chords form a valid chain. The algorithm may unluckily produce the following Eulerian tour*

$$c'_k, c''_k, c''_{2k}, c''_{k-1}, c''_{2k-1}, \cdots, c''_1, c''_{k+1}, c''_0,$$
$$c_0, c'_0, c_{k+1}, c'_{k+1}, c_1, c'_1, c_{k+2}, c'_{k+2}, \cdots,$$
$$c_{k-1}, c'_{k-1}, c_{2k}, c'_{2k}, c_k.$$

At Step 3, the following valid chains are generated:

- *one open valid 3-chain* $\{c'_k, c''_k, c''_{2k}\}$,
- $k-1$ *open valid 2-chains* $\{\{c''_i, c''_{i+k}\} \mid 1 \le i < k\}$,
- *one closed 3-chain* $\{c''_0, c_0, c'_0\}$,
- $2k-1$ *open valid 2-chains* $\{\{c_i, c'_i\} \mid 1 \le i \le 2k, i \ne k\}$.
- *one open valid 1-chain* $\{c_k\}$.

So totally $3k$ *open chains and one closed chain are obtained. These chains costs*

$$|C| + 3k = 3(2k+1) + 3k = 9k + 3 = \frac{3}{2} \cdot opt - \frac{3}{2}.$$

Thus the approximation ratio of ***ETD*** *is at least* $\frac{3}{2}$.

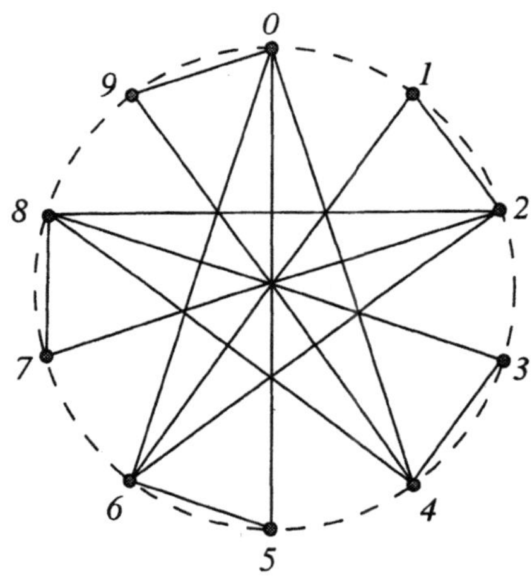

Figure 10.15. An instance used in Example 10.34 with $n = 10$.

From Lemma 10.33 and Example 10.34, we have the following theorem.

THEOREM 10.35 *The approximation ratio of* ***ETD*** *is exactly* 1.5.

7.2 ITERATIVE MATCHING

In this subsection, we modify the heuristics **Iterative Matching (IM)** and **Preprocessed Iterative Matching (PIM)** described in Subsection 5.3 such that they are applicable to chords. The **IM** for chords also maintains a set of valid chains of chords $\mathcal{P}$ throughout its execution. Initially $\mathcal{P}$ consists of 1-chains of single chord in C. While the fit graph $F(\mathcal{P})$ of $\mathcal{P}$ has non-empty edge set, we find a maximum matching M in $F(\mathcal{P})$ and then merge each matched pair of chains (of chords) in M into a larger chain (of chords). When $F(\mathcal{P})$ has empty edge set, $\mathcal{P}$ is output as the solution.

By using the same argument used in Lemma 10.20, one can prove the approximation ratio of **IM** for chords is at most $\frac{5}{3}$. On the other hand, replacing each arc in Example 10.21 by a chord between its two endpoints leads to an instance (see Figure 10.16) for which the **IM** for chords may produce a solution of $\frac{3}{2}$ times the optimum. So we have the following theorem.

THEOREM 10.36 *The approximation ratio of **IM** for chords is between $\frac{3}{2}$ and $\frac{5}{3}$.*

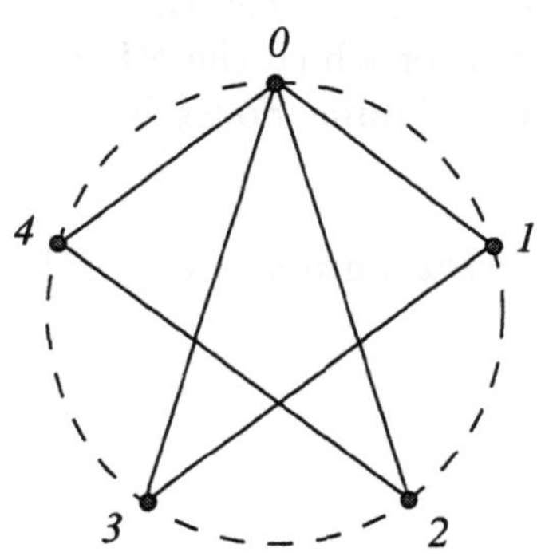

Figure 10.16. The instance obtained by replacing each arc in Example 10.21 with a chord between the same endpoints.

PIM for chords is adapted from the PIM algorithm for arcs. The chord-version PIM also runs in two phases:

1. Preprocessing Phase: repeatedly take valid closed chains out of the remaining chords until no more closed chain can be obtained from the remaining chords.

2. Matching Phase: apply the algorithm **IM** to the remaining chords.

The following procedure is used by the Preprocessing Phase to obtain a closed valid chain containing a specified chord c, if there is any, from a set of chords S. Let c^- be the circular arc between the two endpoints of c that passes through the link between node $n-1$ and node 0, and let c^+ be the arc complementary to c^-. Let S_c^+ (S_c^- respectively) be the set of chords in $S - \{c\}$ whose two endpoints are both in c^+ (c^- respectively). Let G_c^+ (G_c^- respectively) be the directed graph with the nodes in c^+ (c^- respectively) as its vertices and directed edges obtained from orienting the chords in S_c^+ to not use the link between node $n-1$ and node 0, (orienting the chords in S_c^- to use the link between node $n-1$ and node 0, respectively). There is a closed valid chain in S that contains c if and only if either there is a path between the two endpoints of c in G_c^+, or there is a path between the two endpoints of c in G_c^-. After constructing G_c^+ and G_c^-, such a path, if there is any, can be found by breadth-first search. Once this path is obtained, we add c to it to obtain a closed valid chain.

Following the argument in Lemma 10.23, we can prove that the approximation ratio of **PIM** for chords is at most 1.5. Replacing each arc in Example 10.19 with a chord between the same endpoints leads to an instance (see Figure 10.17) for which the **PIM** for chords may produce a solution of $\frac{4}{3}$ times the optimum. So we have the following theorem.

THEOREM 10.37 *The approximation ratio of **PIM** for chords is between* $\frac{4}{3}$ *and* $\frac{3}{2}$.

8. VALID CHAIN GENERATION FROM SPLITABLE CHORDS

We begin with an example to illustrate the benefit of splitting chords. Consider the following example depicted in Figure 10.18: let $n = 5$ and $C = \{c_i : 0 \leq i \leq 4\}$ with $c_0 = (0,2), c_1 = (1,3), c_2 = (2,4), c_3 =$

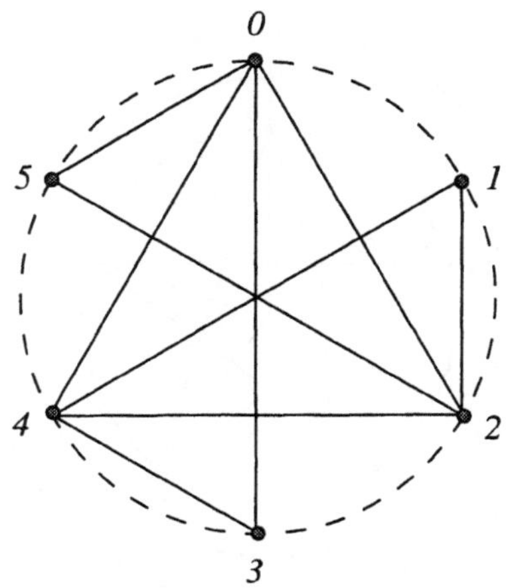

Figure 10.17. The instance obtained by replacing each arc in Example 10.19 with a chord between the same endpoints.

$(3,0), c_4 = (4,1)$. The optimum without splits is 8, while with splits a solution of cost 6 can be obtained by splitting c_3 into two chords

$$c_3' = (3,4), c_3'' = (4,0),$$

and then c_0, c_2, c_3'' form one valid chain and c_1, c_3', c_4 form another valid chain. Thus the optimum with splits can be 25% lower.

Now we describe our algorithm which is based on Eulerian rounding developed in Section 6.:

- The input is a set of chords C.
- Add a set of fake chords H such that $|H| = d(C)$ and $d(C \cup H) = 0$. This can be easily done by adding one by one fake chords in between nodes of odd degree.
- Find an Eulerian cycle EC of $C \cup H$. EC can be traversed in two directions, and each direction transforms the chords of $C \cup H$ into arcs. We call A' and A'' the sets of arcs obtained from C this way. If $b(A') > b(A'')$, let A be A'', otherwise let $A = A'$. C is oriented to obtain A.
- Do the Eulerian rounding of A.

For a chord $c \in C$, we call $a_c' \in A'$ and $a_c'' \in A''$ the corresponding arcs obtained by orienting c in opposite directions. Exactly one of a_c'

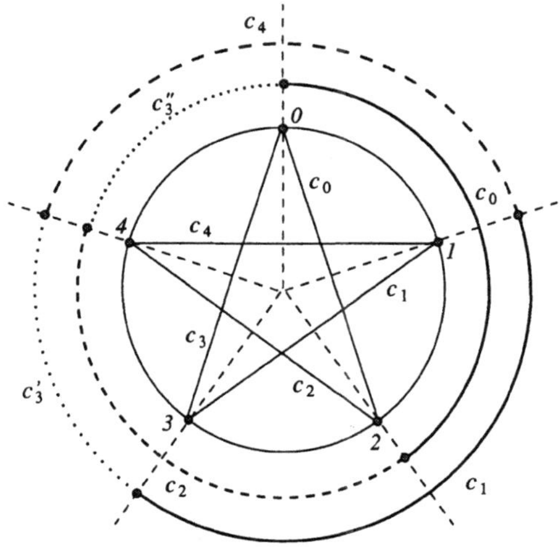

Figure 10.18. Chord c_3 is split into c_3' and c_3'' to obtain a solution with splits of cost 6.

and $a_c'' \in A''$ is blue, and therefore $b(A') + b(A'') = |C|$. We conclude that $b(A) \leq |C|/2$. Note that $d(A) = |H| = d(C)$. By lemma 10.25, with Eulerian rounding of A we obtain a solution of cost at most

$$|A| + b(A) + d(A) \leq \frac{3}{2}|C| + d(C).$$

Thus the above algorithm has approximation ratio at most $\frac{3}{2}$. The next theorem shows that its approximation ratio is exactly $\frac{3}{2}$.

THEOREM 10.38 *The above algorithm has approximation ratio exactly* $\frac{3}{2}$.

Proof. The fact that the approximation ration is at most $\frac{3}{2}$ follows from the discussion above. The following example shows that the approximation ratio of the algorithm is at least $\frac{3}{2}$.

Let n, the number of nodes on the ring, be odd. Consider the following set of chords:

$$C = \{c_0, c_1, \cdots, c_{n-3}, c', c''\},$$

where for $0 \leq i \leq n-3$, the endpoints of c_i are the nodes i and $i+2$, the endpoints of c' are 0 and 1, and the endpoints of c'' are $n-2$ and $n-1$.

The Eulerian tour selected by the algorithm is

$$c_{n-3}, c_{n-5}, \cdots, c_0, c', c_1, c_3, \cdots, c_{n-4}, c''.$$

The chords

$$c', c_1, c_3, \cdots, c_{n-4}, c''$$

are oriented clockwise, and the chords

$$c_{n-3}, c_{n-5}, \cdots, c_0$$

are oriented counterclockwise. For Eulerian rounding, node 0 is selected. Then all the arcs

$$c_{n-3}, c_{n-5}, \cdots, c_2$$

are going to be split, obtaining a solution of cost $n + \frac{n-3}{2}$, since there are n chords and $\frac{n-3}{2}$ of them are split.

However, an optimum of cost $n+2$ exists: orient all the chords clockwise and produce two open chains:

$$c_0, c_2, \cdots, c_{n-3}$$

and

$$c', c_1, c_3, \cdots, c_{n-4}, c''.$$

Therefore, the cost ratio of the output of the algorithm to the optimum is $\frac{n+(n-3)/2}{n+2}$. If we let n large enough, this ratio gets arbitrarily close to $\frac{3}{2}$.

In conclusion, the approximation ratio of the algorithm presented in this section is exactly $\frac{3}{2}$. ■

An obvious practical improvement is: once an orientation has been selected, run the algorithm from Section 6. on the given set of arcs. Another practical improvement would be the equivalent of Phases 4.1 and 4.3 of the algorithm from Section 6. applied directly to chords, before Eulerian rounding. That is, before an orientation is selected, find cycles in C such that, when oriented in the best of the two possible ways, they have small blue number (recall that a closed chain with blue number one is valid). Then do an Eulerian rounding on the cycle, and remove the cycle from C.

9. GROOMING OF PRIMITIVE RINGS

We assume that the ring network consists of n nodes, labeled by $0, 1, \cdots, n-1$. Then each primitive ring can be represented by a set from the universe $\{0, 1, \cdots, n-1\}$. Thus an instance I of Ring Grooming consists of m sets $A_1, A_2, \cdots, A_m$ from the universe $\{0, 1, \cdots, n-1\}$ and a constant parameter g representing the grooming granularity. Let P be any subset of $\{1, 2, \cdots, m\}$. Its *size* is $|P|$, and it is called as a d-group if $|P| = d$. The cost of P is given by $\left|\bigcup_{r\in P} A_r\right|$, and the savings of P is given by $\sum_{r\in P} |A_r| - \left|\bigcup_{r\in P} A_r\right|$. Let $\pi = \{P_1, P_2, \cdots, P_k\}$ be any partition of the indices $\{1, 2, \cdots, m\}$. It is called as a d-grouping if $|P_i| = d$ for all $1 \le i \le k$. The cost of π is sum of the costs of all groups in π, i.e., $\sum_{i=1}^{k} \left|\bigcup_{r\in P_i} A_r\right|$, and the savings of π is the sum of the savings of all groups in π, i.e.,

$$\sum_{i=1}^{k} \left(\sum_{r\in P_i} |A_r| - \left| \bigcup_{r\in P_i} A_r \right| \right) = \sum_{r=1}^{m} |A_r| - \sum_{i=1}^{k} \left| \bigcup_{r\in P_i} A_r \right| .$$

We distinguish two variations of Ring Grooming. In *Unrestricted Ring Grooming*, a solution is a partition of $\{1, 2, \cdots, m\}$ into groups of sizes at most g. The term "unrestricted" indicates that there is no restriction on the number of groups in a partition. Such partition is referred to as an *unrestricted partition*. In *Restricted Ring Grooming*, a solution is a partition of $\{1, 2, \cdots, m\}$ into $\left\lceil \frac{m}{g} \right\rceil$ groups of sizes at most g. The term "restricted" specifies that the number of groups in a partition has to be $\left\lceil \frac{m}{g} \right\rceil$, the smallest possible. Such partition is referred to as a *restricted partition*. In either variation, the objective is to find a solution (a partition) with minimum cost.

Although no restriction is put on the number of groups in Unrestricted Ring Grooming, one can always convert any partition into one with at most $\left\lfloor \frac{2m}{g+1} \right\rfloor$ groups without increasing the cost. Indeed, we observe that any two groups whose total size is at most g can be merged into a larger group and the cost of resulting partition is not increased. By repeatedly performing this merging, we can reach a partition in which the total size of any pair of groups is at least $g+1$. Let k be the number of groups and $d_1, d_2, \cdots, d_k$ be their sizes. Then we have $\binom{k}{2}$ inequalities

$$d_i + d_j \ge g + 1.$$

Sum up these inequalities and notice that each term d_i appears in exactly $k-1$ of these inequalities. We have

$$(k-1)\sum_{i=1}^{k} d_i \geq \binom{k}{2}(g+1) = \frac{k(k-1)}{2}(g+1).$$

This implies that

$$k \leq \frac{2\sum_{i=1}^{k} d_i}{g+1} = \frac{2m}{g+1}$$

and thus $k \leq \left\lfloor \frac{2m}{g+1} \right\rfloor$. In particular, when $g=2$, the obtained partition is also a restricted partition. Thus when $g=2$, an optimal restricted partition is also an optimal unrestricted partition.

In addition, one can always convert any partition into one with exactly $\left\lceil \frac{m}{g} \right\rceil$ groups and at most twice the cost, as follows. Let $\{P_1, P_2, \cdots, P_k\}$ be a solution to Unrestricted Ring Grooming. Reorder the indices $1, 2, \cdots, m$ such that for any $1 \leq i \leq k$, P_i consists of consecutive indices. We construct $\left\{Q_1, Q_2, \cdots, Q_{\left\lceil \frac{m}{g} \right\rceil}\right\}$, a solution to Restricted Ring Grooming, by setting

$$Q_j = \{g(j-1)+1, g(j-1)+2, \cdots, gj\},$$

for $1 \leq j \leq \left\lfloor \frac{m}{g} \right\rfloor$, and if $m \bmod g > 0$,

$$Q_{\left\lceil \frac{m}{g} \right\rceil} = \left\{g\left\lfloor \frac{m}{g} \right\rfloor + 1, g\left\lfloor \frac{m}{g} \right\rfloor + 2, \cdots, m\right\}.$$

This construction is illustrated in Figure 10.19. As $|P_i| \leq g$ for all $1 \leq i \leq k$, each group P_i is entirely contained in either one group Q_j, or two consecutive groups Q_j and Q_{j+1}. For $1 \leq j \leq \left\lceil \frac{m}{g} \right\rceil$, let j' be the index such that $g(j-1)+1 \in P_{j'}$, and let j'' be the index such that $gj \in P_{j''}$ if $j \leq \left\lfloor \frac{m}{g} \right\rfloor$ or k if $m \bmod g > 0$ and $j = \left\lceil \frac{m}{g} \right\rceil$. Then for $1 \leq j \leq \left\lceil \frac{m}{g} \right\rceil$, $Q_j \subseteq \bigcup_{i=j'}^{j''} P_i$, and therefore

$$\left|\bigcup_{r \in Q_j} A_r\right| \leq \left|\bigcup_{r \in \bigcup_{i=j'}^{j''} P_i} A_r\right| \leq \sum_{i=j'}^{j''} \left|\bigcup_{r \in P_i} A_r\right|.$$

As each term inside the righthand side summation can contribute to either only $\left|\bigcup_{r\in Q_j} A_r\right|$ or both $\left|\bigcup_{r\in Q_j} A_r\right|$ and $\left|\bigcup_{r\in Q_{j+1}} A_r\right|$,

$$\sum_{j=1}^{\left\lceil \frac{m}{g} \right\rceil} \left| \bigcup_{r\in Q_j} A_r \right| \leq 2 \sum_{i=1}^{k} \left| \bigcup_{r\in P_i} A_r \right| .$$

Thus the cost of

$$\left\{ Q_1, Q_2, \cdots, Q_{\left\lceil \frac{m}{g} \right\rceil} \right\}$$

is at most twice the cost of

$$\{P_1, P_2, \cdots, P_k\} .$$

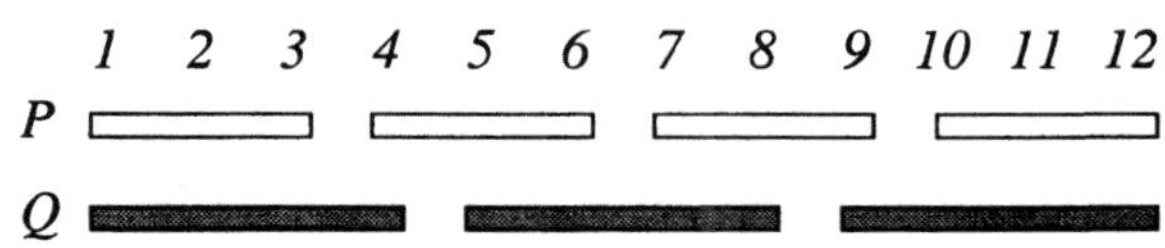

Figure 10.19. Conversion from a solution to Unrestricted Ring Grooming into a solution to Restricted Ring Grooming.

The following lemma shows that Restricted Ring Grooming is as least as hard to approximate as Unrestricted Ring Grooming, but not much harder.

LEMMA 10.39 *Any polynomial-time α-approximation algorithm for Restricted Ring Grooming can lead to a polynomial-time α-approximation algorithm for Unrestricted Ring Grooming; and any polynomial-time α-approximation algorithm for Unrestricted Ring Grooming is also 2α-approximation algorithm for Restricted Ring Grooming.*

Proof. Fix an instance I of Ring Grooming, let $opt\_u(I)$ and $opt\_r(I)$ to denote its minimum unrestricted cost and restricted cost respectively.

Assume that the instance I contains m sets. Then there is an optimal unrestricted partition with at most $\left\lfloor \frac{2m}{g+1} \right\rfloor$ groups. Let I' be the instance obtained from I by adding $\left\lfloor \frac{2m}{g+1} \right\rfloor g - m$ dummy empty sets. Obviously,

$$opt\_u(I) = opt\_u(I') = opt\_r(I') \leq opt\_r(I) \leq 2 \cdot opt\_u(I) .$$

Given any polynomial-time α-approximation algorithm for Restricted Ring Grooming, we apply it to the instance I' and let π' denote the output partition. Then the cost of π' is at most

$$\alpha \cdot opt\_r\,(I') = \alpha \cdot opt\_u\,(I') = \alpha \cdot opt\_u\,(I)\,.$$

Let π denote the partition obtained from π' by removing the indices of the dummy empty sets from the groups of π'. Then π is a solution to the instance I, and it has the same cost as π'. So the cost of π is at most $\alpha \cdot opt\_u\,(I)$.

Given any polynomial-time α-approximation algorithm for Unrestricted Ring Grooming, we apply it to the instance I and let π denote the output partition. Then the cost of π is at most $\alpha \cdot opt\_u\,(I) \leq \alpha \cdot opt\_r\,(I)$. Using the conversion described above, we convert π to a restricted partition of at most twice the cost of π. Then the cost of the obtained restricted partition at most $2\alpha \cdot opt\_r\,(I)$. ■

In this section, we will study the computational complexity of both variations and propose heuristic solutions. In Subsection 9.1, we provide optimal algorithms when $g = 2$ and prove the NP-hardness of both variations for any fixed constant $g > 2$. Therefore, for the case $g > 2$, we search for polynomial-time approximation algorithms. In Subsections 9.2 and 9.3, we develop some graph-theoretical results on edge-connectivity and matchings, which play critical roles in the later analysis of the approximation algorithms. In Subsection 9.4 we propose an approximation algorithm based on iterative matching for Restricted Ring Grooming with g a power of two. A trivial upper bound on its approximation ratio is $\frac{g}{2}$. We attempt to give tighter analysis of its approximation ratio. When $g = 4$, its approximation ratio is exactly 1.5. When $g = 8$, its approximation ratio is at most 2.5. Following our general analyzing framework, we conjecture that its approximation ratio is at most $\frac{g}{4} + \frac{1}{2}$. We also present an extension of the iterative matching algorithm for Unrestricted Ring Grooming with arbitrary g. In Subsection 9.5, we present a set-cover based algorithm for Unrestricted Ring Grooming, whose approximation ratio is at most $1 + \log g$, and running-time is polynomial in m^g where m is the number of primitive rings. From the results of Subsection 9.1, the same algorithm can be used to obtain a $2(1 + \log g)$ approximation ratio for Restricted Ring Grooming.

9.1 COMPUTATIONAL COMPLEXITY

When $g = 2$, simple polynomial-time algorithm exists for Restricted Ring Grooming, which also outputs an optimal solution to Unrestricted

Ring Grooming. By adding one dummy empty set if necessary, we assume that m is even. Then any solution to Restricted Ring Grooming is a *perfect* matching of $\{1, 2, \cdots, m\}$ and vice versa. To establish the relation between Ring Grooming and matching, we define a weighted complete graph K_m with vertex set $\{1, 2, \cdots, m\}$ by setting the weight of any edge $e = (i, j)$ equal to

$$w(e) = |A_i \cap A_j|.$$

Using the fact that

$$|A_1| + |A_2| - |A_1 \cup A_2| = |A_1 \cap A_2|,$$

the weight of any perfect matching is equal to its savings as a solution to Restricted Ring Grooming. Thus a maximum-weighted perfect matching gives rise to an optimal solution to Ring Grooming, Restricted or Unrestricted.

However when $g > 2$, both Unrestricted Ring Grooming and Restricted Ring Grooming are NP-hard, as shown in the next theorem.

THEOREM 10.40 *For any* fixed *constant $g > 2$, both Unrestricted Ring Grooming and Restricted Ring Grooming are NP-hard.*

Proof. The reduction is made from Partition Into Isomorphic Subgraphs [62]. Partition Into Isomorphic Subgraphs is the following problem: given two graphs G and H with $|V(G)| = r|V(H)|$ for some positive integer r, determine whether there is a partition of $V(G)$ into r disjoint sets $V_1, V_2, \cdots, V_r$ such that for all $1 \le i \le r$ the subgraph of G induced by V_i is isomorphic to H. The problem remains NP-Complete for any fixed H with at least three vertices, even if the maximum degree of G is bounded by a constant depending on $|V(H)|$ [88]. We are going to use as H the clique with g vertices to prove the NP-hardness of both optimal ring grooming and optimal restricted ring grooming for any fixed $g \ge 3$, even if all sets are bounded by a constant depending on g.

Given a graph G with $|V(G)| = rg$, construct an instance of the Ring Grooming as follows: the universe $\{0, 1, \cdots, n-1\}$ is $E(G)$, and the sets $\{A_v \mid v \in V(G)\}$ are defined by

$$A_v = \{e \in E(G) \mid e \text{ incident to } v\},$$

for all $v \in V(G)$. For any $U \subseteq V(G)$, let $G[U]$ denote the subgraph of G induced by U, and A_U denote the group of sets $\{A_v \mid v \in U\}$. Then the total savings of the group A_U is $|E(G[U])|$ since each edge of $G[U]$ gives a savings of one. Notice that

$$|E(G[U])| \leq \frac{|U|(|U|-1)}{2},$$

and the equality holds if and only $G[U]$ is a clique.

Consider any partition of $V(G)$ into k disjoint sets

$$\pi = \{V_1, V_2, \cdots, V_k\}$$

with $|V_i| \leq g$ for all $1 \leq i \leq k$. Then the total savings of π is at most

$$\sum_{i=1}^{k} \frac{|V_i|(|V_i|-1)}{2} = \frac{1}{2}\sum_{i=1}^{k} |V_i|^2 - \frac{1}{2}|V(G)|,$$

and it is achievable if and only if for all $1 \leq i \leq k$ the graph $G[V_i]$ is a clique. As $|V_i| \leq g$ for all $1 \leq i \leq k$, it is easy to verify that

$$\sum_{i=1}^{k} |V_i|^2 \leq rg^2,$$

and the equality holds if and only if $k = r$ and $|V_i| = g$ for all $1 \leq i \leq k = r$. Thus the total savings of π is $\frac{g(g-1)}{2}r$ if and only if $k = r$ and for all $1 \leq i \leq k$ the graph $G[V_i]$ is a clique of size g. It follows that the Ring Grooming instance has a solution of cost at most $|E(G)| - \frac{g(g-1)}{2}r$ if and only if there is a partition of $V(G)$ into r disjoint sets $V_1, V_2, \cdots, V_r$ such that for all $1 \leq i \leq r$ the subgraph of G induced by V_i is isomorphic to H. This completes the proof. ■

Because the NP-hardness of the Ring Grooming when $g > 2$, we will seek polynomial-time approximation algorithms. In the next two sections, we develop some graph-theoretical results which will be used later for the design and analysis of our approximation algorithms.

9.2 EDGE-CONNECTIVITY

In this section, we develop some graph-theoretic results on edge connectivities, which play critical roles in the later analysis of the approximation algorithms.

For any graph G, we use $V(G)$ and $E(G)$ to denote the vertex set and edge set respectively. For any $v \in V(G)$, its degree is denoted by

$\deg_G(v)$. Although we are interested mainly on loopless multigraph, we remark on that a loop contributes two to the degree of the node it is adjacent to. For any $U \subseteq V(G)$, we use $E_G(U)$ to denote the set of edges with both endpoints in U, and $\Delta_G(U)$ denote the *cut* between U and $V(G) \backslash U$, i.e., the set of edges with exactly one endpoint in U. A cut containing c edges is referred to as a c-cut. For any two vertices u and v of G, the *connectivity* between u and v, denote by $c_G(u,v)$ is the maximum number of edge-disjoint paths in G from u to v. The subscript G may be omitted if it can be understood from the context. Note that removing all loops does not affect any cut or pairwise edge-connectivity. It is well known that

$$|E_G(U)| = \frac{1}{2}\left(\sum_{u\in U} \deg_G(u) - |\Delta_G(U)|\right), \forall U \subseteq V$$

$$c_G(u,v) = \min_{U\subseteq V(G), u\in U, v\notin U} |\Delta_G(U)|.$$

Thus if the degree of each node in G is even, so are all cuts and all pairwise edge connectivities. Given any positive integer c, we can partition V into classes such that for any two vertices in the same class, their connectivity is at least c in G. We call one class a *c-cluster*. An *odd* (*even*) c-cluster has an odd (even respectively) number of vertices.

Consider any two edges $e_1, e_2 \in E$. Let v_1 and v_2 be the two endpoints of e_1, and v_3 and v_4 be the two endpoints of e_2. Let e_1', e_2', e_1'' and e_2'' be four new edges between v_1 and v_3, v_2 and v_4, v_1 and v_4, and v_2 and v_3 respectively. A *switch* of e_1 and e_2 refers to the replacing of e_1 and e_2 either by e_1' and e_2', or by e_1'' and e_2'' (see Figure 10.20). A switch is said to be *permissible* if no pairwise edge connectivity drops after the switch.

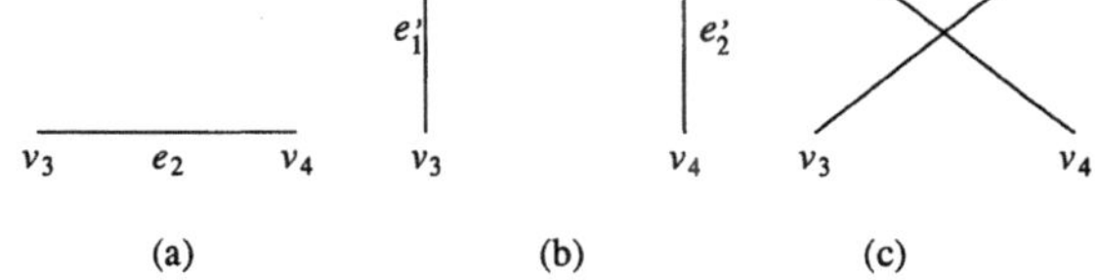

Figure 10.20. The two possible switches of two edges.

LEMMA 10.41 *Let $G = (V, E)$ be any multigraph in which all nodes have even degrees. Then any pair of its edges allows a permissible switch.*

Proof. We prove the lemma by contradiction. Assume that there exist two edges $e_1, e_2 \in E$ such that after either switch, some pairwise edge connectivity drops. Let v_i for $1 \le i \le 4$, e'_i and e''_i for $1 \le i \le 2$ be the same as above. Let G' (G'') be the graph obtained from G by replacing e_1 and e_2 by e'_1 and e'_2 (e''_1 and e''_2 respectively). Note that in both G' and G'' all nodes have even degree.

We first claim that all v_i's must be distinct. In fact, first of all, e_1 and e_2 are not adjacent to any common vertex; for otherwise either G' or G'' or both are isomorphic to G and thus the lemma is true. Second, e_1 and e_2 can not both be loops; for otherwise both G' and G'' would contain the subgraph $G - \{e_1, e_2\}$ which preserves the pairwise edge-connectivities of G. At last, neither of e_1 and e_2 can be a loop; for otherwise, say e_1 is a loop and e_2 is not, replacing e_2 by the concatenation of e'_1 and e'_2 (or e''_1 and e''_2 in G'') in any path going through e_2 will at least maintain the same number of edge-disjoint paths between any pair of nodes.

From the set of pairs of vertices u, v with $\min(c_{G'}(u,v), c_{G''}(u,v)) < c_G(u,v)$, pick one with minimum $c_G(u,v)$. Assume without loss of generality that $c_{G'}(u,v) \le c_{G''}(u,v)$. Then, since in both G and G' pairwise edge-connectivity is even, we have:

$$c_{G'}(u,v) = c_G(u,v) - 2.$$

We first observe that the edge connectivity between u and v in the graph $G - \{e_1, e_2\}$ is exactly is $c_G(u,v) - 2$, since on one hand it is at least $c_G(u,v) - 2$, and on the other hand it is at most $c_{G'}(u,v)$. Let U be any subset of V such that $u \in U$, $v \notin U$, and exactly $c_G(u,v) - 2$ edges across between U and $V - U$ in $G - \{e_1, e_2\}$. Then both e_1 and e_2 are in $\Delta_G(U)$; for otherwise $|\Delta_G(U)|$ is at most $c_G(u,v) - 1$, which is impossible, as it would contradict the max-flow min-cut theorem. Let e_k for $1 \le k \le c_G(u,v)$ denote the $c_G(u,v)$ edges in $\Delta_G(U)$. Consider $p_1, p_2, \dots, p_{c_G(u,v)}$, $c_G(u,v)$ edge-disjoint paths from u to v in G. Then each path p_k goes through a unique edge in $\Delta_G(U)$. By proper relabeling, assume that path p_k passes through edge e_k for $1 \le k \le c_G(u,v)$. The edge e_k breaks the path p_k into three parts: the subpath p'_k in U, e_k itself, and the subpath p''_k in $V - U$, as illustrated in Figure 10.21.

Without loss of generality, assume that $v_1 \in U$. Then $v_2 \notin U$. We prove by contradiction that $v_3 \in U$ and thus $v_4 \notin U$. Suppose to the contrary. Then there are still $c_G(u,v)$ edge-disjoint paths between u and v in G': one is p'_1, followed by e'_1 and then p''_2; the second is p'_2, followed by e'_2 and then p''_1; and the other $c_G(u,v) - 2$ paths are p_k for $3 \le k \le c_G(u,v)$ (see Figure 10.22). All these paths are edge-disjoint, so $c_{G'}(u,v) = c_G(u,v)$, and this is a contradiction.

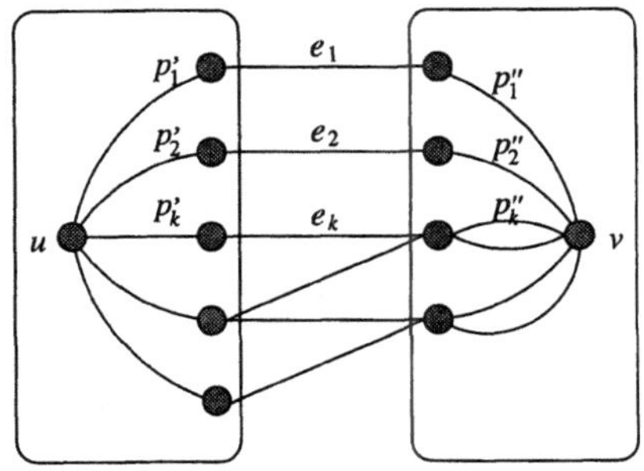

Figure 10.21. The breaking-up of the edge-disjoint paths.

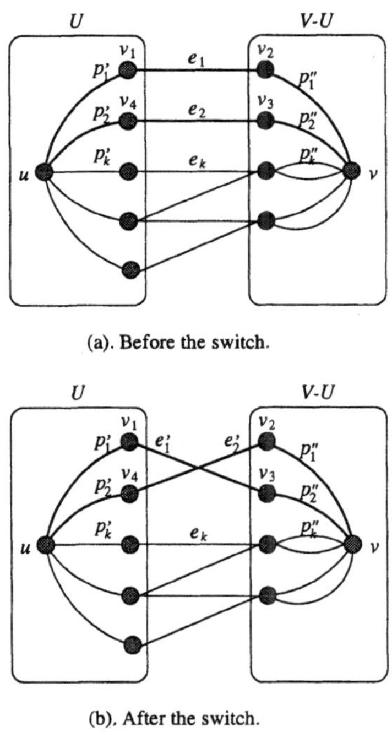

Figure 10.22. Replacing e_1 and e_2 by e_1' and e_2' does not decrease the edge-connectivity between u and v.

We will prove by contradiction that in G'' any pairwise edge connectivity does not decrease compared to G. Suppose to the contrary that there exist two vertices u' and v' with $c_{G''}(u', v') < c_G(u', v')$. From the selection of u and v, we have

$$c_G(u',v') \geq c_G(u,v).$$

Let C_u and C_v denote the connected components of u and v, respectively, in $G - \Delta_G(U)$. Then both u' and v' must be in $C_u \cup C_v$, for otherwise their connectivity is not affected. Without loss of generality, assume that $u' \in C_u$. If $v' \in C_v$, then $c_G(u',v')$ is at most $c_G(u,v)$ and thus must be equal to $c_G(u,v)$. This implies that for any set of $c_G(u',v')$ edge-disjoint paths between u' and v' in G, each of them passes through one unique edge in $\Delta_G(U)$. Following the similar argument as in Figure 10.22, we can show that $c_{G''}(u',v') = c_G(u',v')$, which is a contradiction. So v' must be in C_u as well. We first observe that among any set of $c_G(u',v')$ edge-disjoint paths between u' and v' in G, exactly two pass through e_1 and e_2 respectively, for otherwise $c_{G''}(u',v')$ would be at most one less than $c_G(u',v')$, hence be equal to $c_G(u',v')$ exactly. Second, we observe that for any set of $c_G(u',v')$ edge-disjoint paths between u' and v' in G, those paths across between C_u and C_v can be converted such that the part of each these paths within C_v is a concatenation of some unique pair of p''_i and p''_j (in which cases all these paths within C_v pass trough v - see Figure 10.23). Consider any such set of $c_G(u',v')$ edge-disjoint paths between u' and v' in G, and let q_1 and q_2 be the two paths among them which pass through e_1 and e_2 respectively. If both e_1 and e_2 are between u' and v in q_1 and q_2 respectively, once again the similar argument as in Figure 10.22 would lead to the conclusion that $c_{G''}(u',v') = c_G(u',v')$ (see Figure 10.24). So without loss of generality, assume that e_1 is between u' and v in $q_1$and e_2 is between v' and v in q_2. However, in this case we can construct a new path between u' and v' which does not pass through e_1 or e_2 as follows: the subpath of q_2 between u' and v, followed by the subpath of q_1 between v' and v (see Figure 10.25). This new path is edge-disjoint from all the other $C_G(u',v') - 2$ paths between u' and v'. Thus $c_{G''}(u',v')$ is at least $c_G(u',v') - 1$, and therefore must be equal to $c_G(u',v')$, again a contradiction. This completes the proof.

■

We now describe the cactus representation of all minimum cuts of a graph, introduced by Dinits, Karzanov, and Lomonosov [49]. In the literature, a cactus is a *simple* graph with the property that any two-*vertex*-connected component is an induced cycle (in other words, no cycle has chords). To clarify notation, a cycle cannot repeat vertices, but can consist of exactly two parallel edges in a graph. We use a slightly modified version: a *cactus* is a multigraph such that each edge appears in exactly one cycle. After removing one edge from each couple of parallel

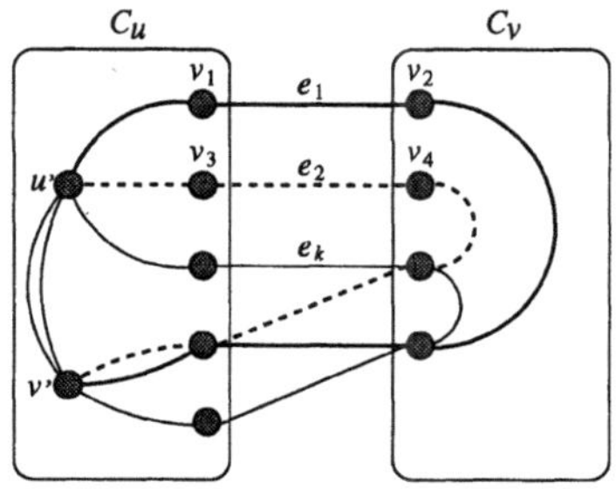

(a). Before the replacement.

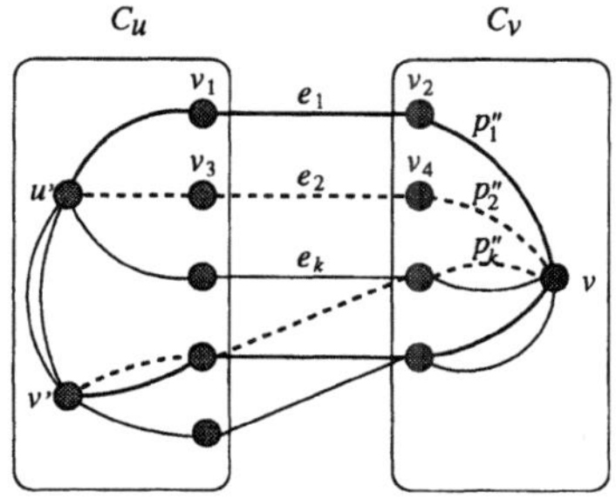

(b). After the replacement.

Figure 10.23. Replacing each subpath within C_v by a concatenation of some unique pair of p_i'' and p_j''.

edges from a cactus as defined by us, one obtains a cactus as defined in the literature. See Figure 10.26 for the comparison. Two important properties of a cactus is that any two cycles are either disjoint or touching at one vertex, and any pair of vertices in a connected component have exactly two edge-disjoint paths in between.

Suppose that G is a two-*edge*-connected multigraph. Its *reduced representation*, denoted by G^*, is defined as follows: the vertices of G^* are the 3-clusters of G; each edge of G whose endpoints are in different 3-clusters is added to the edge set of G^*. G^* can have parallel edges. From [49] we know that the reduced representation of G^* is a cactus, and it is called in the literature the cactus representation. In addition, all the cuts of size two of G correspond to all the cuts of size two of G^*. As the original reference is in Russian, and we only need a special case, we provide a proof for the sake of completeness.

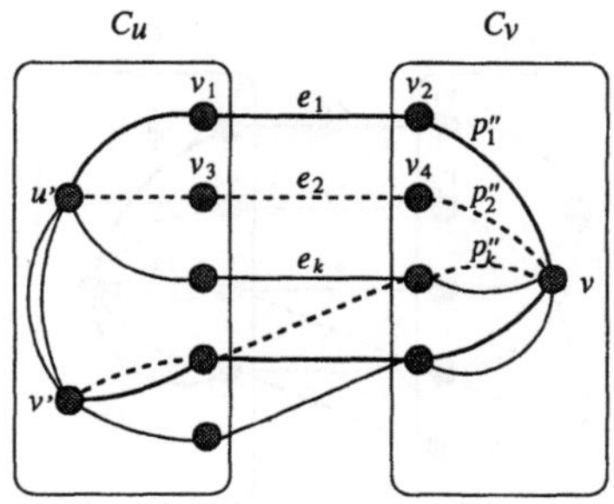

(a). Before the switch.

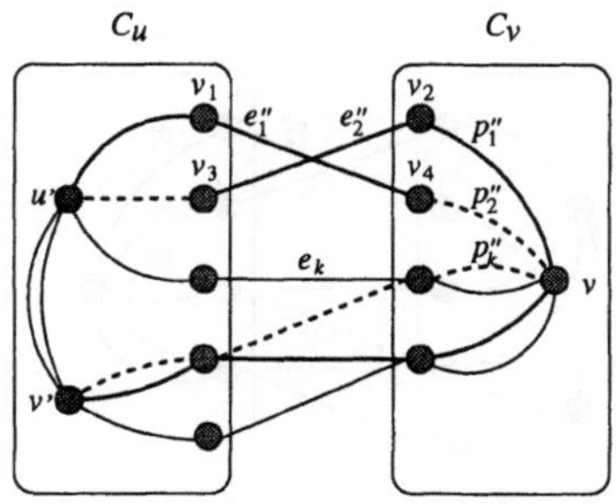

(b). After the switch.

Figure 10.24. If both e_1 and e_2 are between u' and v, replacing e_1 and e_2 by e_1'' and e_2'' also does not decrease the edge-connectivity between u' and v'.

THEOREM 10.42 *[49] Suppose that G is a two-edge-connected multigraph. Then its reduced representation G^* is a cactus, and the 2-cuts of G correspond to the 2-cuts of G^*.*

Proof. Let $(U, \overline{U})$ be a 2-cut of G. A 3-cluster of G is either contained in U or disjoint from U. Therefore $(U, \overline{U})$ induces a partition of $V(G^*)$ and this partition defines a 2-cut of G^*. From a 2-cut of G^* it is immediate to obtain a 2-cut of G: a partition of $V(G^*)$ induces a partition of $V(G)$.

A 1-cut of G^* does not exist since it will induce a 1-cut of G. Therefore each edge of G^* is in a cycle. Assume by contradiction that there are two distinct cycles C_1 and C_2 in G^* which contain a common edge. Walking along C_1 from the common edge in an arbitrary direction, let c be the first vertex from which C_1 and C_2 branch, and c' be the first vertex after c where C_1 and C_2 meet again. Note that c' is different from c. Then, in G^*, we have three edge disjoint paths from c to c': two on C_2 and one on C_1 that is in the previous walk from c to c' (see Figure 10.27). Let u be a vertex of G in the 3-cluster represented by c and v be a vertex of

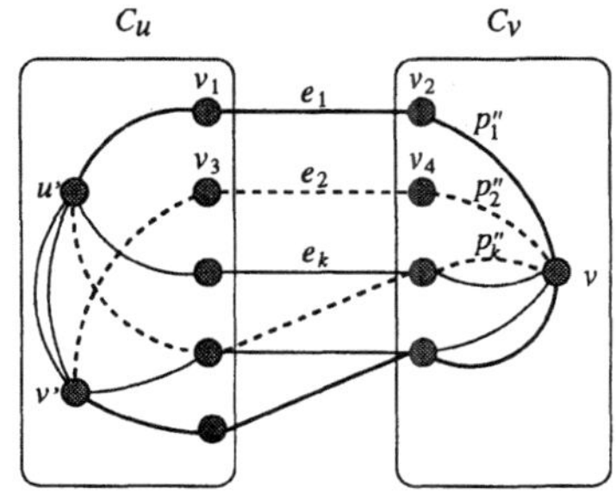

(a). Before the switch.

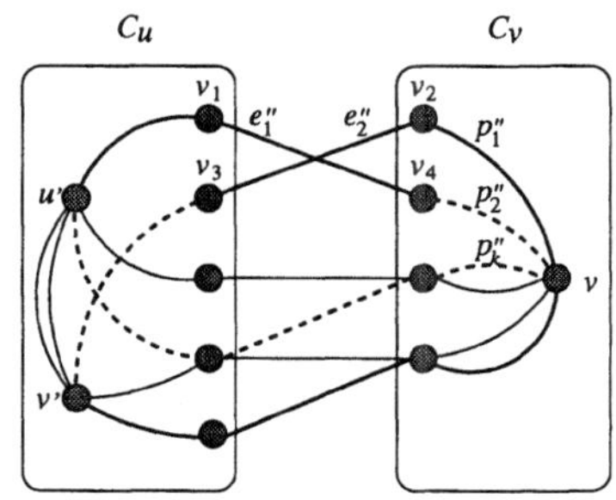

(b). After the switch.

Figure 10.25. If e_1 is between u' and v and e_2 between v' and v, replacing e_1 and e_2 by e_1'' and e_2'' also does not decrease the edge-connectivity between u' and v'.

G in the 3-cluster represented by c'. There is no 2-cut in G separating u from v, since there is no 2-cut in G^* separating c from c'. Therefore $c_G(u,v) > 2$. We obtained a contradiction to the fact that u and v are in different 3-clusters. In conclusion, no edge of G^* appears in two distinct cycles of G^*, and therefore G^* is a cactus. ■

Next, we investigate what happens to the cactus representation when we validly switch two edges. For any vertex $v \in V$, let v^* be the 3-cluster contains v. Let e be any edge of G with endpoints u and v. If $u^* = v^*$, i.e., u and v are in the same 3-cluster, then e^* is not defined, otherwise, let e^* denote the edge of G^* obtained from e. We say e is *related* to a 3-cluster c if either e^* is not defined and $u^* = v^* = c$, or e^* is defined and c is on the unique cycle of the G^* containing e^*. Then any edge of G is related to some vertex of the cactus. The following lemma shows that under certain circumstances the switch leads to an "improved" G^*:

LEMMA 10.43 *Suppose that G is a two-edge-connected multigraph and it has two edges e_1 and e_2 which allow a permissible switch. If there is*

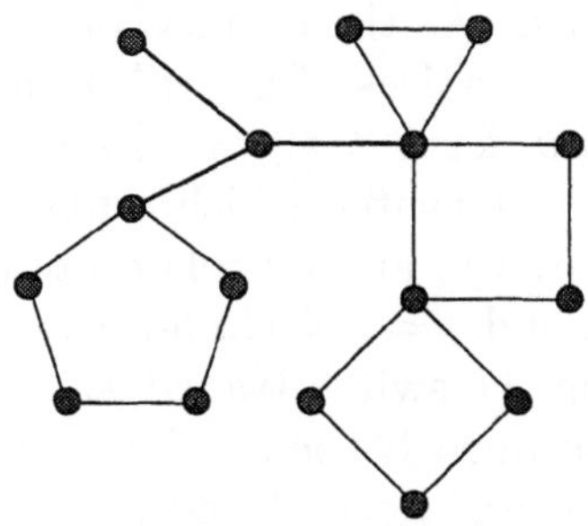

(a). Cactus defined in the literature.

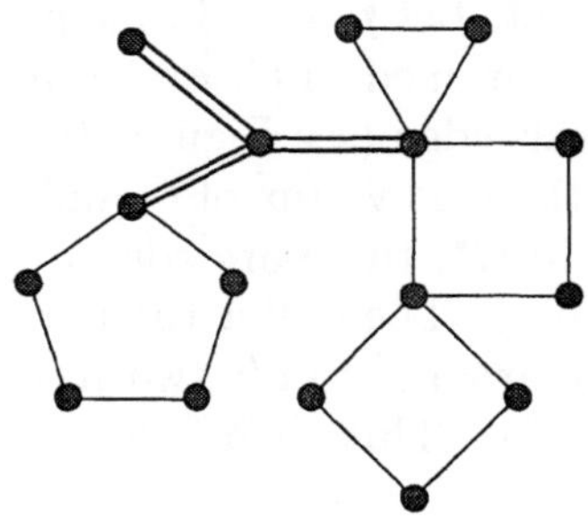

(b). Cactus defined by us.

Figure 10.26. The comparison between the cactus defined inthe literature and the cactus defined by us.

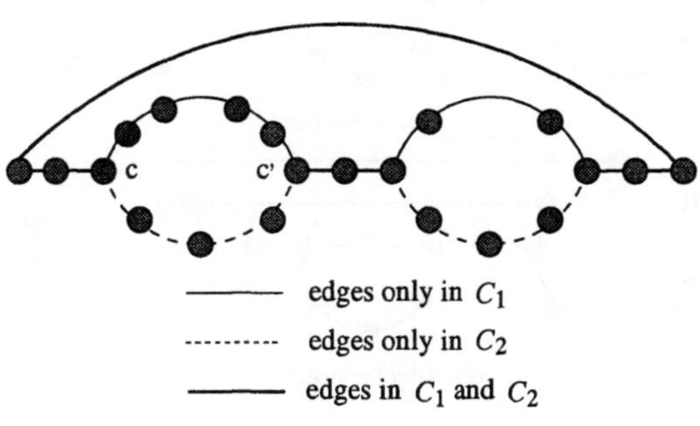

Figure 10.27. Three edge-disjoint paths between c and c'.

no vertex of G^* to which e_1 and e_2 are both related, then any permissible switch of e_1 and e_2 either increases the number of the edges of G^*, or decreases the number of vertices of G^*.

Proof. Assume by contradiction that after a permissible switch, neither does the number of vertices of G^* decrease, nor does the number of the edges of G^* increase. As the permissible switch can not split any 3-clusters, the number of vertices of G^* will remain the same after the permissible switch. Let v_i for $1 \leq i \leq 4$, e'_i and e''_i for $1 \leq i \leq 2$ be as above, and assume that the switch which replaces e_1 and e_2 by e'_1 and e'_2 is permissible. As e_1 and e_2 are not related to any common 3-cluster, $v_i^* \neq v_j^*$ for any $i = 1, 2$ and $j = 3, 4$. In addition, $v_1^* \neq v_2^*$ and $v_3^* \neq v_4^*$; for otherwise after doing the switch two edges are added while at most one is removed, and consequently the number of edges of G^* is increased. Thus all v_i^*'s are distinct, and both e_1^* and e_2^* are edges of G^*. For $i = 1$ and 2, let C_i be the unique cycle of G^* containing e_i^*. From the hypothesis, C_1 and C_2 are disjoint. Let p be a simple path in G^* from v_1^* to v_3^*. Pick any edge in p not contained in C_1 or C_2, and let C be the unique cycle of G^* containing this edge (see Figure 10.28). As we walk along p from v_1^* to v_3^*, let c' be the first vertex of C and c'' the last vertex of C. From p we obtain that in G^*, there are edge disjoint paths p_1, from c' to v_1^*, and p_2, from c'' to v_3^*. Then after the permissible switch we have three edge-disjoint paths from c' to c'': two in C, and one obtained by concatenating p_1, $e_1'^*$ and p_2. This means that G^* is no longer a cactus, a contradiction. ■

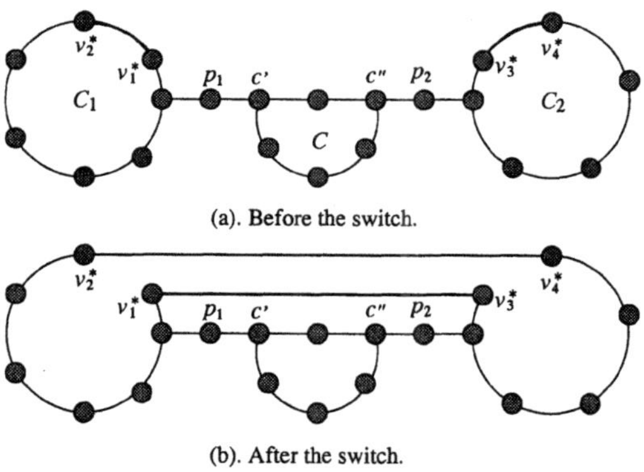

Figure 10.28. After the permissible switch, there are three edge-disjoint paths from c' to c''.

It can be easily verified that if a set of edges are *pairwise* related to a common 3-cluster, then they are *all* related to a common 3-cluster. Consequently, the vertices of these edges are either in clusters on a single cycle of G^*, or in clusters on cycles of G^* containing a common vertex. After applying a finite number of times permissible switches in two-edge-connected graph G as in the above lemma, we eventually reach the situation in which, any pair of edges in G which allow a permissible switch are related to a common vertex of G^*.

9.3 INSIDE MATCHINGS AND VALID MATCHINGS

Let $G = (V, E)$ be any loopless graph. A *fractional matching* is a function x: $E \longrightarrow Q$ satisfying that for any $v \in V$,

$$\sum_{e:e \text{ incident to } v} x(e) \leq 1.$$

If the equality holds for $v \in V$, then x is said to be perfect. A fractional matching x is a normal matching if $x(e) \in \{0, 1\}$ for any $e \in E$. Let w be any edge-weight function. The weight of a fractional matching x is defined as

$$w(x) = \sum_{e \in E} w(e)\, x(e).$$

The following is a well-known theorem of Edmonds (for a reference, see [69], pages 201-206) regarding fractional matchings.

THEOREM 10.44 *(Edmonds) If the fractional matching x satisfies:*

$$\sum_{e \in E(U)} x(e) \leq \left\lfloor \frac{|U|}{2} \right\rfloor, \quad \forall U \subseteq V$$

then it can be converted to a normal matching of at least the same weight.

In the remaining of this section, all matchings are meant to be normal and perfect unless otherwise specified explicitly.

Now we consider a *complete* weighted graph K_m with even m. The number of perfect matchings in K_m is

$$(m-1)!! = (m-1) \cdot (m-3) \cdot \cdots \cdot 3 \cdot 1.$$

As each edge is contained in $(m-3)!!$ different perfect matchings, the total weight of all these perfect matchings is $(m-3)!! \sum_{e \in E} w(e)$. Another useful fact is that for any perfect matching x in K_m, there are exactly $(m-2)!!$ different normal perfect matchings in K_m, each of which together with x form a cycle containing of all nodes in K_m. Let $\pi = \{V_1, V_2, \cdots, V_k\}$ be a partition of $V(K_m)$. For any $v \in V(K_m)$, let $\pi(v)$ be the unique value i with $v \in V_i$.

An π-*inside matching* of K_m is a perfect matching of K_m satisfying that the endpoints of each edge of the matching are within the same group of π. For an example, see Figure 10.29. A necessary and sufficient condition for the existence of a π-inside perfect matching is that the size of any group is even. Notice that any inside matching can be generated by picking up a perfect matching of each group and put them together. Thus the total number of inside matchings is $\prod_{i=1}^{k}(|V_i|-1)!!$.

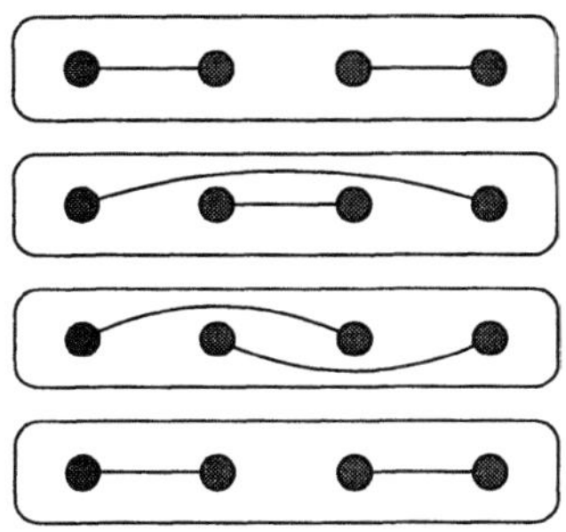

Figure 10.29. The dots represent the vertices, and the partition π is given by the rectangles. The matching in the figure is a π-inside matching, as each edge has both endpoints in the same group of π.

A matching x is said to be π-*valid*, or simply *valid* if π is understood from the context, if the endpoints of each edge of x are within different groups of π. For an example, see Figure 10.30. A necessary condition for the existence of a π-valid perfect matching is that the size of any group is at most $\frac{m}{2}$. This also turns to be a sufficient condition, which is proved below.

Let $|V_i| = d_i$ for $1 \leq i \leq k$. We use $f(d_1, d_2, \cdots, d_k)$ to denote the number of π-valid matchings. We observe that merging any two groups into one group will not increase the number of valid matchings. So we only need to consider the case when the total size of any two groups is greater than $\frac{m}{2}$. Then k must be at most three. If $k = 2$, there are exactly $\left(\frac{m}{2}\right)!$ π-valid perfect matchings. So we assume that $k = 3$. In this case, any π-valid perfect matching can be generated as follows. Split

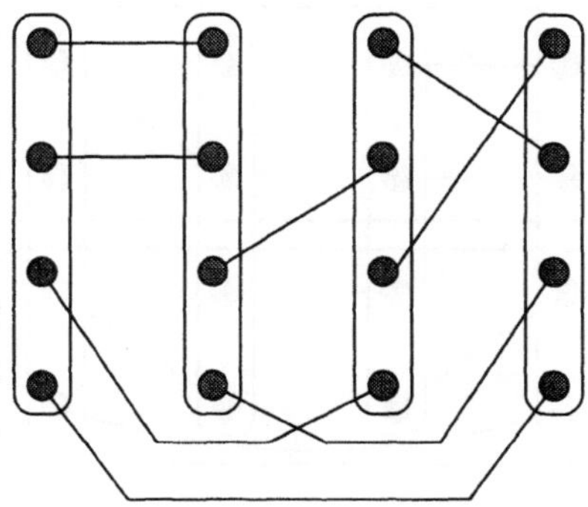

Figure 10.30. The dots represent the vertices and the partition π is given by the rectangles. The matching in the figure is a π-valid matching, as no edge has both endpoints in the same group of π.

V_1 into two parts V_{12} and V_{13}, V_2 into two parts V_{21} and V_{23}, V_3 into two parts V_{31} and V_{32} such that

$$|V_{12}| = |V_{21}|, |V_{13}| = |V_{31}|, |V_{23}| = |V_{32}|.$$

This is always possible and indeed the split must satisfy that

$$\begin{aligned} |V_{12}| = |V_{21}| &= \frac{m}{2} - d_3, \\ |V_{13}| = |V_{31}| &= \frac{m}{2} - d_2, \\ |V_{23}| = |V_{32}| &= \frac{m}{2} - d_1. \end{aligned}$$

Then for any pair of i and j, generate a bipartite matching between V_{ij} and V_{ji}. Combine the three generated bipartite matchings to obtain a π-valid perfect matching. Thus a π-valid perfect matching always exists.

Now consider two partitions of $V(K_m)$:

$$\begin{aligned} \pi &= \{V_1, V_2, \cdots, V_k\}, \\ \sigma &= \{U_1, U_2, \cdots, U_l\}. \end{aligned}$$

Notice that partition π induces a partition $\pi|_{U_j}$ of U_j. If $|U_j|$ is even and is at least $2|V_i|$, for $1 \leq i \leq k$, then there is a $\pi|_{U_j}$-valid perfect matching of U_j. Therefore if for all $1 \leq j \leq l$, $|U_j|$ is even and is at least $2|V_i|$ for $1 \leq i \leq k$, then there exists a π-valid σ-inside matching. Any such matching can be generated by picking up a $\pi|_{U_j}$-valid perfect matching of U_j and combining them together. For an example, see Figure 10.31.

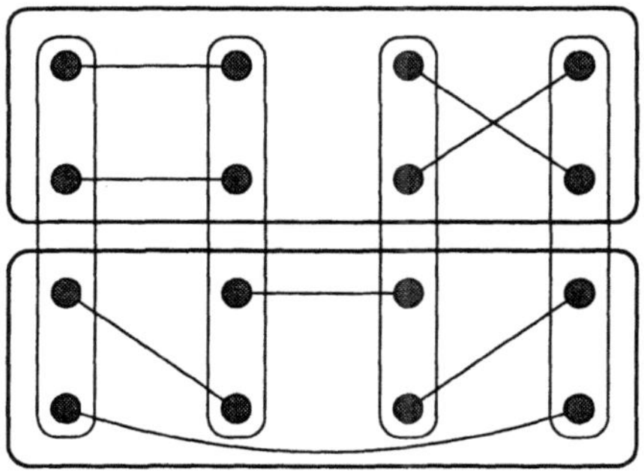

Figure 10.31. The dots represent the vertices, the partition π is given by the vertical rectangles, and the partition σ is given by the horizontal think solid rectangles. The matching in the figure is a π-valid σ-inside matching, as no edge has both endpoints in the same group of π, but every edge has both endpoints in the same group of σ.

Let x be any perfect matching of K_m. We construct a weighted multigraph $H(\pi, x)$ as follows: the vertex set is $\{1, 2, \cdots, k\}$; for any edge $e = (u, v)$ in x, add an edge between $\pi(u)$ and $\pi(v)$ in $H(\pi, x)$ and set its weight equal to $w(e)$. Note that if $|V_i| = d$ for all $1 \leq i \leq k$, then $H(\pi, x)$ is d-regular. If x is π-valid, then the graph $H(\pi, x)$ has no loop.

LEMMA 10.45 *Let $\pi = \{V_1, V_2, \cdots, V_k\}$ be a partition of $V(K_m)$ with $|V_i| = d$ for all $1 \leq i \leq k$. Let x be a π-valid perfect matching of K_m. Assume that for any odd set $U \subseteq V(H(\pi, x))$, $\left|\Delta_{H(\pi,x)}(U\right| \geq d$. Then there is a matching in $H(\pi, x)$ of weight at least $\frac{1}{d} w(x)$.*

Proof. Note that $H(\pi, x)$ is d-regular and loopless. Let y be the fractional perfect matching of $H(\pi, x)$ given by $y(e) = \frac{1}{d}$ for any edge

e. Then the weight of y is $\frac{1}{d}w(x)$. For any set $U \subseteq V(H(\pi, x))$,

$$\begin{aligned}
&\sum_{e \in E_{H(\pi,x)}(U)} y(e) \\
&= \frac{1}{d}|E_{H(\pi,x)}(U)| \\
&= \frac{1}{2d}\left(\sum_{u \in U} \deg_{H(\pi,x)}(u) - |\Delta_{H(\pi,x)}(U)|\right) \\
&= \frac{1}{2d}(d|U| - |\Delta_{H(\pi,x)}(U)|) \\
&= \frac{|U|}{2} - \frac{1}{2d}|\Delta_{H(\pi,x)}(U)|.
\end{aligned}$$

Thus if U is odd,

$$\sum_{e \in E_{H(\pi,x)}(U)} y(e) \leq \frac{|U|}{2} - \frac{1}{2d}d = \frac{|U|-1}{2} = \left\lfloor \frac{|U|}{2} \right\rfloor.$$

The lemma then follows from Theorem 10.44. ■

Let $\pi = \{V_1, V_2, \cdots, V_k\}$ be a partition of $V(K_m)$with $|V_i| = d$ for all $1 \leq i \leq k$. Let x be a π-valid perfect matching of K_m. Let (v_1, v_2) and (v_3, v_4) be two edges of x. Let x' (x'') be the matching obtained from x by replacing these two edges with (v_1, v_3) and (v_2, v_4) ((v_1, v_4) and (v_2, v_3) respectively). If $\pi(v_1) = \pi(v_3)$ or $\pi(v_2) = \pi(v_4)$, then x'' is π-valid and $H(\pi, x'')$ is isomorphic to $H(\pi, x)$. Similarly, if $\pi(v_1) = \pi(v_4)$ or $\pi(v_2) = \pi(v_3)$, then x' is π-valid and $H(\pi, x')$ is isomorphic to $H(\pi, x)$. If the four $\pi(v_i)$'s are all distinct, then both x' and x'' are π-valid and from Lemma 10.41, either $H(\pi, x')$ or $H(\pi, x'')$ does not decrease any pairwise edge connectivity compared to $H(\pi, x)$. Thus in either case, either x' is π-valid and $H(\pi, x')$ does not decrease any pairwise edge connectivity compared to $H(\pi, x)$, and/or x'' is π-valid and $H(\pi, x'')$ does not decrease any pairwise edge connectivity compared to $H(\pi, x)$. If x' (x'' respectively) is π-valid and $H(\pi, x')$ ($H(\pi, x'')$ respectively) does not decrease any pairwise edge connectivity compared to $H(\pi, x)$, we call the replacing of (v_1, v_2) and (v_3, v_4) with (v_1, v_3) and (v_2, v_4) ((v_1, v_4) and (v_2, v_3) respectively) as a *permissible switch* of (v_1, v_2) and (v_3, v_4). Then any pair of edges in a π-valid matching allows a permissible switch.

9.4 THE ITERATIVE MATCHING ALGORITHM

We define a weighted complete graph K_m with vertex set $\{1, 2, \cdots, m\}$ by setting the weight of any edge $e = (i, j)$ equal to

$$w(e) = |A_i \cap A_j|.$$

Let $\pi = \{V_1, V_2, \cdots, V_k\}$ be a partition of $V = \{1, 2, \cdots, m\}$. The *intersection graph* of π, denoted by $K(\pi)$, is a weighted complete graph defined as follows: its vertex set is $\{1, 2, \cdots, k\}$, for any $e = (i, j)$ its weight is

$$\left| \left(\bigcup_{r \in V_i} A_r \right) \cap \left(\bigcup_{r \in V_j} A_r \right) \right|.$$

In this section, we assume that g is a power of two and $m \bmod g = 0$. The iterative matching produces a solution to Restricted Ring Grooming. An initial partition π_0 consists of m groups, each one with one index. The algorithm runs in $\log g$ iterations. For any $1 \leq i \leq \log g$, the i-th iteration finds a maximum-weighted perfect matching of $K(\pi_{i-1})$, and then construct a new partition π_i as follows: for each edge (j_1, j_2) of the obtained matching, a group of π_i is generated by merging the j_1-th group and the j_1-th group. Note that each π_i is a 2^i-grouping, and the partition $\pi_{\log g}$ is a solution to Restricted Ring Grooming.

The analysis of the iterative matching algorithm is based on savings. Let S_i be the weight of any maximum-weighted perfect matching of $K(\pi_{i-1})$. S_i is referred to as the savings of the i-th iteration. Then it is easy to verify that the total savings of $\pi_{\log g}$ is $\sum_{j=1}^{\log g} S_j$. Recall that the cost of the output is $\sum_{i=1}^{m} |A_i| - \sum_{j=1}^{\log g} S_j$.

Fix an optimal restricted ring grooming OPT_g which consists of $\frac{m}{g}$ groups $D_1, \cdots, D_{\frac{m}{g}}$ where for any $1 \leq k \leq \frac{m}{g}$,

$$D_k = \{k_i \mid 1 \leq i \leq g\}.$$

Let opt_j denote the minimum cost of a grooming of $A_1, A_2, \ldots, A_m$ with grooming granularity j. One trivial upper bound on the approximation ratio of the iterative matching is $\frac{g}{2}$. In fact, the cost of the iterative matching is

$$\begin{aligned}
&\sum_{r=1}^{m}|A_r| - \sum_{j=1}^{\log g} S_j \\
&= \left(\sum_{r=1}^{m}|A_r| - S_1\right) - \sum_{j=2}^{\log g} S_j \\
&= opt_2 - \sum_{j=2}^{\log g} S_j \le opt_2 \\
&\le \sum_{k=1}^{\frac{m}{g}} \sum_{j=1}^{\frac{g}{2}} \left|A_{k_{2j-1}} \cup A_{k_{2j}}\right| \\
&\le \frac{g}{2} \cdot \sum_{k=1}^{\frac{m}{g}} \left|\bigcup_{i=1}^{g} A_{k_i}\right| = \frac{g}{2} \cdot opt_g.
\end{aligned}$$

In the remaining of this section, we will provide a tighter analysis of the approximation ratio of the iterative matching.

Any OPT_g-inside perfect matching is simply referred to as an *inside matching*, i.e., the term inside is always with respect to the optimal partition OPT_g. From the discussion in Subsection 9.3, the total number of inside matchings is $((g-1)!!)^{\frac{m}{g}}$. We are particularly interested in the following $(g-1)!!$ inside matchings. For any $1 \le k \le \frac{m}{g}$, arrange the $(g-1)!!$ perfect matchings inside D_k in the *non-decreasing* order of their weights and let $M_j(D_k)$ denote the weight of the j-th perfect matching. For $1 \le j \le (g-1)!!$, let $M_j = \sum_{k=1}^{\frac{m}{g}} M_j(D_k)$. The sum of these $(g-1)!!$ weights is

$$(g-3)!! \sum_{k=1}^{\frac{m}{g}} \sum_{1 \le i < j \le g} \left|A_{k_i} \cap A_{k_j}\right|.$$

For any $i \ge 1$, let $\ell(i) = \left(\frac{g}{2^{i-1}} - 1\right) \cdot (g-3)!!$. The next lemma is the key to our analysis of the performance of the iterative matching.

LEMMA 10.46 *If* $S_i \ge \frac{M_{\ell(i)}}{2^{i-1}}$ *for any* $1 \le i \le \log g$, *then the approximation ratio of the iterative matching is at most* $\frac{g}{4} + \frac{1}{2}$.

Proof. Suppose that $S_i \geq \frac{M_{\ell(i)}}{2^{i-1}}$ for any $1 \leq i \leq \log g$. Since

$$\ell(i) - \ell(i+1) = \frac{g \cdot (g-3)!!}{2^i},$$

the total savings of $\pi_{\log g}$ is

$$\begin{aligned}
\sum_{i=1}^{\log g} S_i \geq \sum_{i=1}^{\log g} \frac{M_{\ell(i)}}{2^{i-1}} &= \frac{2}{g \cdot (g-3)!!} \sum_{i=1}^{\log g} \frac{g \cdot (g-3)!!}{2^i} M_{\ell(i)} \\
&\geq \frac{2}{g \cdot (g-3)!!} \sum_{i=1}^{\log g} \sum_{j=\ell(i+1)+1}^{\ell(i)} M_j \\
&= \frac{2}{g \cdot (g-3)!!} \sum_{j=1}^{(g-1)!!} M_j \\
&= \frac{2}{g} \sum_{k=1}^{\frac{m}{g}} \sum_{1 \leq i < j \leq g} \left| A_{k_i} \cap A_{k_j} \right|.
\end{aligned}$$

So the total cost of $\pi_{\log g}$ is at most

$$\begin{aligned}
&\sum_{i=1}^{m} |A_i| - \frac{2}{g} \sum_{k=1}^{\frac{m}{g}} \sum_{1 \leq i < j \leq g} |A_{k_i} \cap A_{k_j}| \\
&= \sum_{k=1}^{\frac{m}{g}} \left(\sum_{i=1}^{g} |A_{k_i}| - \frac{2}{g} \sum_{1 \leq i < j \leq g} |A_{k_i} \cap A_{k_j}| \right) \\
&\leq \left(\frac{g}{4} + \frac{1}{2} \right) \sum_{k=1}^{\frac{m}{g}} \left| \bigcup_{i=1}^{g} A_{k_i} \right| = \left(\frac{g}{4} + \frac{1}{2} \right) \cdot opt_g.
\end{aligned}$$

The above inequality follows from this inequality:

$$\sum_{i=1}^{g} |A_{k_i}| - \frac{2}{g} \sum_{1 \leq i < j \leq g} |A_{k_i} \cap A_{k_j}| \leq \left(\frac{g}{4} + \frac{1}{2} \right) \left| \bigcup_{i=1}^{g} A_{k_i} \right|$$

which can be proved as follows. Consider any $a \in \bigcup_{i=1}^{g} A_{k_i}$. Suppose that it appears in exactly r sets among $A_{k_1}, A_{k_2}, \cdots, A_{k_g}$. Then it appears in exactly $\binom{r}{2}$ $A_{k_i} \cap A_{k_j}$'s. Thus the total contribution of a in the left-hand side is

$$r - \frac{2}{g} \binom{r}{2} \leq \frac{g}{4} + \frac{1}{2},$$

where the previous inequality is proved as follows: we have $(r - \frac{g+1}{2})^2 \geq 0$, and therefore $r^2 - (g+1)r + \frac{g^2}{4} + \frac{g}{2} + \frac{1}{4} \geq 0$. Since g is even, $r^2 - (g+1)r + \frac{g^2}{4} + \frac{g}{2}$ is an integer, and therefore nonnegative. The above inequality follows by algebraic manipulation. This completes the proof. ∎

Note that $S_1 \geq M_{\ell(1)}$ is true. In general, for any π_{i-1}-valid perfect matching x, S_i is at least the weight of any matching of $H(\pi_{i-1}, x)$. Notice that each such graph $H(\pi_{i-1}, x)$ is a 2^{i-1}-regular graph. From Lemma 10.45, for any $2 \leq i \leq \log g$, $S_i \geq \frac{M_{\ell(i)}}{2^{i-1}}$ if there is a π_{i-1}-valid perfect matching x satisfying that

- **Weight Condition**: $w(x) \geq M_{\ell(i)}$;
- **Odd Condition**: $\left|\Delta_{H(\pi_{i-1},x)}(U)\right| \geq 2^{i-1}$ for any odd set $U \subseteq V(H(\pi_{i-1}, x))$.

For the ease of argument for **Weight Condition**, we restrict our attention to inside matchings. In the next we will show the existence of the π_{i-1}-valid inside matching that meets both **Weight Condition** and **Odd Condition** when $g = 4$ and 8.

9.4.1 THE CASE $G = 4$

We only have to consider the second iteration. As any valid inside matching has weight at least M_1, the **Weight Condition** is met trivially. So we only have to consider the **Odd Condition**. For any π_1-valid inside matching x, $H(\pi_1, x)$ is 2-regular, and thus consists of a collection of disjoint cycles. As the number of edges crossing any cut of $H(\pi_1, x)$ is even, the odd condition thus translates into the fact that all cycles in $H(\pi_1, x)$ are even. By applying valid switchings of pairs of edges inside the same optimal group, we can ensure that each connected component of $H(\pi_1, x)$ contains all or none of

$$\pi_1(k_1), \pi_1(k_2), \pi_1(k_3), \pi_1(k_4)$$

for any 4-group $\{k_1, k_2, k_3, k_4\}$ of OPT_4, and therefore will have an even number of vertices. Therefore, the approximation ratio of the iteration matching is at most 1.5.

In the next, we will use an example to show that the above analysis is tight, that is the ratio of the iterative matching is exactly 1.5 when

$g = 4$. Let $G = (V, E)$ be 4-regular graph and with girth (the size of a smallest cycle) at least five, whose existence follows from [109] (Problem 10.12). Let $n = 2|V| = |E|$. Assign to every edge e of G one index from $\{0, 1, \cdots, n-1\}$, and to every vertex v of G two indices from $\{0, 1, \cdots, n-1\}$, such that no index is assigned twice. Let v be a vertex of G, such that v is assigned indices i_1 and i_4, and the four edges incident to it have assigned the indices i_2, i_3, i_5, i_6. Define four sets

$$A_{v1} = \{i_1, i_2, i_3\},$$
$$A_{v2} = \{i_3, i_4, i_5\},$$
$$A_{v3} = \{i_5, i_6, i_1\},$$
$$A_{v4} = \{i_2, i_4, i_6\}.$$

Repeat for all vertices v, obtaining $4|V|$ sets; this gives the instance of Ring Grooming. An (optimum) solution of cost $6|V|$ exists: any 4-group consists of the four sets defined by any vertex. Note that for $i \neq j$,

$$|A_{vi} \cap A_{vj}| = 1.$$

But suppose A_{v1} picks up i_2 and matches, in the first iteration, with whoever picks i_2 from the other side (i_2 is assigned to an edge). Similarly, A_{v2} picks i_3, A_{v3} picks i_5, and A_{v4} picks i_6, and each matches with an A_{ui} with $u \neq v$. The savings in this first iteration is $2|V|$: each two sets who match save exactly one. In the second iteration, consider any two 2-groups $\{A_{va}, A_{ub}\}$ and $\{A_{xc}, A_{yd}\}$, where $\{a, b, c, d\} \subseteq \{1, 2, 3, 4\}$. Note that and xy are edges in G. Let $B_i = A_{va} \cup A_{ub}$ and $B_j = A_{xc} \cup A_{yd}$. Note that $|B_i| = 5$ and specifically B_i consists of the labels assigned to three edges and one each from the labels assigned to two vertices (each vertex gets assigned two labels). Moreover, these three edges and two vertices form a path in G. B_j has the same properties. Then we will prove:

$$|B_i \cap B_j| \leq 1. \tag{10.1}$$

We have two cases. First, assume u, v, x, y are distinct vertices of G. Assume by symmetry that $A_{ub} \cap A_{xc} \neq \emptyset$ and therefore u and x are adjacent in G. As G has no cycles of length three or four, u is not adjacent to y, x is not adjacent to v, and y is not adjacent to v. In conclusion, $A_{ub} \cap A_{yd} = A_{xc} \cap A_{va} = A_{va} \cap A_{yd} = \emptyset$, and therefore $|B_i \cap B_j| = |A_{ub} \cap A_{xc}| = 1$.

In the second case, assume $x = u$, and therefore $b \neq c$. As A_{ub} and A_{xc} pick up different edges of G incident to $x = u$, $v \neq y$. Note that $|A_{ub} \cap A_{xc}| = 1$. Since G has no cycle of length three, v and y are not adjacent, and therefore $A_{va} \cap A_{yd} = \emptyset$. Also $A_{va} \cap A_{xc}$, if nonempty, consists of the label of the edge of G xv, and since $x = u$, $A_{va} \cap A_{xc} \subseteq A_{ub} \cap A_{xc}$. Similarly, $A_{ub} \cap A_{yd} \subseteq A_{ub} \cap A_{xc}$, and therefore $B_i \cap B_j = A_{ub} \cap A_{xc}$. In conclusion inequatity 10.1 holds.

Therefore, the total savings in the second iteration cannot exceed V, and hence iterative matching produces a solution of cost at least

$$12|V| - 2|V| - |V| = 9|V| = \frac{3}{2}opt.$$

Combining the example above, and the discussion which precedes it with Lemma 10.46 we have the following theorem.

THEOREM 10.47 *For* $g = 4$, *the performance ratio of the iteration matching algorithm is exactly* 1.5.

9.4.2 THE CASE $G = 8$

We begin with the second iteration. Let x be any π_1-valid inside matching. Now we change x, if necessary, to ensure that $H(\pi_1, x)$ does not have any odd cycle, and x has weight at least M_{45}. The odd condition is met if each cycle of $H(\pi_1, x)$ has an even number of edges from each 8-group. So we now concentrate on one 8-group

$$D_k = \{k_1, k_2, \cdots, k_8\}$$

of OPT_8. Let x' be the matching obtained from x by removing the four edges inside D_k. Then there exactly four paths in $H(\pi_1, x')$, with some possibly consisting of only one vertex. Permute the vertices such that we can assume the paths are: from $\pi_1(k_1)$ to $\pi_1(k_2)$, from $\pi_1(k_3)$ to $\pi_1(k_4)$, from $\pi_1(k_5)$ to $\pi_1(k_6)$, and from $\pi_1(k_7)$ to $\pi_1(k_8)$. Now consider a set of inside matchings of D_k obtained as follows: k_1 is matched to some k_j with $j > 2$. We have six choices. Let k_r be such $\pi_1(k_j)$ and $\pi_1(k_r)$ are connected in by one of the paths above. Match k_r to k_s, where $s \notin \{1, 2, j, r\}$. We have four choices. Let k_t be such that $\pi_1(k_s)$ and $\pi_1(k_t)$ are connected by one of the paths above. Match k_t to k_q, where $q \notin \{1, 2, j, r, s, t\}$. We have two choices. For the last edge, we pick what is left. We note that all these 48 matchings are valid, as whenever we match k_j to k_q, $\pi_1(k_j)$ and $\pi_1(k_q)$ are in different components of $H(\pi_1, x')$

and therefore are distinct. We also note that any of these 48 matchings links the four paths above into one cycle. The best among these 48 matching beats $M_{45}(D_k)$. Add the best one to x', and let y be the resulting matching. Then y is a perfect π_1-valid inside matching, and any cycle of $H(\pi_1, y)$ has an even number of edges inside D_k. Moreover, the edge connectivity between any pair of nodes is not decreased compared to $H(\pi_1, x)$. We repeat this procedure for all 8-groups of OPT_8. At the end, we obtain a desired perfect π_1-valid inside matching.

We proceed to the third iteration. Let x be any perfect π_2-valid inside matching. This time $H = H(\pi_2, x)$ is a 4-regular graph. We first repeatedly perform permissible switchings of pairs of edges inside the same 8-group of OPT_8, until each connected component of H contains either all four or none of the edges of a perfect inside matching of any 8-group of OPT_8. Recall that permissible switchings do not decrease any pairwise edge connectivity, and therefore while arranging one 8-group of OPT_8 we do not damage any other 8-group. At this time, each connected component of H has an even number of vertices and is a 4-regular 2-edge-connected graph. This allows us to work separately on the connected components of H. So from now on we assume H is 2-edge-connected. If there is a pair of edges inside the same 8-group of OPT_8 which are not both related to some 4-cluster of H, we execute a permissible switch, thus by Lemma 10.43 either increasing the number of edges of H^*, or decreasing the number of edges of H^*. We repeat the process above a finite number of times, until any two edges of H from any 8-group are both related to some 4-cluster of H. The next lemma shows that at this time no 4-cluster can be odd.

LEMMA 10.48 *Assume that for any two edges of H from any 8-group, there is a 4-cluster of H to which both edges are related. Then no 4-cluster of H can be odd.*

Proof. For intuition, we refer to Figure 10.32. Let C be a fixed simple cycle of H^*, the cactus representation of H. For all $c \in V(C)$, define S_c (a subcactus) to be the cactus induced by those cactus vertices c' such that any path from c' to $V(C) \setminus \{c\}$ goes through c. Note that $c \in V(S_c)$. Let $V_c \subseteq V(H)$ be the union of the 4-clusters of H represented by vertices of S_c. Since any two edges from the same 8-group are related to some common vertex of the cactus, if one edge from an 8-group has both endpoints in V_c, then any of the 4 edges of the 8-group has to be either with both endpoints in V_c or with endpoints in two distinct vertices of

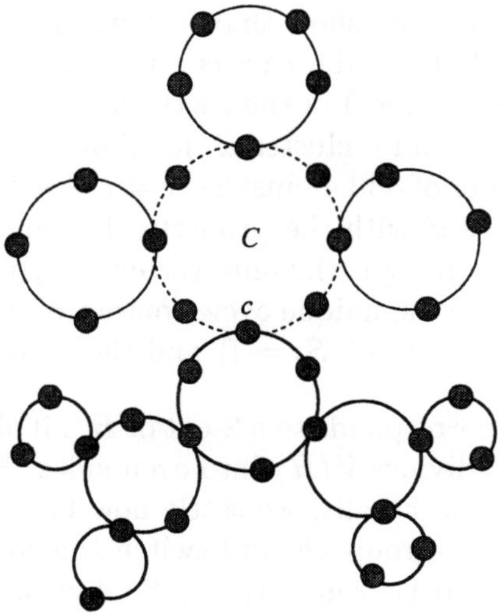

Figure 10.32. **The subcactus S_c represented by the collection of thick solid circiles.**

C. We say that an 8-group *belongs* to c if at least one of the four edges of H given by the 8-group has both endpoints in V_c. If an 8-group belongs to c and it has $4 - j$ edges with both endpoints in V_c, then j edges of C have to come from this 8-group. We say c has a *deficit* of size j if j edges of C have to be from 8-groups belonging to c. Note that if $c' \neq c$ is another vertex of C, the 8-groups belonging to c' are all distinct from the 8-groups belonging to c. So the sum of the deficits of the vertices of C cannot exceed $|C|$.

As a vertex of H is made from four original sets, the vertices of V_c are made from $4|V_c|$ original sets. Two of these original sets are linked to original sets outside V_c: the two edges of C incident with c. These two original sets are both matched outside V_c. Excluding them, we are left with $4|V_c| - 2$ original sets in V_c, matched two by two. If $|V_c|$ is even, there must be an 8-group which belongs to c and without all four inside edges (as edges of H) having both endpoints in V_c, and thus c has a deficit of at least one. If $|V_c|$ is odd, then either a) there is an 8-group with exactly two matched original sets in V_c, b) there is one 8-group with exactly four matched original sets in V_c and another 8-group with exactly six matched original sets in V_c, or c) there are three distinct 8-groups each with exactly six matched original sets in V_c. In all cases c has a deficit of at least three. As the sum of the deficits of the vertices of C cannot exceed $|C|$, we conclude that for each $c \in V(C)$, c has a deficit of exactly one. In conclusion $|V_c|$ is even.

To complete the proof, we show that if there are odd 4-clusters, then there is a simple cycle C of the cactus and some $c \in V(C)$ with $|V_c|$ odd. Pick some vertex r (root) of the cactus and let c be a vertex of the cactus representing an odd 4-cluster at maximum distance from r. As there is an even number of odd 4-clusters, $r \neq c$. Let T_c be the subcactus defined by the vertices c' with the property that any path from c' to r passes trough c. Note that c is the only vertex of T_c which represents an odd 4-cluster. Let C be the unique cycle containing c and not contained in T_c. Then with respect to C, $S_c = T_c$ and therefore $|V_c|$ is odd. ∎

As any 2-cut of H corresponds to a 2-cut of H^*, if all the 4-clusters are even, any 2-cut of H divides $V(H)$ into even sides. So at this time the **Odd Condition** holds. Finally, we settle now beating M_{15} argument. We concentrate on an 8-group D_k and switch the edges. Any permissible switch can only increase connectivity. So 4-clusters either merge or remain the same, and in both cases remain all of even size. Assume the current inside matching is

$$(k_1, k_2), (k_3, k_4), (k_5, k_6), (k_7, k_8).$$

We produce another 15 matchings as follows: three by switching (k_1, k_2) with any of the other three edges; four by first switching (k_3, k_4) with (k_5, k_6) and then switching (k_1, k_2) with any of the other three edges; four by first switching (k_3, k_4) with (k_7, k_8) and then switching(k_1, k_2) with any of the other three edges; and four by first switching (k_5, k_6) with (k_7, k_8) and then switching (k_1, k_2) with any of the other three edges. Pick up the heaviest one among these 16 matchings and we beat $M_{15}(D_k)$. By repeating this procedure for all 8-groups of OPT_8, we conclude that there is an π_2-inside matching of weight at least M_{15}, and such that each odd set $S \subset V(H)$ has $|\Delta(S)| \geq 4$. As both **Weight Condition** and **Odd Condition** are met, we have the following theorem.

THEOREM 10.49 *For* $g = 8$, *the performance ratio of the iteration match is at most* 2.5.

9.4.3 DISCUSSION

When g is greater than 8, we make a the following conjecture on the approximation ratio of the **Iterative Matching**.

CONJECTURE 10.50 *For any $g = 2^k$ with $k \geq 4$, the approximation ratio of the Iterative Matching for Restricted Ring Grooming is at most $\frac{g}{4} + \frac{1}{2}$.*

The iterative matching can be modified to generate a solution to Unrestricted Grooming with arbitrary value of g. Let $\pi = \{V_1, V_2, \cdots, V_k\}$ be a partition of $V = \{1, 2, \cdots, m\}$ with $|V_i| \leq g$ for $1 \leq i \leq k$. Let $K'(\pi)$ be the subgraph of $K(\pi)$ which excludes all edges (i, j) in $K(\pi)$ with either $|V_i| + |V_j| > g$ or $w(i, j) = 0$. An initial partition π_0 consists of m groups, each one with one index. The l-th iteration finds a maximum-weighted matching of $K'(\pi_{l-1})$, and then construct a new partition π_l as follows: for each edge (j_1, j_2) of the obtained matching, a group of π_l is generated by merging the j_1-th group and the j_2-th group. The algorithm repeats this iteration until for some l, the graph $K'(\pi_{l-1})$ has no edge. At this moment, no further improvement can be made. However, the number of groups can be potentially further reduced by applying approximation algorithms for bin-packing problem [40].

Following the same argument as in the beginning of this subsection, we can get a loose upper bound of $\lceil \frac{g}{2} \rceil$ on the approximation ratio of the proposed general iterative matching algorithm. However, its exact ratio remains open.

9.5 A SET COVER APPROACH

We can model the Unrestricted Ring Grooming problem as a Weighted Set Cover problem as follows: the elements are $\{1, 2, \cdots, m\}$, the sets are the

$$\left(\binom{m}{g}\right) = \binom{m}{0} + \binom{m}{1} + \cdots + \binom{m}{g}$$

subsets of $\{1, 2, \cdots, m\}$ of size *at most* g, and each set P has cost $\left|\bigcup_{r \in P} A_r\right|$. A partition $\{P_1, P_2, \cdots, P_k\}$, solution to Unrestricted Ring Grooming, corresponds to picking the sets $P_1, P_2, \cdots, P_k$ in the set cover problem we defined, while from a Set Cover solution we can construct easily an Unrestricted Ring Grooming solution (which is a partition, not just a collection of sets) without increasing the cost. Thus the minimum cost of the defined Weighted Set Cover problem is *equal to* the minimum cost of the original Unrestricted Ring Grooming problem. Chvatal's algorithm [39] approximates within a factor of $1 + \log g$ the set cover problem. Using $\left(\binom{m}{g}\right) \leq gm^g$, we conclude that there is an algorithm for Unrestricted Ring Grooming with running time polynomial in m^g and performance ratio $1 + \log g$.

From Lemma 10.39, the above algorithm can be used to obtain a $2(1 + \log g)$ approximation ratio for Restricted Ring Grooming.

10. SUMMARY

The minimum ADM problem is NP-hard for any constant $g \geq 1$. It is decomposed into two subproblems: the primitive ring generation and the primitive ring grooming. The primitive ring generation can be reduced to valid chain generation. A number of algorithms for various types of optimal valid channel generation without splits have been proposed and their performances are summarized in Table 10.2 and Table 10.3.

Heuristic	Lower Bound	Upper Bound
CM	1.75	1.75
PCM	1.5	$\frac{5}{3}$
ET-TS	1.75	1.75
MCC-TS	1.5	1.6
IM	1.5	$\frac{5}{3}$
PIM	$\frac{4}{3}$	1.5

Table 10.2. Bounds on the approximation ratios of heuristics for optimal valid channel generation from arcs without splits.

Heuristic	Lower Bound	Upper Bound
ETD	1.5	1.5
IM	1.5	$\frac{5}{3}$
PIM	$\frac{4}{3}$	1.5

Table 10.3. Bounds on the approximation ratios of heuristics for optimal valid channel generation from chords without splits.

In addition, we have proposed approximation algorithms based on Eulerian Rounding for optimal valid chain generation with splits. For the arc-version, the approximation ratio is between $\frac{10}{9}$ and $\frac{4}{3}$. For the chord-version, the approximation ratio is exactly 1.5. For most of these algorithms, there is a gap between the lower bound and the upper bound for most heuristics. It will be interesting to reduce the gap, and ideally to get the exact values of the approximation ratios.

Two variations of Ring Grooming are addressed in this chapter: Unrestricted Ring Grooming and Restricted Ring Grooming. When $g = 2$, both have the same minimum cost and can be solved in polynomial time. However, when $g > 2$, both are NP-hard. Any polynomial-time α-approximation algorithm for Restricted Ring Grooming can lead to a polynomial-time α-approximation algorithm for Unrestricted Ring Grooming; and any polynomial-time α-approximation algorithm for Unrestricted Ring Grooming is also a 2α-approximation algorithm for Restricted Ring Grooming. When g is a power of two, an iterative matching algorithm is proposed for Restricted Ring Grooming. Its approximation ratio is exactly 1.5 when $g = 4$, at most 2.5 when $g = 8$, and at most $\frac{g}{2}$ in general while it is conjectured to be at most $\frac{g}{4} + \frac{1}{2}$. The iterative algorithm is also extended for Unrestricted Ring Grooming with arbitrary g, and a loose upper bound on its approximation ratio is $\lceil \frac{g}{2} \rceil$. In addition, set-cover based approximation algorithms have been proposed for both Unrestricted Ring Grooming and Restricted Ring Grooming. They have approximation ratios of at most $1 + \log g$, and respectively at most $2(1 + \log g)$, but running time polynomial in m^g, which may be prohibitively large for a practical implementation.

Chapter 11

OPTIMAL GROOMING OF UNIFORM TRAFFIC IN WDM BLSRS

This chapter studies the optimal grooming of all-to-all uniform traffic in WDM BLSRs with single line-speed. By scaling the line-speed of the wavelength channels, we assume that each traffic stream has unital demand. The routing of each traffic stream is not predetermined. In addition, the two opposite traffic streams between any pair of nodes are not required to be routed along the same path in opposite directions. Thus the BLSR are represented as two counter-rotated rings over the same set of nodes and each traffic stream is represented as a directed chord. Although the algorithms presented in the previous chapter can be readily adopted for traffic grooming of all-to-all uniform traffic, special geometric and combinatorial structures can be exploited to develop simpler but potentially better solutions.

The general framework of the construction of the grooming is still the two-phased approach, but different algorithms will be used in each phase. The optimal generation of primitive rings is equivalent to the optimal grooming of all-to-all full-wavelength traffic demands. Various groomings have the minimum ADM cost and use minimal wavelengths at the same time. But most of them are recursively constructed and the resulting primitive rings lack regular structural properties. The irregular primitive rings not only make the ring grooming difficult to handle, but may also lead to higher ADM cost. The grooming of the uniform all-to-all full-wavelength traffic demands presented in this chapter is extremely simple and geometrically intuitive. The primitive rings produced by this grooming exhibit elegant structural properties. These primitive rings allow an unified approach to the construction of ring grooming. Indeed, the ring grooming can be reduced to a graph decomposition problem. By applying some well-known combinatorial designs, we have been able to construct optimal or suboptimal groomings for various traffic granularities.

This chapter proceeds as follows. In Section 1., we construct an optimal grooming of all-to-all full-wavelength traffic demands geometrically. Such grooming is also used to generate primitve rings. In Section 2., we present a simple optimal grooming of all-to-all half-wavelength traffic demands. In Section 3., we construct an grooming of all-to-all quarter-wavelength traffic demands combinatorially. Finally we discuss the general approaches in Section 4..

1. ALL-TO-ALL FULL-WAVELENGTH TRAFFIC

Throughout of this chapter, we assume the WDM BLSR consists of n nodes, which are numbered by $0, 1, \cdots, n-1$ clockwise. Each traffic demand is represented by a directed chord. The length of a directed chord from node i to node j is defined as

$$\min \{|j-i| \bmod n, n-|j-i| \bmod n\}.$$

When n is even, those chords of length $\frac{n}{2}$ are called *diameters*. For each $0 \leq i < n$, the diameter between from node i to node $\left(i+\frac{n}{2}\right) \bmod n$ is referred to as the i-th diameter. When n is odd, those chords of length $\left\lfloor\frac{n}{2}\right\rfloor$ are called *subdiameters*. For each $0 \leq i < n$, the diameter between from node i to node $\left(i+\left\lfloor\frac{n}{2}\right\rfloor\right) \bmod n$ is referred to as the i-th diameter.

The all-to-all full-wavelength traffic over an n-node WDM BLSR requires at least $n(n-1)$ SONET ADMs and at least $\left\lceil\frac{n^2-1}{8}\right\rceil$ wavelengths. Many groomings can have this minimum ADM cost. For example, if we do not care the number of wavelengths to be used, we can use $\left\lceil\frac{n(n-1)}{4}\right\rceil$ clockwise logical SONET rings and $\left\lfloor\frac{n(n-1)}{4}\right\rfloor$ counterclockwise logical SONET rings, each of which carries two directed chords between a pair of nodes. Obviously, this grooming uses $n(n-1)$ SONET ADMs and $\left\lceil\frac{n(n-1)}{4}\right\rceil$ working wavelengths. In order to have a minimal wavelength cost, the chords should be routed properly:

- When n is odd, all chords *must* be routed in the short way.

- When $n \bmod 4 = 0$, all chords which are not diameters *must* be routed in the short way; any pair of opposite diameters *must* be routed in the same direction, and exactly $\frac{n}{4}$ pairs of opposite diameters *must* be routed in each direction.

- When $n \bmod 4 = 2$, we *may* route all chords which are not diameters in the short way, $\lceil \frac{n}{4} \rceil$ pairs of opposite diameters in clockwise direction, and the remaining $\lfloor \frac{n}{4} \rfloor$ pairs of opposite diameters in counterclockwise direction. However some other routings can also lead to the minimal wavelength cost.

The groomings based on the routing and wavelength assignment algorithms [23] [56] [165] all have both minimal ADM cost and minimal wavelength cost. All these algorithms have iterative nature. The grooming in a ring of n nodes is built upon a grooming in a ring of $n-1$ or $n-2$ nodes. The SONET rings produced by these groomings vary dramatically on the ring size (or ADM cost). Their structures are hard or even impossible to characterize. If these SONET rings are primitive rings, no special advantages can be taken in developing simpler or better groomings. Because of these drawbacks, we provide yet another optimal grooming that also uses minimal ADMs and minimal wavelengths simultaneously. This algorithm is similar to but slightly different from to the one proposed in [127].

When n is odd, the grooming is constructed as follows (see Figure 11.1). For any $0 \le i < j < \lfloor \frac{n}{2} \rfloor$, the four endpoints of the i-th subdiameter and the j-th subdiameter form two opposite SONET rings. These two SONET rings are said to be generated by two subdiameters i and j. Both are denoted by (i, j) and their directions are either specified explicitly or inferred from the context. For each $0 \le i < \frac{n}{2}$, the two endpoints of the i-th subdiameter and node $n-1$ form two opposite SONET rings. These two SONET rings are said to be generated by a single diameter i. Both are denoted by (i, i) and their directions are also either specified explicitly or inferred from the context. The total number of SONET rings is

$$2\binom{\lfloor \frac{n}{2} \rfloor}{2} + 2\left\lfloor \frac{n}{2} \right\rfloor = \frac{n^2-1}{4},$$

and the number of working wavelengths used in each direction is

$$\binom{\lfloor \frac{n}{2} \rfloor}{2} + \left\lfloor \frac{n}{2} \right\rfloor = \frac{n^2-1}{8}.$$

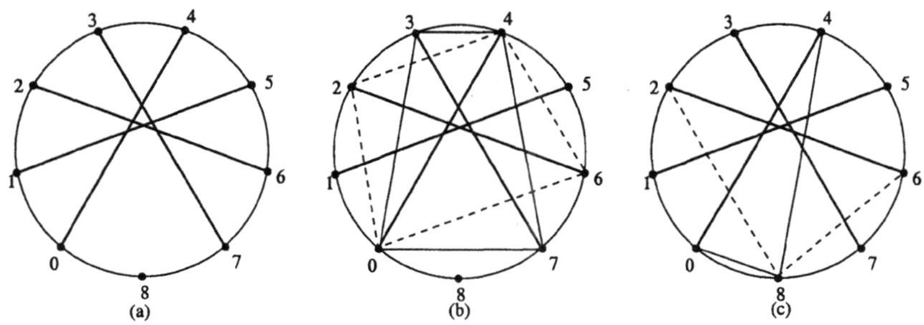

Figure 11.1. Generation of primitive rings in odd ring.

When n is even, the grooming is carried out as follows (see Figure 11.2). For any $0 \leq i < j < \frac{n}{2}$, the four endpoints of the i-th diameter and the j-th diameter form two opposite SONET rings. These two SONET rings are said to be generated by two diameters i and j. Both are denoted by (i, j) and their directions are either specified explicitly or inferred from the context. For each $0 \leq i < \frac{n}{2}$, the two endpoints of the i-th diameter form one SONET ring. This SONET ring is said to be generated by a single diameter i, and is denoted by (i, i). Among these $\frac{n}{2}$ SONET rings generated by single diameters, $\lceil \frac{n}{4} \rceil$ of them are routed clockwise, and the remaining $\lfloor \frac{n}{4} \rfloor$ rings are routed counterclockwise. The total number of SONET rings is

$$2\binom{\frac{n}{2}}{2} + \frac{n}{2} = \frac{n^2}{4},$$

and the number of working wavelengths used in each direction is

$$\binom{\frac{n}{2}}{2} + \left\lceil \frac{n}{4} \right\rceil = \left\lceil \frac{n^2}{8} \right\rceil.$$

The SONET rings constructed as above exhibit regular and balanced structure. Each ring generated by two (sub)diameters consists of four nodes, and each ring generated by single (sub)diameter consists of three nodes if n is odd or two nodes if n is even. When n is odd, any two SONET rings in the same direction share either two nodes if they have a common subdiameter, or exactly one node $n-1$ if they are both formed by one subdiameter, or no node otherwise. When n is even, any two SONET rings in the same direction either share two nodes if they have a common diameter, or do not share any nodes otherwise.

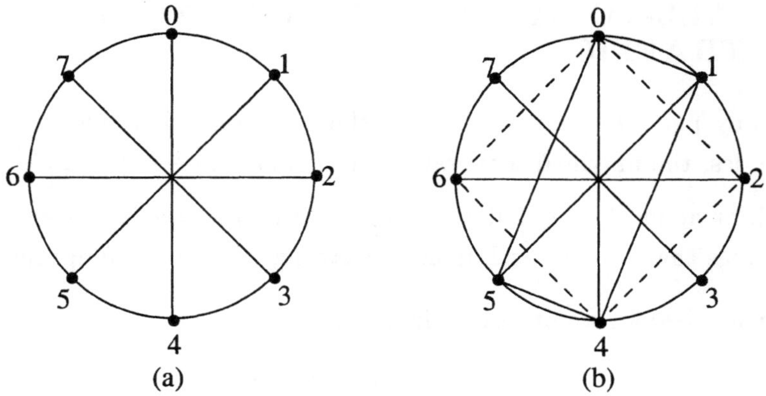

Figure 11.2. Generation of primitive rings in even ring.

In addition, when n is odd, the SONET rings in the two directions are completely symmetric naturally: each ring and its reversal are routed in opposite directions. When n is even, the same symmetry holds for rings generated by two diameters. For rings generated by single diameters, we can artificially pair up them. We use the notation $((i,i);(j,j))$ to denote a pairing between a clockwise ring (i,i) and a counterclockwise ring (j,j).

Because of these elegant properties, we will these SONET rings as the primitive rings for traffic grooming when the traffic demands are less than the wavelength channel's line-speed. The union of a set of primitve rings in the same direction is referred to as an aggregated ring. We will use the operator $\sqcup$ to union the primitive rings. Due to the symmetry between the two opposite directions, the grooming in most situation is also symmetric unless explicitly pointed out. For examples, the notation $(0,1) \sqcup (0,2)$ means that the clockwise primitive rings $(0,1)$ and the $(0,2)$ form a clockwise aggregated ring, and the counterclockwise primitive rings $(0,1)$ and the $(0,2)$ form a counterclockwise aggregated ring. The notation $(0,1) \sqcup ((0,0);(1,1))$ represents that

- the primitive ring $(0,0)$ is placed clockwise and it forms a clockwise aggregated ring together with the clockwise primitive ring $(0,1)$;
- the primitive ring $(1,1)$ is placed counterclockwise and it forms a counterclockwise aggregated ring together with the counterclockwise primitive ring $(0,1)$.

2. ALL-TO-ALL HALF-WAVELENGTH TRAFFIC

As any pair of primitive rings in the same direction share at most two nodes, the maximal total savings is at most $2\left\lfloor\frac{n^2-1}{8}\right\rfloor = 2\left\lfloor\frac{n^2}{8}\right\rfloor$ if n is odd, or at most $2\left\lfloor\frac{n^2}{8}\right\rfloor$ if n is even. In the following we present the groomings that achieve such maximal savings and thus are optimal.

When n is odd, we first sort the primitive rings

$$\left\{(i,j) : 0 \leq i \leq j < \left\lfloor\frac{n}{2}\right\rfloor\right\}$$

in the increasing order of i and those primitive rings with the same i are further sorted in the decreasing order of j. With this ordering, every two consecutive primitive rings share exactly two nodes. Therefore, the sequential pairing according to this order is optimal.

When $n \mod 4 = 0$, the grooming given below has savings of $2\left\lfloor\frac{n^2}{8}\right\rfloor$ and thus is optimal.

$$\begin{aligned}&\left\{(2k, 2k+1) \sqcup ((2k, 2k); (2k+1, 2k+1)) \mid 0 \leq k < \frac{n}{4}\right\}\\ \cup &\left\{(2k, i) \sqcup (2k+1, i) \mid 0 \leq k < \frac{n}{4} - 1, 2k+2 \leq i < \frac{n}{2}\right\}.\end{aligned}$$

When $n \mod 4 = 2$, the grooming given below has savings of $2\left\lfloor\frac{n^2}{8}\right\rfloor$ and thus is optimal.

$$\begin{aligned}&\left\{(2k, 2k+1) \sqcup ((2k, 2k); (2k+1, 2k+1)) \mid 0 \leq k < \left\lfloor\frac{n}{4}\right\rfloor\right\}\\ \cup &\left\{(2k, i) \sqcup (2k+1, i) \mid 0 \leq k < \left\lfloor\frac{n}{4}\right\rfloor, 2k+2 \leq i < \frac{n}{2}\right\}\\ \cup &\left(\frac{n}{2} - 1, \frac{n}{2} - 1\right).\end{aligned}$$

The ring grooming given above is optimal and the total ADM cost of our solution is

$$n(n-1) - 2\left\lfloor\frac{n^2}{8}\right\rfloor.$$

In the next, we investigate the impact of the generation of primitive rings on the overall ADM cost. We first consider the primitive rings generated by algorithms in [23] [56] [165]. We use the notation $\{i_1, i_2, \cdots, i_k\}$ to denotes the primitive ring formed by the nodes $i_1, i_2, \cdots, i_k$. Assume that $n = 7$. Due to the symmetry, only clockwise savings are presented. The algorithms in [23] [56] [165] generate the following set of primitive rings respectively:

$$\{0,3,6\}, \{0,2,5\}, \{0,1,4\}, \{1,3,5\}, \{1,2,4,6\}, \{2,3,4,5,6\};$$
$$\{0,3,4\}, \{1,3,5\}, \{2,3,6\}, \{0,1,4,5\}, \{0,2,5,6\}, \{1,2,4,6\};$$
$$\{0,3,6\}, \{1,4,6\}, \{2,5,6\}, \{0,2,4\}, \{1,3,4,5\}, \{0,1,2,3,5\}.$$

The weighted graphs corresponding to them are shown in Figure 11.3, Figure 11.4 and Figure 11.5 respectively. The weight of dashed edges is one; the weights of thin solid edges are two; and the weight of thick sold line is three. It's easy to verify that the weight of any maximum-weighted matching in each graph is 5. Thereby the maximal savings is 5. This is less than the savings of the grooming presented in this subsection, which is $\left\lfloor \frac{7^2}{8} \right\rfloor = 6$.

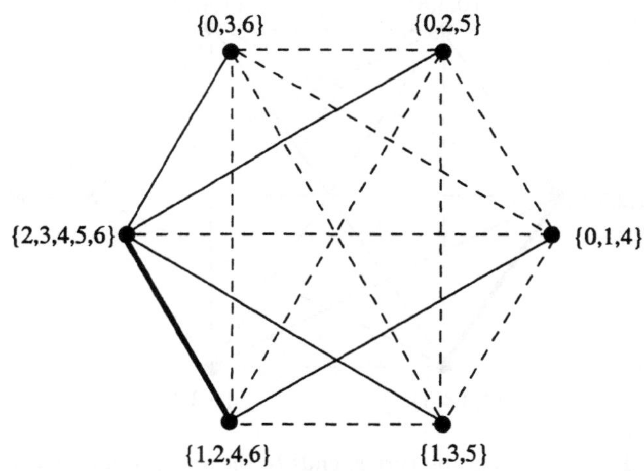

Figure 11.3. **The weighted graph corresponds to the set of primitive rings generated by optimal route and wavelength assignment algorithm given in [165].**

On the other hand, when $n = 11$ the construction [139] in Table 11.1 uses 40 ADMs in each direction, one less than the ADMs used by our constructions. It thus becomes an interesting open problem to find how far our construction is from optimum.

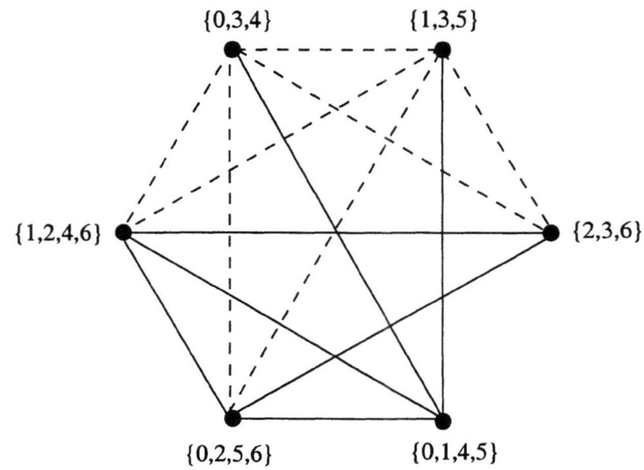

Figure 11.4. The weighted graph corresponds to the set of primitive rings generated by optimal route and wavelength assignment algorithm given in [56].

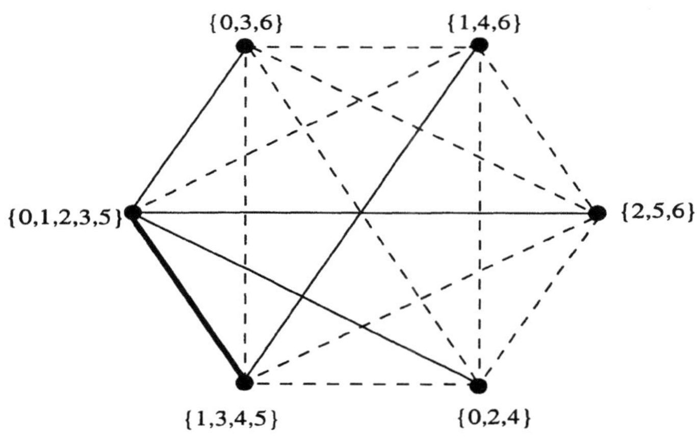

Figure 11.5. The weighted graph corresponds to the set of primitive rings generated by optimal route and wavelength assignment algorithm given in [23].

3. ALL-TO-ALL QUARTER-WAVELENGTH TRAFFIC

When n is small, the minimum ADM cost can be easily found. When $n = 2, 3$, and 4, the minimal number of ADMs is 2, 3 and 7 respectively.

$\{0,4,5,7,9\} \sqcup \{0,5,6,7\}$
$\{1,2,4,9\} \sqcup \{4,6,9,10\}$
$\{2,5,8,9\} \sqcup \{3,5,9\}$
$\{3,7,8,10\} \sqcup \{2,7,10\}$
$\{0,3,6,10\} \sqcup \{0,1,6\}$
$\{0,2,6,8\} \sqcup \{2,3,8\}$
$\{1,3,4,7\} \sqcup \{1,4,8\}$
$\{1,5,10\}$

Table 11.1. A better grooming when $n = 11$ and $g = 2$.

In the following, we study the optimal grooming with $g = 4$ when $n \geq 5$. We first study how much savings that an individual aggregated ring can have and when an aggregated ring can have the maximal savings.

LEMMA 11.1 *The maximal savings of an aggregated ring consisting of g primitive rings is $4g - 2m$ where $m = \min\left\{\ell : \binom{\ell}{2} \geq g\right\}$. The maximal savings is achieved if and only if either all primitive rings are generated by two (sub)diameters from a set of m (sub)diameters, or $g = 1 + \binom{m-1}{2}$ and all but one primitive rings are generated by two (sub)diameters from a set of $m - 1$ (sub)diameters.*

Proof. (1). Assume that n is even. Let x be the number of primitive rings which are generated by single diameters. Since the maximal number of primitive rings that can be generated by ℓ diameters is at most

$$\binom{\ell}{2} + \binom{\ell}{1} = \binom{\ell+1}{2}$$

and

$$\binom{m}{2} \geq g > \binom{m-1}{2},$$

there are at least $m - 1$ diameters contained in the g primitive rings.

If there are exactly $m - 1$ diameters, the total size is $2(m-1)$ and

$$x \geq g - \binom{m-1}{2} \geq 1.$$

Thus the total savings is

$$\begin{aligned} &4(g - x) + 2x - 2(m-1) \\ &= 4g - 2m - 2(x-1) \\ &\leq 4g - 2m \end{aligned}$$

and the equality holds if and only if $g = 1 + \binom{m-1}{2}$ and all but one primitive rings are generated by two diameters.

If there are exactly m diameters, the total size is $2m$. Suppose that there are x primitive rings which are generated by a single diameter. Then the total savings is

$$\begin{aligned} &4(g - x) + 2x - 2m \\ &= 4g - 2x - 2m \\ &\leq 4g - 2m \end{aligned}$$

and the equality holds if and only if all primitive rings are generated by two diameters.

If there are $\ell > m$ diameters, the total size is 2ℓ. Suppose that there are x primitive rings which are generated by a single diameter. Then the total savings is

$$\begin{aligned} &4(g - x) + 2x - 2\ell \\ &= 4g - 2x - 2\ell \\ &\leq 4g - 2\ell \\ &\leq 4g - 2m - 2. \end{aligned}$$

(2). Assume that n is odd. We first assume that all primitive rings are generated by two diameters. Then the number of subdiameters ℓ contained in the g primitive rings is at least m. Thus the maximal saving is at most

$$4g - 2\ell \leq 4g - 2m$$

and the equality holds if and only if there are exactly m subdiameters.

In the next we assume that not all primitive rings are generated by two diameters. Let x be the number of primitive rings which are generated by single diameters. As the maximal number of primitive rings that can be generated by ℓ subdiameters is at most

$$\binom{\ell}{2} + \binom{\ell}{1} = \binom{\ell+1}{2}$$

and

$$\binom{m}{2} \geq g > \binom{m-1}{2},$$

there are at least $m - 1$ subdiameters contained in the g primitive rings.

If there are exactly $m-1$ subdiameters, the total size is $2(m-1)+1$ and

$$x \geq g - \binom{m-1}{2} \geq 1.$$

Thus the total savings is

$$\begin{aligned} & 4(g-x) + 3x - 2(m-1) - 1 \\ & = 4g - 2m - (x-1) \\ & \leq 4g - 2m \end{aligned}$$

and the equality is achieved if and only if $g = 1 + \binom{m-1}{2}$ and all but one primitive rings are generated by two subdiameters.

If there are $\ell \geq m$ subdiameters, the total size is $2\ell + 1$ and the total saving is

$$\begin{aligned} & 4(g-x) + 3x - 2\ell - 1 \\ & = 4g - x - 2\ell - 1 \\ & \leq 4g - 2\ell - 1 \\ & \leq 4g - 2m - 1. \end{aligned}$$

Therefore the lemma is true for both even rings and odd rings. ∎

Table 11.2 lists the maximal savings of the aggregated rings consisting of $g \leq 8$ primitive rings. From the proof of Lemma 11.1, if there are x primitive rings that are generated by single diameters (subdiameters respectively), then the total saving is $2(x-1)$ ($x-1$ respectively) less than the upper bound $4g - 2m$.

g	1	2	3	4	5	6	7	8
savings	0	2	6	8	12	16	18	22

Table 11.2. The maximal savings of individual aggregated rings.

To simplify the presentation, for different integers i, j, k and ℓ, the aggregated ring $(i,j) \sqcup (i,k) \sqcup (j,k)$ is denoted by a triple (i,j,k), and the aggregated ring $(i,k) \sqcup (i,\ell) \sqcup (j,k) \sqcup (j,\ell)$ is denoted by $\{i,j\} \times \{k,\ell\}$. Then from Lemma 11.1, when $g = 3$ an aggregated ring has the maximal savings (of six) if and only if its configuration is (i,j,k). If $g = 4$, an aggregated ring has the maximal savings (of eight) if and only if its configuration is either $(i,j,k) \sqcup (i,i)$, or $(i,j,k) \sqcup (i,\ell)$ or $\{i,j\} \times \{k,\ell\}$.

The next lemma study how much savings that optimal grooming of a set of primitive rings with $g = 4$ can achieve.

LEMMA 11.2 *Suppose that $g = 4$. The total savings by any grooming of p primitive rings is no more than $8\left\lceil\frac{p}{4}\right\rceil, 8\left\lceil\frac{p}{4}\right\rceil - 6, 8\left\lceil\frac{p}{4}\right\rceil - 4$ and $8\left\lceil\frac{p}{4}\right\rceil - 2$ respectively if $p \bmod 4 = 0, 1, 2$ and 3. Furthermore, these savings are achievable only if the configuration of each aggregated ring is either (i, j, k), or $(i, j, k) \sqcup (i, i)$, or $(i, j, k) \sqcup (i, \ell)$ or $\{i, j\} \times \{k, \ell\}$.*

Proof. We prove the lemma by induction on p. When $p = 1, 2, 3$ and 4, the lemma follows immediately from Lemma 11.1. Now assume that the lemma is true if the number of primitives is less than p. We show that it is also true when there are p primitive rings. Obviously there must exist an optimal grooming in which some aggregated ring consists of at least three primitive rings. We first assume that there is an aggregated ring consisting of 4 primitive rings. Its saving is at most 8, and is achievable only if its configuration is either $(i, j, k) \sqcup (i, i)$, or $(i, j, k) \sqcup (i, \ell)$ or $\{i, j\} \times \{k, \ell\}$. As

$$(p - 4) \bmod 4 = p \bmod 4$$

and

$$\left\lceil\frac{p-4}{4}\right\rceil = \left\lceil\frac{p}{4}\right\rceil - 1,$$

from the induction hypothesis, the total saving is at most $8\left\lceil\frac{p}{4}\right\rceil, 8\left\lceil\frac{p}{4}\right\rceil - 6, 8\left\lceil\frac{p}{4}\right\rceil - 4$ and $8\left\lceil\frac{p}{4}\right\rceil - 2$ respectively if $p \bmod 4 = 0, 1, 2$ and 3; and these savings are achievable only if the configuration of each aggregated ring is either (i, j, k), or $(i, j, k) \sqcup (i, i)$, or $(i, j, k) \sqcup (i, \ell)$ or $\{i, j\} \times \{k, \ell\}$.

Now we assume that there is no aggregated ring consisting of 4 primitive rings. Consider any aggregated ring consisting of three primitive rings. Its savings is at most 6, and is achievable only if its configuration is (i, j, k). The savings of the remaining $p - 3$ primitive rings depends on $p \bmod 4$. We consider the following four cases.

If $p \bmod 4 = 0$, then $(p - 3) \bmod 4 = 1$. Thus the total savings is at most

$$6 + 8\left\lceil\frac{p-3}{4}\right\rceil - 6 = 8\left\lceil\frac{p-3}{4}\right\rceil = 8\left\lceil\frac{p}{4}\right\rceil$$

and it is achievable only if the configuration of each aggregated ring is either (i, j, k), or $(i, j, k) \sqcup (i, i)$, or $(i, j, k) \sqcup (i, \ell)$ or $\{i, j\} \times \{k, \ell\}$.

If $p \bmod 4 = 1$, then $(p-3) \bmod 4 = 2$. Thus the total saving is at most

$$6 + 8\left\lceil \frac{p-3}{4} \right\rceil - 4 = 8\left\lceil \frac{p-3}{4} \right\rceil + 2 = 8\left\lceil \frac{p}{4} \right\rceil - 6$$

and it is achievable only if the configuration of each aggregated ring is either (i,j,k), or $(i,j,k) \sqcup (i,i)$, or $(i,j,k) \sqcup (i,\ell)$ or $\{i,j\} \times \{k,\ell\}$.

If $p \bmod 4 = 2$, then $(p-3) \bmod 4 = 3$. Thus the total saving is at most

$$6 + 8\left\lceil \frac{p-3}{4} \right\rceil - 2 = 8\left\lceil \frac{p-3}{4} \right\rceil + 4 = 8\left\lceil \frac{p}{4} \right\rceil - 4$$

and it is achievable only if the configuration of each aggregated ring is either (i,j,k), or $(i,j,k) \sqcup (i,i)$, or $(i,j,k) \sqcup (i,\ell)$ or $\{i,j\} \times \{k,\ell\}$.

If $p \bmod 4 = 3$, then $(p-3) \bmod 4 = 0$. Thus the total saving is at most

$$6 + 8\left\lceil \frac{p-3}{4} \right\rceil = 8\left\lceil \frac{p}{4} \right\rceil - 2$$

and it is achievable only if the configuration of each aggregated ring is either (i,j,k), or $(i,j,k) \sqcup (i,i)$, or $(i,j,k) \sqcup (i,\ell)$ or $\{i,j\} \times \{k,\ell\}$.

In either cases, the lemma is true by the induction principle. ■

From Lemma 11.2, one particular necessary condition to achieve the maximal savings specified in Lemma 11.2 is that the primitive rings generated by single (sub)diameters must be grouped with the triples, which further implies that the number of triples should be no less than the single (sub)diameters generating the rings. If the number of primitive rings generated by single diameters (subdiameters respectively) is x more than the maximal number of triples, then the total saving is $2x$ (x respectively) less than the upper bound given in Lemma 11.2.

From the above discussion, the triples play an important role in the grooming and we would like to have as many triples as possible. The maximal number of triples has been well-studied in the combinatorial design [41]. Given a point set V, a *triple system* of V is a set of triples such that any pair of points in V appear in at most one triple. The size of a triple system is the number of triples in it. A *maximal triple system* is a triple system with maximum size. If a triple system covers all pair of elements, it is called as a *Steiner triple system* (STS), otherwise it is called as *partial triple system* (PTS). It's well known that the minimum pairs not covered by a triple system is

$$\ell(v) = \begin{cases} \frac{v}{2} & \text{if } v \equiv 0, 2 \bmod 6 \\ 0 & \text{if } v \equiv 1, 3 \bmod 6 \\ \frac{v}{2}+1 & \text{if } v \equiv 4 \bmod 6 \\ 4 & \text{if } v \equiv 5 \bmod 6 \end{cases}$$

where $v = |V|$. The size of any maximal triples system is thus

$$\frac{\frac{v(v-1)}{2} - \ell(v)}{3}.$$

Various constructions of maximal triple systems can be found in [41].

3.1 GROOMING WHEN N MOD $4 = 2$

When n is even, there are totally $\frac{n^2}{4}$ primitive rings, among which $\frac{n^2}{8} - \frac{n}{4}$ primitive rings each generated by two diameters must be routed clockwise, other $\frac{n^2}{8} - \frac{n}{4}$ primitive rings each generated by two diameters must be routed counterclockwise, and the remaining $\frac{n}{2}$ primitive rings each generated by single diameter can be routed in either direction to minimize the total cost.

When $n \bmod 4 = 2$, there are totally $\frac{n^2}{4}$ primitive rings. As in this case $\frac{n^2}{4} \bmod 4 = 1$, from Lemma 11.2, the total saving is at most

$$8\left\lceil\frac{n^2}{16}\right\rceil - 6 = \frac{n^2}{2},$$

and this total saving is achievable only if the configuration of each aggregated ring is either (i,j,k), or $(i,j,k) \sqcup (i,i)$,or $(i,j,k) \sqcup (i,\ell)$ or $\{i,j\} \times \{k,\ell\}$.

When $n = 6$ or 10, the number of triples is one or two respectively in each direction, which is less than the total number of diameters. Thereby, the total savings is at most $\frac{n^2}{2} - 2$. Such savings is achieved by the groomings listed in Table 11.3,Table 11.4 for $n = 6$ and 10 respectively. When $n = 6$ and 10, the minimum ADM cost is 14 and 42 respectfully.

$(0,1,2) \sqcup ((0,0);(1,1))$	$(2,2)$

Table 11.3. Optimal grooming when $n = 6$.

$(0,1,2) \sqcup ((0,0);(1,1))$	$\{0,1\} \times \{3,4\}$
$(2,3,4) \sqcup ((2,2);(3,3))$	$(4,4)$

Table 11.4. Optimal grooming when $n = 10$.

0	1	2	$\cdots$	$\frac{n}{2}-7$	$\frac{n}{2}-6$	$\cdots$	$\frac{n}{2}-1$
1		2	$\cdots$	$\frac{n}{2}-7$	$\frac{n}{2}-6$	$\cdots$	$\frac{n}{2}-1$
$\vdots$			$\ddots$	$\vdots$	$\vdots$	$\vdots$	$\vdots$
$\frac{n}{2}-8$				$\frac{n}{2}-7$	$\frac{n}{2}-6$	$\cdots$	$\frac{n}{2}-1$
$\frac{n}{2}-7$					$\frac{n}{2}-6$	$\cdots$	$\frac{n}{2}-1$
$\vdots$						$\ddots$	$\frac{n}{2}-1$
$\frac{n}{2}-2$							$\frac{n}{2}-1$

Table 11.5. Optimal grooming when $n \bmod 4 = 2$ and $n \geq 14$.

When $n \geq 14$ and $n \bmod 4 = 2$, the optimal grooming is illustrated in Table 11.5. We group the primitive rings involving diameters $0, 1, \cdots, \frac{n}{2} - 8$ as follows:

$$\left\{(2i, 2i+1, 2i+2) \sqcup ((2i, 2i);(2i+1, 2i+1)) \mid 0 \leq i < \left\lfloor \frac{n}{4} \right\rfloor - 3\right\} \cup$$
$$\left\{\{2i, 2i+1\} \times \{2j-1, 2j\} \mid 0 \leq i < \left\lfloor \frac{n}{4} \right\rfloor - 3, i+2 \leq j \leq \left\lfloor \frac{n}{4} \right\rfloor\right\}$$

The remaining clockwise primitive rings are groomed as shown in Table 11.6, and the remaining counterclockwise primitive rings are groomed as shown in Table 11.7.

$\left(\frac{n}{2}-7, \frac{n}{2}-6, \frac{n}{2}-5\right) \sqcup \left(\frac{n}{2}-7, \frac{n}{2}-7\right)$
$\left(\frac{n}{2}-7, \frac{n}{2}-4, \frac{n}{2}-3\right) \sqcup \left(\frac{n}{2}-3, \frac{n}{2}-3\right)$
$\left(\frac{n}{2}-7, \frac{n}{2}-2, \frac{n}{2}-1\right) \sqcup \left(\frac{n}{2}-1, \frac{n}{2}-1\right)$
$\left(\frac{n}{2}-5, \frac{n}{2}-4, \frac{n}{2}-1\right) \sqcup \left(\frac{n}{2}-5, \frac{n}{2}-5\right)$
$\left(\frac{n}{2}-6, \frac{n}{2}-4, \frac{n}{2}-2\right)$
$\left(\frac{n}{2}-6, \frac{n}{2}-3, \frac{n}{2}-1\right)$
$\left(\frac{n}{2}-5, \frac{n}{2}-3, \frac{n}{2}-2\right)$

Table 11.6. Optimal grooming of the remaining clockwise primitive rings.

The total savings of the above grooming is

$$8 \cdot \frac{\frac{n^2}{4} - 9}{4} + 6 \cdot 3 = \frac{n^2}{2}$$

and thus is optimal. So when $n \bmod 4 = 2$ and $n \geq 14$, the ADM cost is

$$n(n-1) - \frac{n^2}{2} = \frac{(n-1)^2 - 1}{2}.$$

$(\frac{n}{2}-7, \frac{n}{2}-6, \frac{n}{2}-5) \sqcup (\frac{n}{2}-6, \frac{n}{2}-6)$
$(\frac{n}{2}-7, \frac{n}{2}-4, \frac{n}{2}-3) \sqcup (\frac{n}{2}-4, \frac{n}{2}-4)$
$(\frac{n}{2}-7, \frac{n}{2}-2, \frac{n}{2}-1) \sqcup (\frac{n}{2}-2, \frac{n}{2}-2)$
$(\frac{n}{2}-5, \frac{n}{2}-4, \frac{n}{2}-1) \sqcup (\frac{n}{2}-5, \frac{n}{2}-3)$
$(\frac{n}{2}-6, \frac{n}{2}-4, \frac{n}{2}-2) \sqcup (\frac{n}{2}-5, \frac{n}{2}-2)$
$(\frac{n}{2}-6, \frac{n}{2}-3, \frac{n}{2}-1) \sqcup (\frac{n}{2}-3, \frac{n}{2}-2)$

Table 11.7. Optimal grooming of the remaining counterclockwise primitive rings.

When $n = 6, 10$, the ADM cost is $14, 42$ respectfully.

3.2 GROOMING WHEN N MOD $4 = 0$

When n mod $4 = 0$, there are totally $\frac{n^2}{4}$ primitive rings. As in this case $\frac{n^2}{4} \bmod 4 = 0$, the total savings is at most

$$8\left\lceil \frac{n^2}{16} \right\rceil = \frac{n^2}{2}$$

from Lemma 11.2, and this total saving is achievable only if the configuration of each aggregated ring is either (i,j,k), or $(i,j,k) \sqcup (i,i)$,or $(i,j,k) \sqcup (i,\ell)$ or $\{i,j\} \times \{k,\ell\}$.

When $n = 8$ or 12, one can verify that the maximal total savings is $\frac{n^2}{2} - 4$. Table 11.8 and Table 11.9 list the two optimal groomings for $n = 8$ and 12 respectively. Thus when $n = 8$ and 12, the minimum total ADM cost is 28 and 64 respectfully.

$(0,1,2) \sqcup ((0,0);(1,1))$
$(0,3) \sqcup (1,3) \sqcup (2,3) \sqcup ((2,2);(3,3))$

Table 11.8. Optimal grooming when $n = 8$.

$(0,1,2) \sqcup ((1,1);(2,2))$	$(1,3,5) \sqcup (1,4)$
$(0,3,4) \sqcup ((3,3);(4,4))$	$(2,4,5) \sqcup (2,3)$
$(0,5) \sqcup ((0,0);(5,5))$	

Table 11.9. Optimal grooming when $n = 12$.

When $n = 16$, Table 11.10 lists an optimal grooming whose total saving is $64 \cdot 2 = 128 = \frac{n^2}{2}$.

$(0,1,2) \sqcup ((0,0);(1,1))$	$(0,5,6) \sqcup (0,7)$
$(2,4,5) \sqcup ((2,2);(5,5))$	$(1,4,7) \sqcup (1,5)$
$(0,3,4) \sqcup ((3,3);(4,4))$	$(3,5,7) \sqcup (2,3)$
$(2,6,7) \sqcup ((6,6);(7,7))$	$(1,3,6) \sqcup (4,6)$

Table 11.10. Optimal grooming when $n = 16$.

When $n = 20$, Table 11.11 lists an optimal grooming, which is obtained in the following way. There is a maximal partial triple system (PTS) over the set $\{0, 1, \cdots, 9\}$ consisting of 13 triples. Six pairs are not covered by such PTS. The clockwise grooming is based on a matching between the 13 triples, and 6 uncovered pairs corresponding to 6 primitive rings together with 7 primitive rings generated by single diameters. The counterclockwise grooming is based on a matching between 12 triples in the PTS, and the three pairs broken from the remaining triple, the six unconvered pairs together with three single diameters. The savings is $\frac{n^2}{2}$ and thus is optimal.

Clockwise Grooming

$(0,1,3) \sqcup (0,0)$	$(2,5,8) \sqcup (0,2)$
$(2,4,7) \sqcup (2,2)$	$(1,4,8) \sqcup (1,2)$
$(0,4,9) \sqcup (4,4)$	$(3,4,6) \sqcup (2,3)$
$(0,5,6) \sqcup (6,6)$	$(1,5,9) \sqcup (4,5)$
$(0,7,8) \sqcup (8,8)$	$(1,6,7) \sqcup (6,8)$
$(3,5,7) \sqcup (7,7)$	$(2,6,9) \sqcup (7,9)$
$(3,8,9) \sqcup (9,9)$	

Counterclockwise Grooming

$(0,1,3) \sqcup (1,1)$	$(2,5,8) \sqcup (0,2)$
$(2,4,7) \sqcup (4,9)$	$(1,4,8) \sqcup (1,2)$
$(0,5,6) \sqcup (5,5)$	$(3,4,6) \sqcup (2,3)$
$(0,7,8) \sqcup (0,4)$	$(1,5,9) \sqcup (4,5)$
$(3,5,7) \sqcup (3,3)$	$(1,6,7) \sqcup (6,8)$
$(3,8,9) \sqcup (0,9)$	$(2,6,9) \sqcup (7,9)$

Table 11.11. Optimal grooming when $n = 20$.

When $n = 24$, Table 11.12 lists an optimal grooming obtained as follows. Over the set $\{0, 1, \cdots, 11\}$ there is a PTS consisting of 20 triples with 6 pairs not covered. In each direction, the grooming is based on a matching between 18 triples in the PTS, and the 6 pairs broken from the remaining two triples, the 6 uncovered pairs together with 6 single diameters. The savings is $\frac{n^2}{2}$ and thus is optimal.

When $n \geq 28$ and $n \bmod 4 = 0$, an optimal grooming is illustrated in Table 11.13. We first groom all the primitive rings containing diameters 0 through 7 as follows: the primitive rings exclusively generated by diameters 0 through 7 are grouped in the same way as in Table 11.10; the others are grouped as follows.

$$\left\{\{2i, 2i+1\} \times \{2j, 2j+1\} \mid 0 \leq i \leq 3, 4 \leq j < \frac{n}{4}\right\}$$

$(0,1,4) \sqcup ((0,0);(1,1))$	$(5,7,0) \sqcup (0,2)$	$(3,4,7) \sqcup (0,3)$
$(2,3,6) \sqcup ((2,2);(3,3))$	$(11,0,6) \sqcup (0,8)$	$(1,2,5) \sqcup (1,7)$
$(4,5,8) \sqcup ((4,4);(5,5))$	$(2,4,10) \sqcup (2,8)$	$(10,11,1) \sqcup (2,11)$
$(6,7,10) \sqcup ((6,6);(7,7))$	$(6,8,1) \sqcup (1,3)$	$(9,11,4) \sqcup (4,6)$
$(9,10,0) \sqcup ((9,9);(10,10))$	$(7,9,2) \sqcup (1,9)$	$(5,6,9) \sqcup (5,10)$
$(7,8,11) \sqcup ((8,8);(11,11))$	$(3,5,11) \sqcup (3,9)$	$(8,10,3) \sqcup (8,9)$

Table 11.12. Optimal grooming when $n = 24$.

0	1	$\cdots$	7	8	9	$\cdots$	$\frac{n}{2}-2$	$\frac{n}{2}-1$
$\vdots$		$\ddots$	$\vdots$	8	9	$\cdots$	$\frac{n}{2}-2$	$\frac{n}{2}-1$
6			7	8	9	$\cdots$	$\frac{n}{2}-2$	$\frac{n}{2}-1$
7				8	9	$\cdots$	$\frac{n}{2}-2$	$\frac{n}{2}-1$
8					9	$\cdots$	$\frac{n}{2}-2$	$\frac{n}{2}-1$
$\vdots$						$\ddots$	$\vdots$	$\vdots$
$\frac{n}{2}-3$							$\frac{n}{2}-2$	$\frac{n}{2}-1$
$\frac{n}{2}-2$								$\frac{n}{2}-1$

Table 11.13. Optimal recursive grooming when $n \bmod 4 = 0$ and $n \geq 28$.

After the first step, the remaining primitive rings are exclusively generated by diameters 8 through $\frac{n}{2} - 1$. We then repeat the grooming approach in the first step until the remaining primitive rings are generated by either 8, 10 or 12 diameters, which are then grouped in the same way we did for $n = 16, 20$, and 24.

Therefore, when $n \geq 16$ and $n \bmod 4 = 0$, the total saving of our grooming is $\frac{n^2}{2}$ and the total ADM cost is

$$n(n-1) - \frac{n^2}{2} = \frac{(n-1)^2 - 1}{2}.$$

When $n = 8, 12$, the total ADM cost is 28, 64 respectfully.

3.3 GROOMING WHEN N MOD $4 = 3$

When n is odd, we present only clockwise grooming as the counter-clockwise grooming can be done in the same way.

Table 11.14 through Table 11.19 list the groomings for n = 7, 11, 15, 19, 23, 27 respectively. When n =7, 11, the listed groomings can be proved to be optimal. When n = 15, 19, 23, 27, the listed groomings all have savings equal to the upper bound given in Lemma 11.2 and thus are optimal.

$(0,1,2) \sqcup (0,0)$	$(1,1) \sqcup (2,2)$

Table 11.14. Optimal grooming when $n = 7$.

$(0,0) \sqcup (4,4) \sqcup (0,4) \sqcup (1,4)$	$(2,3,4) \sqcup (2,2)$
$(1,1) \sqcup (3,3) \sqcup (1,3) \sqcup (0,3)$	$(0,1,2)$

Table 11.15. Optimal grooming when $n = 11$.

$(0,1,3) \sqcup (0,0)$	$(1,2,4) \sqcup (1,1)$
$(2,3,5) \sqcup (2,2)$	$(3,4,6) \sqcup (3,3)$
$(4,5,0) \sqcup (4,4)$	$(5,6,1) \sqcup (5,5)$
$(6,0,2) \sqcup (6,6)$	

Table 11.16. Optimal grooming when $n = 15$.

$(0,1,2) \sqcup (0,0)$	$(1,3,6) \sqcup (1,1)$
$(2,4,5) \sqcup (2,2)$	$(0,3,4) \sqcup (3,3)$
$(1,4,7) \sqcup (4,4)$	$(0,5,6) \sqcup (5,5)$
$(2,6,7) \sqcup (6,6)$	$(3,5,7) \sqcup (7,7)$
$(0,7,8) \sqcup (8,8)$	$(1,5,8)$
$(2,3,8)$	$(4,6,8)$

Table 11.17. Optimal grooming when $n = 19$.

$(0,3,4) \sqcup (0,1)$	$(4,5,7) \sqcup (5,5)$
$(0,5,6) \sqcup (0,2)$	$(2,6,10) \sqcup (6,6)$
$(1,5,10) \sqcup (1,3)$	$(2,7,8) \sqcup (7,7)$
$(1,2,4) \sqcup (2,3)$	$(0,8,10) \sqcup (8,8)$
$(0,7,9) \sqcup (0,0)$	$(1,8,9) \sqcup (9,9)$
$(1,6,7) \sqcup (1,1)$	$(3,7,10) \sqcup (10,10)$
$(2,5,9) \sqcup (2,2)$	$(3,6,9)$
$(3,5,8) \sqcup (3,3)$	$(4,9,10)$
$(4,6,8) \sqcup (4,4)$	

Table 11.18. Optimal grooming when $n = 23$.

When $n \geq 31$ and $n \bmod 4 = 3$, an optimal grooming is illustrated in Table 11.20. We first groom all the primitive rings containing subdi-

$(0,1,4) \sqcup (0,0)$	$(5,7,0) \sqcup (0,2)$
$(1,2,5) \sqcup (1,1)$	$(11,0,6) \sqcup (0,8)$
$(2,3,6) \sqcup (2,2)$	$(8,10,3) \sqcup (2,8)$
$(3,4,7) \sqcup (3,3)$	$(6,8,1) \sqcup (1,3)$
$(4,5,8) \sqcup (4,4)$	$(9,11,4) \sqcup (1,9)$
$(5,6,9) \sqcup (5,5)$	$(3,5,11) \sqcup (3,9)$
$(6,7,10) \sqcup (6,6)$	$(7,9,2) \sqcup (2,4)$
$(7,8,11) \sqcup (7,7)$	$(10,12,5) \sqcup (2,10)$
$(8,9,12) \sqcup (8,8)$	$(4,6,12) \sqcup (4,10)$
$(9,10,0) \sqcup (9,9)$	$(12,1,7)$
$(10,11,1) \sqcup (10,10)$	
$(11,12,2) \sqcup (11,11)$	
$(12,0,3) \sqcup (12,12)$	

Table 11.19. Optimal grooming when $n = 27$.

0	1	···	6	7	8	9	···	$\lfloor \frac{n}{2} \rfloor - 1$
⋮		⋱	6	7	8	9	···	$\lfloor \frac{n}{2} \rfloor - 1$
5			6	7	8	9	···	$\lfloor \frac{n}{2} \rfloor - 1$
6				7	8	9	···	$\lfloor \frac{n}{2} \rfloor - 1$
7					8	9	···	$\lfloor \frac{n}{2} \rfloor - 1$
8						9	···	$\lfloor \frac{n}{2} \rfloor - 1$
⋮							⋱	⋮
$\lfloor \frac{n}{2} \rfloor - 2$								$\lfloor \frac{n}{2} \rfloor - 1$

Table 11.20. Optimal recursive grooming when $n \bmod 4 = 3$ and $n \geq 31$.

ameters 0 through 7: the primitive rings exclusively generated by subdiameters 0 through 6 are grouped in the same way as in Table 11.16, the others are grouped as follows.

$$\begin{aligned}&\{(6,7,8) \sqcup (7,7)\} \cup \\ &\{\{2i, 2i+1\} \times \{7,8\} \mid 0 \leq i \leq 2\} \cup \\ &\left\{\{2i, 2i+1\} \times \{2j-1, 2j\} \mid 0 \leq i \leq 3, 5 \leq j < \left\lceil \frac{n}{4} \right\rceil\right\}.\end{aligned}$$

After the first step, the remaining primitive rings are generated by subdiameters 8 through $\lfloor \frac{n}{2} \rfloor - 1$. We then repeat the grooming approach in the first step until the remaining primitive rings are generated by either 7, 9, 11 or 13 subdiameters, which are then grouped in the same way we did for $n =$ 15, 19, 23 and 27.

Now we show that when $n \bmod 4 = 3$ and $n \geq 15$, the total ADM cost is $\frac{(n-1)^2}{2}$. When $n \bmod 16 = 15$ and $n \geq 15$, the total number of

ADMs used is

$$\begin{aligned}&2\left(\binom{n}{2}-8\left\lceil\frac{n^2-1}{32}\right\rceil\right)\\&=2\left(\binom{n}{2}-8\cdot\frac{n^2-1}{32}\right)\\&=\frac{(n-1)^2}{2}.\end{aligned}$$

When $n \bmod 16 = 3$ and $n \geq 19$, the total number of ADMs used is

$$\begin{aligned}&2\left(\binom{n}{2}-8\left\lceil\frac{n^2-1}{32}\right\rceil+6\right)\\&=2\left(\binom{n}{2}-8\left(\frac{n^2-1}{32}+\frac{3}{4}\right)+8\right)\\&=\frac{(n-1)^2}{2}.\end{aligned}$$

When $n \bmod 16 = 7$ and $n \geq 23$, the total number of ADMs used is

$$\begin{aligned}&2\left(\binom{n}{2}-8\left\lceil\frac{n^2-1}{32}\right\rceil+4\right)\\&=2\left(\binom{n}{2}-8\left(\frac{n^2-1}{32}+\frac{1}{2}\right)+4\right)\\&=\frac{(n-1)^2}{2}.\end{aligned}$$

When $n \bmod 16 = 11$ and $n \geq 27$, the total number of ADMs used is

$$\begin{aligned}&2\left(\binom{n}{2}-8\left\lceil\frac{n^2-1}{32}\right\rceil+2\right)\\&=2\left(\binom{n}{2}-8\left(\frac{n^2-1}{32}+\frac{1}{4}\right)+2\right)\\&=\frac{(n-1)^2}{2}.\end{aligned}$$

Thus whenever $n \bmod 4 = 3$ and $n \geq 15$, the total number of ADMs used is $\frac{(n-1)^2}{2}$.

When $n = 7, 11$, the total ADM cost is $24, 54$ respectfully.

3.4 GROOMING WHEN N MOD $4 = 1$

Again we only present the clockwise grooming.

Table 11.21 through Table 11.29 list the groomings for n = 5, 9, 13, 17, 21, 25, 29, 33, 37 respectively. When n = 5, 9, 13, the listed groomings can be proved to be optimal. When n = 17, 21, the listed groomings can be proved to be either optimal or use one more than the optimum in each direction. When n = 25, 29, 33, 37, the listed groomings all have savings equal to the upper bound given in Lemma 11.2 and thus are optimal.

$(0,0) \sqcup (1,1) \sqcup (0,1)$

Table 11.21. Optimal grooming when $n = 5$.

$(0,1,2)$	$(0,0) \sqcup (3,3) \sqcup (0,3)$
	$(1,3) \sqcup (2,3) \sqcup (1,1) \sqcup (2,2)$

Table 11.22. Optimal grooming when $n = 9$.

$(0,1,2) \sqcup (0,6)$	$(0,3,4)$
$(1,3,5) \sqcup (5,6)$	$(1,4,6)$
$(2,4,5) \sqcup (0,5)$	$(2,3,6)$

Table 11.23. Optimal grooming when $n = 13$.

$(1,3,6) \sqcup (1,1)$	$(0,1,2) \sqcup (1,5)$
$(3,5,7) \sqcup (2,2)$	$(0,3,4) \sqcup (2,3)$
$(2,4,5) \sqcup (3,3)$	$(2,6,7) \sqcup (4,6)$
$(1,4,7) \sqcup (4,4)$	$(0,0) \sqcup (6,6) \sqcup (7,7) \sqcup (0,7)$
$(0,5,6) \sqcup (5,5)$	

Table 11.24. Optimal grooming when $n = 17$.

When $n \geq 41$ and n mod $4 = 1$, an optimal grooming is illustrated in Table 11.30. We first groom all the primitive rings containing subdiameters 0 through 15: the primitive rings exclusively generated by subdiameters. $0, 1, \cdots, 15$ are grouped in the same way as we did with $n = 33$; the others are grouped as follows.

$$\left\{ \{2i, 2i+1\} \times \{2j, 2j+1\} \mid 0 \leq i \leq 7, 8 \leq j < \left\lceil \frac{n}{4} \right\rceil - 1 \right\}.$$

$(0,1,3) \sqcup (1,1)$	$(3,8,9) \sqcup (9,9)$
$(3,5,7) \sqcup (3,3)$	$(1,5,9) \sqcup (1,2)$
$(0,4,9) \sqcup (4,4)$	$(3,4,6) \sqcup (2,3)$
$(2,5,8) \sqcup (5,5)$	$(0,5,6) \sqcup (4,5)$
$(2,6,9) \sqcup (6,6)$	$(0,7,8) \sqcup (6,8)$
$(2,4,7) \sqcup (7,7)$	$(1,6,7) \sqcup (7,9)$
$(1,4,8) \sqcup (8,8)$	$(0,0) \sqcup (2,2) \sqcup (0,2)$

Table 11.25. Optimal grooming when $n = 21$.

$(0,1,4) \sqcup (0,0)$	$(10,11,1) \sqcup (10,10)$
$(1,2,5) \sqcup (1,1)$	$(11,0,6) \sqcup (11,11)$
$(2,3,6) \sqcup (2,2)$	$(0,2,8) \sqcup (0,3)$
$(3,4,7) \sqcup (3,3)$	$(1,3,9) \sqcup (1,7)$
$(4,5,8) \sqcup (4,4)$	$(2,4,10) \sqcup (2,11)$
$(5,6,9) \sqcup (5,5)$	$(6,8,1) \sqcup (4,6)$
$(6,7,10) \sqcup (6,6)$	$(5,7,0) \sqcup (5,10)$
$(7,8,11) \sqcup (7,7)$	$(7,9,2) \sqcup (8,9)$
$(8,10,3) \sqcup (8,8)$	$(3,5,11)$
$(9,10,0) \sqcup (9,9)$	$(9,11,4)$

Table 11.26. Optimal grooming when $n = 25$.

$(0,1,4) \sqcup (0,0)$	$(0,2,8) \sqcup (0,3)$
$(1,2,5) \sqcup (1,1)$	$(1,3,9) \sqcup (1,7)$
$(2,3,6) \sqcup (2,2)$	$(2,4,10) \sqcup (2,13)$
$(3,4,7) \sqcup (3,3)$	$(4,6,12) \sqcup (4,9)$
$(4,5,8) \sqcup (4,4)$	$(5,7,13) \sqcup (5,12)$
$(5,6,9) \sqcup (5,5)$	$(8,10,1) \sqcup (6,8)$
$(6,7,10) \sqcup (6,6)$	$(3,5,11) \sqcup (10,11)$
$(7,8,11) \sqcup (7,7)$	$(7,9,0) \sqcup (2,7)$
$(8,9,12) \sqcup (8,8)$	$(9,11,2) \sqcup (2,12)$
$(9,10,13) \sqcup (9,9)$	$(10,12,3) \sqcup (7,12)$
$(10,0,5) \sqcup (10,10)$	$(11,13,4)$
$(11,12,0) \sqcup (11,11)$	$(13,0,6)$
$(12,13,1) \sqcup (12,12)$	$(1,6,11)$
$(13,3,8) \sqcup (13,13)$	

Table 11.27. Optimal grooming when $n = 29$.

After the first step, the remaining primitive rings are exclusively generated by subdiameters 16 through $\lfloor \frac{n}{2} \rfloor - 1$. We then repeat the grooming approach in the first step until the remaining primitive rings are exclusively generated by either $12, 14, 16$ or 18 subdiameters, which are then grouped in the same way we did for $n = 25, 29, 33$, and 37.

$(0,4,8) \sqcup (0,0)$	$(0,6,12) \sqcup (12,12)$	$(5,7,9) \sqcup (9,10)$
$(1,6,8) \sqcup (1,1)$	$(1,13,15) \sqcup (13,13)$	$(0,7,13) \sqcup (0,2)$
$(2,5,12) \sqcup (2,2)$	$(0,9,14) \sqcup (14,14)$	$(1,4,5) \sqcup (1,2)$
$(3,10,12) \sqcup (3,3)$	$(0,10,15) \sqcup (15,15)$	$(1,7,10) \sqcup (2,10)$
$(4,11,14) \sqcup (4,4)$	$(0,3,5) \sqcup (0,1)$	$(2,14,15) \sqcup (3,15)$
$(5,6,15) \sqcup (5,5)$	$(2,7,11) \sqcup (0,11)$	$(1,9,12) \sqcup (4,12)$
$(6,11,13) \sqcup (6,6)$	$(1,3,14) \sqcup (1,11)$	$(3,8,11) \sqcup (5,11)$
$(7,8,15) \sqcup (7,7)$	$(2,4,6) \sqcup (2,3)$	$(3,4,7) \sqcup (6,7)$
$(8,12,13) \sqcup (8,8)$	$(2,8,9) \sqcup (2,13)$	$(7,12,14) \sqcup (8,14)$
$(9,10,11) \sqcup (9,9)$	$(5,13,14) \sqcup (3,13)$	$(3,6,9) \sqcup (9,13)$
$(6,10,14) \sqcup (10,10)$	$(4,10,13) \sqcup (4,9)$	
$(11,12,15) \sqcup (11,11)$	$(5,8,10) \sqcup (4,10)$	

Table 11.28. Optimal grooming when $n = 33$.

$(0,1,8) \sqcup (0,0)$	$(15,16,4) \sqcup (15,15)$	$(4,6,9) \sqcup (4,8)$
$(1,2,9) \sqcup (1,1)$	$(16,17,5) \sqcup (16,16)$	$(9,11,14) \sqcup (4,14)$
$(2,3,10) \sqcup (2,2)$	$(17,0,3) \sqcup (17,17)$	$(14,16,0) \sqcup (8,14)$
$(3,4,11) \sqcup (3,3)$	$(0,2,5) \sqcup (0,4)$	$(9,13,0) \sqcup (0,7)$
$(4,5,12) \sqcup (4,4)$	$(5,7,10) \sqcup (0,10)$	$(16,1,7) \sqcup (1,4)$
$(5,6,13) \sqcup (5,5)$	$(10,12,15) \sqcup (4,10)$	$(11,15,2) \sqcup (2,16)$
$(6,7,14) \sqcup (6,6)$	$(1,3,6) \sqcup (1,5)$	$(12,16,3) \sqcup (3,9)$
$(7,8,15) \sqcup (7,7)$	$(6,8,11) \sqcup (1,11)$	$(5,9,15) \sqcup (5,14)$
$(8,9,16) \sqcup (8,8)$	$(11,13,16) \sqcup (5,11)$	$(15,0,6) \sqcup (6,17)$
$(9,10,17) \sqcup (9,9)$	$(2,4,7) \sqcup (2,6)$	$(17,2,8) \sqcup (8,12)$
$(10,14,1) \sqcup (10,10)$	$(7,9,12) \sqcup (2,12)$	$(6,10,16) \sqcup (10,11)$
$(11,12,0) \sqcup (11,11)$	$(12,14,17) \sqcup (6,12)$	$(15,17,1) \sqcup (13,15)$
$(12,13,1) \sqcup (12,12)$	$(3,5,8) \sqcup (3,7)$	$(7,11,17)$
$(13,14,2) \sqcup (13,13)$	$(8,10,13) \sqcup (3,13)$	
$(14,15,3) \sqcup (14,14)$	$(13,17,4) \sqcup (7,13)$	

Table 11.29. Optimal grooming when $n = 37$.

0	1	$\cdots$	15	16	17	$\cdots$	$\lfloor \frac{n}{2} \rfloor - 2$	$\lfloor \frac{n}{2} \rfloor - 1$
$\vdots$		$\ddots$	$\vdots$	16	17	$\cdots$	$\lfloor \frac{n}{2} \rfloor - 2$	$\lfloor \frac{n}{2} \rfloor - 1$
14			15	16	17	$\cdots$	$\lfloor \frac{n}{2} \rfloor - 2$	$\lfloor \frac{n}{2} \rfloor - 1$
15				16	17	$\cdots$	$\lfloor \frac{n}{2} \rfloor - 2$	$\lfloor \frac{n}{2} \rfloor - 1$
16					17	$\cdots$	$\lfloor \frac{n}{2} \rfloor - 2$	$\lfloor \frac{n}{2} \rfloor - 1$
$\vdots$						$\ddots$	$\vdots$	$\vdots$
$\lfloor \frac{n}{2} \rfloor - 3$							$\lfloor \frac{n}{2} \rfloor - 2$	$\lfloor \frac{n}{2} \rfloor - 1$
$\lfloor \frac{n}{2} \rfloor - 2$								$\lfloor \frac{n}{2} \rfloor - 1$

Table 11.30. Optimal recursive grooming when $n \bmod 4 = 1$ and $n \geq 37$.

Now we show that when $n \bmod 4 = 1$ and $n \geq 25$, the total ADM cost is $\frac{(n-1)^2}{2}$. When $n \bmod 16 = 9$ and $n \geq 25$, the total number of ADMs used is

$$\begin{aligned}
&2\left(\binom{n}{2} - 8\left\lceil\frac{n^2-1}{32}\right\rceil + 4\right)\\
&= 2\left(\binom{n}{2} - 8\left(\frac{n^2-1}{32} + \frac{1}{2}\right) + 4\right)\\
&= \frac{(n-1)^2}{2}.
\end{aligned}$$

When $n \bmod 16 = 13$ and $n \geq 29$, the total number of ADMs used is

$$\begin{aligned}
&2\left(\binom{n}{2} - 8\left\lceil\frac{n^2-1}{32}\right\rceil + 6\right)\\
&= 2\left(\binom{n}{2} - 8\left(\frac{n^2-1}{32} + \frac{3}{4}\right) + 6\right)\\
&= \frac{(n-1)^2}{2}.
\end{aligned}$$

When $n \bmod 16 = 1$ and $n \geq 33$, the total number of ADMs used is

$$\begin{aligned}
&2\left(\binom{n}{2} - 8\left\lceil\frac{n^2-1}{32}\right\rceil\right)\\
&= 2\left(\binom{n}{2} - 8\cdot\frac{n^2-1}{32}\right)\\
&= \frac{(n-1)^2}{2}.
\end{aligned}$$

When $n \bmod 16 = 5$ and $n \geq 37$, the total number of ADMs used is

$$\begin{aligned}
&2\left(\binom{n}{2} - 8\left\lceil\frac{n^2-1}{32}\right\rceil + 2\right)\\
&= 2\left(\binom{n}{2} - 8\left(\frac{n^2-1}{32} + \frac{1}{4}\right) + 2\right)\\
&= \frac{(n-1)^2}{2}.
\end{aligned}$$

Thus whenever $n \bmod 4 = 1$ and $n \geq 25$, the total number of ADMs used is $\frac{(n-1)^2}{2}$.

When $n = 5, 9, 13, 17, 21$, the total ADM cost is 10, 36, 76, 132, 202 respectfully.

4. DISCUSSION ON GENERAL APPROACHES

In this chapter, we have presented efficient groomings of uniform traffic with $g = 2$ and $g = 4$. When $g = 2$, the grooming is very efficient, and its ADM cost is

$$n(n-1) - 2\left\lfloor \frac{n^2}{8} \right\rfloor .$$

When $g = 4$, the grooming heavily uses the combinatorial design of triple systems, and its ADM cost is summarized in Table 11.31. Such combinatorial approach can be generalized to general g as well. In this section, we discuss some general approaches.

n	2	3	4	5	6	7	8	9
ADMs	2	3	7	10	14	24	28	36
n	10	11	12	13	17	21	*others*	
ADMs	42	54	64	76	132	202	$\left\lceil \frac{(n-1)^2}{2} \right\rceil$	

Table 11.31. Summary of the ADMs used.

For any integer m, we use LK_m to denote a complete graph of m vertices with a loop on each vertex. Then the optimal ring grooming problem can be reduced to a graph decomposition problem in LK_m. We first consider the odd rings and consider the clockwise grooming only due to the exact symmetry between two directions. Let $n = 2m + 1$. Associate with each vertex i in LK_m a subdiameter i, with each edge (i, j) the primitive ring generated by subdiameters i and j, and with each loop (i, i) the primitive ring generated by subdiameter i and the node $\left\lfloor \frac{n}{2} \right\rfloor = m$. Then each aggregated ring corresponds to a subgraph of LK_m having at most g edges. When a vertex appears in more than one edge, a savings results. When no loops are chosen in this subgraph, the ADM cost for this subgraph is twice the number of vertices of nonzero degree in the subgraph; when at least one loop is chosen, it is one plus twice the number of vertices of nonzero degree in the subgraph.

When n is even, the situation is fundamentally the same, but some details differ. We examine this next. While for the case of odd n, the primitive rings fall naturally into pairs, the primitive rings generated by single diameters do not naturally pair when n is even. Thus we will consider the grooming in both directions simultaneously. Let $n = 2m$.

Associate with each vertex i in LK_m a diameter i, with each edge (i, j) the two primitive ring generated by diameters i and j, and with each loop (i, i) with the primitive ring generated by diameter i. Then for any subgraph whose number of non-loop edges plus *half* the number of loops does not exceed g, then we can make a valid grooming as follows. On the clockwise ring, we place a primitive ring generated by two diameters,and its reversal on the counterclockwise ring. However, for primitive rings generated by single diameters, we place half on the clockwise ring and half on the counterclockwise ring. Hence in the even case we treat loops differently than when the ring size is odd. While a loop is counted as one edge when the ring size is odd, a loop is counted as a half-edge when the ring size is even. In this case, the cost of a subgraph is always simply twice the number of vertices of nonzero degree in the subgraph.

Our task in both cases has been reduced to a graph decomposition problem. Partition the edges of LK_m into subgraphs, each containing at most g edges (counting loops as edges when the ring size is odd, and as half-edges when the ring size is even) so that the total cost of all subgraphs is minimized. See [41] for results on graph decompositions in general and for further references.

Based on this graph decomposition approach, we have been constructed efficient grooming when $g = 8$ and the ring size is a multiple of four. Due to its lengthy presentation, we do not provide the construction here. The construction in detail can be found in [42].

Chapter 12

MINIMIZING SONET ADMS IN SINGLE-HUB WDM SELF-HEALING RINGS

1. INTRODUCTION

This chapter is devoted to the minimum ADM cost problem in single-hub WDM self-healing rings, in which the all traffic demands are between a central hub and other nodes. Some known results on the traffic grooming in single-hub WDM self-healing rings were obtained in [68] and [115]. The network architecture considered in [115] is the WDM UPSR. Thus the routing is predetermined for each traffic demand. In addition, all wavelength channels are assumed to have the same transmission capacity; all traffic streams are assumed to be simplex (i.e., all are originated from the hub), uniform (i.e., all have the same traffic demand) and sub-wavelength (i.e., less than the transmission capacity of the wavelength). An optimal grooming of these traffic demands in the WDM UPSR was developed in [115]. On the other hand, [68] observed that the cost of ADM does not increase linearly with the line speed. For example, the cost ratio between an OC-$4n$ ADM and an OC-n ADM is 2.5. Thus the total ADM cost might be decreased if the wavelength channels are allowed to have different line-speeds. Based on this observation, [68] proposed a simple approach in a two-speed WDM self-healing rings as follows. Assume that the two line speeds are $g_1$and g_2 with $g_2 = 4g_1$. For each traffic demand with value r, assign $r \bmod g_2$ traffic to the SONET rings with speed g_1 and $r - r \bmod g_2$ traffic to the SONET rings with speed g_2. The performance of this approach comparing to the optimal assignment was not discussed in [68]. As shown later in this chapter, this simple approach is not optimal in many cases.

This chapter is intended to extend the works in [68] and [115] in various aspects. All results in this chapter are applicable to all WDM self-healing ring architectures, to both duplex traffic demands and simplex traffic demands. In addition, both uniform traffic demands and non-uniform traffic demands are addressed. The most important contribution of this chapter is the in-depth study of optimal traffic grooming when multiple line-speeds are allowed. The remaining of this chapter is structured as follows. In Section 2., we prove that in any WDM self-healing ring the minimum ADM cost of any duplex traffic is same as that of the corresponding simplex traffic. In Section 3. we show that the search for optimal grooming can be confined to a narrow subset of groomings, referred to as *canonical groomings.* In Section 4., we construct canonical optimal grooming of uniform traffic demands and canonical suboptimal grooming of non-uniform traffic demands in WDM self-healing rings with single line-speed. In Section 5. we discuss optimal traffic grooming in WDM self-healing rings with two line-speeds. Finally Section 6. summarizes this chapter.

2. DUPLEX VS. SIMPLEX

We consider a single-hub WDM self-healing ring comprising of one hub, numbered with 0, and n access nodes, numbered with $1, \cdots, n$ clockwise. Let R_d denote the set of traffic requests in which the traffic between the hub and node i is fully duplex with demand r_i for all $1 \leq i \leq n$. Let R_s denote the set of traffic requests in which the traffic between the hub and node i is simplex with the hub as the source and has demand r_i for $1 \leq i \leq n$.

Since R_s is a subset of R_d, the minimum ADM cost of R_s is no more than the minimum ADM cost of R_d for any self-healing ring architecture and any configuration of the SONET line-speeds. On the other hand, any grooming of R_s naturally gives rise to a grooming of R_d with the same cost in the following way. Let w be any wavelength used in the grooming of R_s, and let $r_{i,w}$ be the portion of the demand from hub to node i carried in wavelength w. Now consider the following grooming of R_d: we use the same set of wavelengths used in the grooming of R_s, each wavelength w carries $r_{i,w}$ units of demand from hub to node i and $r_{i,w}$ units of demand from node i to hub for all $1 \leq i \leq n$. It's easy to see that such grooming is a valid solution and it uses the same set of ADMs as the grooming of R_s. Thus the minimum ADM cost of R_s is no less than the minimum ADM cost of R_d. So we have the following lemma.

LEMMA 12.1 *For any self-healing ring architecture and any configuration of the SONET line-speeds, the minimum ADM cost of any duplex traffic is same as that of the corresponding simplex traffic.*

So from now on, we will only concentrate the grooming of the simplex traffic. We assume that all traffic are originated from the hub and the traffic demand from the hub to node i is r_i for $1 \leq i \leq n$.

3. CANONICAL GROOMING

In [115], it was proved that in a single-hub WDM UPSR with single line-speed if the traffic demands are uniform and sub-wavelength then there is an optimal grooming in which each demand is carried in exactly one wavelength, i.e., split of a demand into more than one wavelengths is not allowed. In this section, we will generalize this property.

For a WDM self-healing ring with single line-speed g, a grooming is said to be *canonical* if at each node $1 \leq i \leq n$, its demand is carried in $\left\lceil \frac{r_i}{g} \right\rceil$ wavelengths, among which $\left\lfloor \frac{r_i}{g} \right\rfloor$ wavelengths each carries g units of demands to node i, and the remaining one, if there is any, carries $r_i \bmod g$ units of demands to node i.

For a WDM self-healing ring with multiple line-speeds, a grooming can be decomposed into a set of sub-groomings, one corresponding to each line speed. A grooming is then said to be canonical if for each line speed, the corresponding sub-grooming is canonical.

LEMMA 12.2 *For any traffic demands in a single-hub WDM self-healing ring with any line-speed configuration, there is a canonical grooming with minimum ADM cost.*

Proof. We only have to prove the lemma when there is a single line-speed, say g. We transform any given optimal grooming into a canonical grooming with the same cost in a number of steps. The procedure at each step is as follows. Suppose that the current optimal grooming is not canonical. Then at some node i, two portion of its demand $0 < f_2 \leq f_1 < g$ are carried in two wavelengths w_1 and w_2 respectively. We consider two cases.

Case 1: $f_1 + f_2 \leq g$. We use an unused wavelength to carry the two portion of demands f_1 and f_2. Then in the new wavelength two ADMs are used. But the two ADMs used in the wavelengths w_1 and w_2 are removed. So the ADM cost remains unchanged.

Case 2: $f_1+f_2 > g$. We swap all traffic except f_1 carried in wavelength w_1 with the $g - f_1$ portion within f_2 in wavelength w_2, as illustrated in Figure 12.1. In the resulting grooming, wavelength $w_1$carries the full g units of demands to node i , and wavelength w_2 carries $f_1 + f_2 - g$ units of demands to node i . The total ADM cost remains the same.

It's easy to see that one canonical grooming can be reached after a finite number of such procedures. The resulting canonical grooming has the minimum ADM cost and thus is optimal. ■

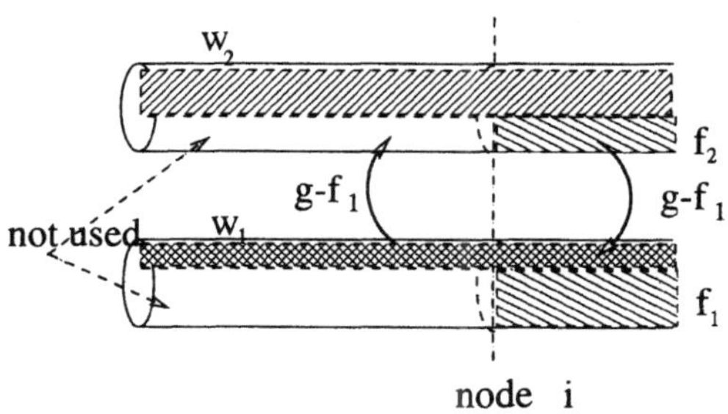

(a) Before the exchange of assignment

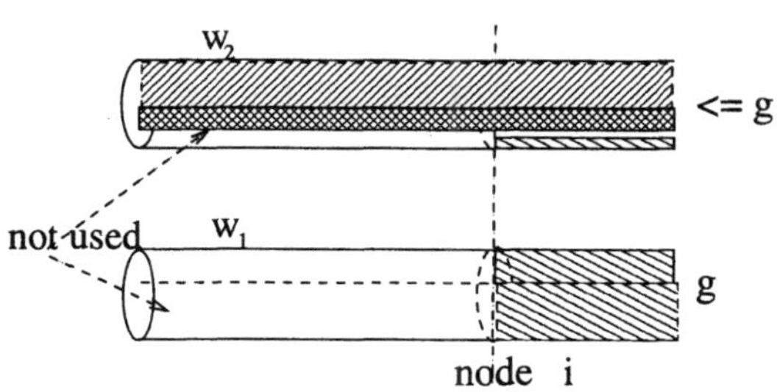

(b) After the exchange of assignment

Figure 12.1. The exchange of wavelengths assignment.

From Lemma 12.2 we will focus on the canonical optimal grooming for any set of traffic demands. It should be noted that a canonical optimal grooming may not use the minimum number of wavelengths.

4. SINGLE LINE-SPEED

In this section, we present canonical optimal grooming of both uniform traffic demands and non-uniform demands in single-hub WDM self-healing rings with single line-speed. We begin with uniform traffic demands. Assume that the traffic demand from the hub to each node is r. [115] essentially gave the canonical optimal grooming in single-hub WDM UPSR when $r < g$. A general canonical optimal grooming can be obtained as follows.

If $r \bmod g = 0$, then the canonical optimal grooming is unique in the sense that each wavelength carry g units of demands exclusively from the hub to some node. Thus each node contributes $2 \cdot \frac{r}{g} = \frac{2r}{g}$ ADMs, half at the node itself and half at the hub. So the total ADM cost is $n \cdot \frac{2r}{g} = \frac{2nr}{g}$.

If $r \bmod g > 0$. In any canonical grooming, at each node $r - r \bmod g$ portion of demands are carried in $\left\lfloor \frac{r}{g} \right\rfloor$ wavelengths exclusively. These demands use $2n \left\lfloor \frac{r}{g} \right\rfloor$ ADMs. In any canonical optimal grooming, the remaining demands at each node, referred to as *residue demands*, must use a minimum number of ADMs. This can be achieved in the same way as in [115]. We partition the n nodes into $\left\lceil \frac{n}{\left\lfloor \frac{g}{r \bmod g} \right\rfloor} \right\rceil$ groups of at most $\left\lfloor \frac{g}{r \bmod g} \right\rfloor$ nodes. The residue demands of the nodes in each group are carried in a single wavelength. These residue demands totally require $n + \left\lceil \frac{n}{\left\lfloor \frac{g}{r \bmod g} \right\rfloor} \right\rceil$ ADMs, one ADM at each of the n nodes and $\left\lceil \frac{n}{\left\lfloor \frac{g}{r \bmod g} \right\rfloor} \right\rceil$ ADMs at the hub. Thus the total ADMs used in the working fiber is

$$2n \left\lfloor \frac{r}{g} \right\rfloor + n + \left\lceil \frac{n}{\left\lfloor \frac{g}{r \bmod g} \right\rfloor} \right\rceil = n \left\lceil \frac{r}{g} \right\rceil + n \left\lfloor \frac{r}{g} \right\rfloor + \left\lceil \frac{n}{\left\lfloor \frac{g}{r \bmod g} \right\rfloor} \right\rceil.$$

Let

$$F(g,r,n) = \begin{cases} \frac{2nr}{g} & \text{if } r \bmod g = 0, \\ n \left\lceil \frac{r}{g} \right\rceil + n \left\lfloor \frac{r}{g} \right\rfloor + \left\lceil \frac{n}{\left\lfloor \frac{g}{r \bmod g} \right\rfloor} \right\rceil & \text{otherwise.} \end{cases}$$

Then we have the following theorem.

THEOREM 12.3 *The minimum ADM cost of uniform traffic demands with rate r in a WDM self-healing ring with line-speed g is $F(g, r, n)$.*

In the next, we consider non-uniform traffic demands. It was proved in [115] that the optimal grooming of non-uniform *sub-wavelength* traffic grooming in WDM UPSR is NP-hard. By applying Lemma 12.2, we can prove that the optimal grooming of any non-uniform traffic demands in any WDM self-healing ring is NP-hard. The reduction is made from the well-known bin-packing problem. In the next, we present approximation algorithms.

Given an instance of traffic demands $r_1, \cdots, r_n$, a canonical grooming is constructed as follows. At each node i, we carry $r_i - r_i \bmod g$ portion of demands in $\left\lfloor \frac{r_i}{g} \right\rfloor$ wavelengths exclusively. Let $\mathcal{A}$ be any approximation algorithm for the bin-packing problem with approximation ratio α. We apply $\mathcal{A}$ to the instance

$$\left\{ \frac{r_i - r_i \bmod g}{g} : 1 \leq i \leq n \right\}.$$

Then each node uses $\left\lceil \frac{r_i}{g} \right\rceil$ ADMs. The number of ADMs at the hub is equal to $\sum_{i=1}^{n} \left\lfloor \frac{r_i}{g} \right\rfloor$ plus the number of bins output by the algorithm $\mathcal{A}$. Let

$$C_{base} = \sum_{i=1}^{n} \left\lceil \frac{r_i}{g} \right\rceil + \sum_{i=1}^{n} \left\lfloor \frac{r_i}{g} \right\rfloor,$$

and let *opt* be the minimum ADM cost. Then the minimum ADM cost at the hub required by the residue demands is

$$opt - C_{base}.$$

Thus the number of bins output by the algorithm $\mathcal{A}$ is at most

$$a \left(opt - C_{base} \right).$$

So the total ADM cost of the grooming constructed in this way is

$$\begin{aligned} & C_{base} + a \left(opt - C_{base} \right) \\ & = a \cdot opt - (a-1) \, C_{base}. \end{aligned}$$

There are a number of bin packing approximation algorithms developed [40]. The well-known First-Fit-Decreasing (FFD) bin packing method first sorts the input objects in the decreasing order, and assign the bins sequentially for objects. The assigned bin is the first bin that still can fit the current object. It gives a 11/9 approximation for the minimal number of bins used [40].

The results obtained in this section can be used as a guide to choose the most cost-effective line speed. On one hand, the higher the line speed, the smaller the number of ADMs. On the other hand, the higher the line speed, the higher the cost of the ADM. However, the cost of ADM does not increase linearly with the line speed. The cost model adopted in [68] assumes that the cost ratio between an OC-$4n$ ADM and an OC-n ADM is 2.5. If the traffic demand is uniform, then the best line speed can be selected by comparing the total ADM costs for different line speeds.

5. TWO LINE-SPEEDS

In this section we consider the optimal traffic grooming in WDM self-healing ring with two line-speeds $g_1$and $g_2 = 4g_1$. We adopt the same cost model used in [68]. We assume that the cost of an ADM of speed g_1 is one, and the cost of an ADM of speed g_2 is 2.5. We call a SONET ring of speed g_1 as a low-speed ring, and a SONET ring of speed g_2 as a high-speed ring without any ambiguity. Similarly, we call a SONET ADM of speed g_1 as a low-speed ADM, and a SONET ADM of speed g_2 as a high-speed ADM. For the simplicity of presentation, g_1 is scaled to one and all demands are scaled accordingly. Thus $g_1 = 1, g_2 = 4$ and all demands are fractional numbers or integers.

When there are two line-speeds, a traffic grooming can be constructed in two phases. The first phase partitions the traffic demands at each node into two portions, one to be carried in the low-speed rings and the other to be carried in the high-speed rings. Such traffic partition can be represented by an n-dimensional vector

$$f = (f_1, \cdots, f_n)$$

where $0 \leq f_i \leq r_i$ is the amount of the traffic at node i that will be placed to a low-speed ring. A simple traffic partition presented in [68] is to let

$$f_i = r_i \bmod g_2$$

for all $1 \leq i \leq n$. The performance of this approach comparing to the optimal assignment was unknown in [68]. In the second phase, we groom the traffic demands to be carried in low-speed rings and the traffic demands to be carried in high-speed rings separately. The grooming in the second phase can apply the algorithms developed in the previous section. The resulting grooming is a canonical grooming. Thus we focus on the optimal traffic partition in the first phase.

5.1 BASIC PROPERTIES

In the subsection, we will present some basic properties of optimal traffic partitions.

LEMMA 12.4 *In any optimal traffic partition* $f = (f_1, \cdots, f_n)$, $f_i < 3$ *for all* $1 \leq i \leq n$, *and there is an optimal solution* $f = (f_1, \cdots, f_n)$ *with* $f_i \leq 2$ *for all* $1 \leq i \leq n$.

Proof. We prove the first part of lemma by contradiction. Let $f = (f_1, \cdots, f_n)$ be any optimal traffic partition with $f_i \geq 3$. Then in a canonical optimal grooming, there at least three low-speed rings devoted exclusively to node i. If we move the traffic carried in any three of these low-speed rings into one-speed ring, we save 6 low-speed ADMs and uses two new high-speed ADMs, and thus decrease the cost by 1. This contradicts to the optimality of $f = (f_1, \cdots, f_n)$.

We now prove the second part of lemma by contradiction. Let $f = (f_1, \cdots, f_n)$ be any optimal traffic partition which contains the least number of entries that are more than two. Suppose $f_i > 2$ for some $1 \leq i \leq n$. Then in a canonical optimal grooming of the traffic demands $\{f_1, \cdots, f_n\}$ into low-speed rings, at least $\lceil f_i \rceil + \lfloor f_i \rfloor$ ADMs are devoted to node i. Now we place such f_i amount of traffic from node i into $\left\lceil \frac{f_i}{4} \right\rceil$ new high-speed rings, i.e. set $f_i = 0$. Then in the new solution, a cost of at least $\lceil f_i \rceil + \lfloor f_i \rfloor$ is saved from the rings of speed g_1 while a cost of $5 \left\lceil \frac{f_i}{4} \right\rceil$ is added to the rings of speed g_2. As

$$\lceil f_i \rceil + \lfloor f_i \rfloor \geq 5 \left\lceil \frac{f_i}{4} \right\rceil$$

when $f_i > 2$, the new solution has no more cost than the solution f but contains one less entries which are more than two. This contradicts to the selection of f. Therefore, the lemma is true. ■

Intuitively, if a traffic can fill a high-speed ring, it should fill fully as many high-speed rings as possible to take advantage of the lower cost per bandwidth of the higher speed rings. The next lemma verifies such intuition.

LEMMA 12.5 *There is an optimal traffic partition $f = (f_1, \cdots, f_n)$ with $f_i \leq r_i \bmod 4$ for all $1 \leq i \leq n$.*

Proof. We prove the lemma by contradiction. Let $f = (f_1, \cdots, f_n)$ be any optimal traffic partition satisfying that $f_i \leq 2$ for all $1 \leq i \leq n$ and the cardinality of the set

$$\{1 \leq i \leq n \mid f_i > r_i \bmod 4\}$$

is the smallest. Assume that $f_i \leq r_i \bmod 4$ for some node i. Then in a canonical optimal grooming of the traffic carried in high-speed rings, in addition to $\left\lfloor \frac{r_i - f_i}{4} \right\rfloor$ high-speed rings which are devoted exclusively to node i, one high-speed ring carries the remaining $4 - f_i + r_i \bmod 4$ amount of traffic from node i. This high-speed ring must also carry traffic from other nodes, for otherwise we can we fill this ring fully with the traffic from node i without any additional cost but the amount of traffic placed in low-speed rings is $r_i \bmod 4$, which contradicts to the selection of $f = (f_1, \cdots, f_n)$. Let $x_i > 0$ be the amount of the traffic carried in this ring from nodes other than node i. Then $x_i > 1$ for otherwise we can decrease the total ADM cost by 0.5 by moving x_i to a dedicated low-speed ring, which again contradicts to the optimality of $f = (f_1, \cdots, f_n)$. As

$$4 - f_i + r_i \bmod 4 + x_i \leq 4,$$

we have

$$1 < x_i \leq r_i \bmod 4 + x_i \leq f_i \leq 2.$$

This implies that x_i is from only one node, say j, for otherwise the portion of the traffic from some node is less than one and again we can decrease the total ADM cost by the moving of it to a dedicated low-speed ring.

Now we look at the f_i amount of traffic from node i carried in low-speed rings. In a canonical optimal grooming, one ring carries the traffic of amount 1 from node i only, another ring carries $f_i - 1$ amount of traffic from node i and may carry additional traffic from other nodes.

Finally we relocate all traffic in these three rings as follows. Fill the high-speed ring fully with the traffic from node i. Fill the first low-speed ring fully with the traffic from node j. In the second low-speed ring, keep the original traffic not from node i, and place $r_i \bmod 4$ amount of traffic from node i and $x_i - 1$ amount of traffic from node j. With this modification, one high-speed ADM is saved but one additional low-speed ADM is used. So the total cost is decreased by $2.5 - 1 = 1.5$, which again contradicts to the optimality of $f = (f_1, \cdots, f_n)$. ∎

From the above lemma, there is an optimal solution in which $\lfloor \frac{r_i}{4} \rfloor$ high-speed rings are dedicated $r_i - r_i \bmod 4$ amount of traffic from node i for all $1 \leq i \leq n$. Thus from now on, we assume that $r_i < 4$ for all node i. For any traffic partition $f = (f_1, \cdots, f_n)$, let

$$S(f) = \{1 \leq i \leq n \mid 0 < f_i < r_i\},$$
$$U(f) = \{1 \leq i \leq n \mid f_i = 0 \text{ or } r_i\}.$$

Thus the traffic from any node in $S(f)$ is carried in both low-speed rings and high-speed rings, and the traffic from any node in $U(f)$ is carried in either low-speed rings or high-speed rings but not both.

The next lemma states that at any node, if the traffic of this node is carried in both types of rings, then the amount of traffic carried in low-speed rings is at most one; and if there is some traffic carried in a high-speed ring, its amount is more than one.

LEMMA 12.6 *Let $f = (f_1, \cdots, f_n)$ be any optimal traffic partition. Then for any $1 \leq i \leq n$, neither $1 < f_i < r_i$ nor $0 < r_i - f_i \leq 1$ is possible.*

Proof. Assume that $1 < f_i < r_i$. Then in a canonical optimal grooming, the total cost of ADMs used by the traffic r_i is at least

$$2 + 1 + 2.5 = 5.5,$$

as at least 2 low-speed ADMs is needed at node i, at least 1 low-speed ADM is needed at the hub, and at least 1 high-speed ADM is required at the node i. But if the traffic r_i is entirely carried by a high-speed ring, the cost of ADMs is at most

$$2.5 + 2.5 = 5 < 5.5,$$

which contradicts to the optimality of $f = (f_1, \cdots, f_n)$.

Now we assume that $0 < r_i - f_i \leq 1$. We remove the $r_i - f_i$ amount of traffic from the high-speed ring and put it in a dedicated low-speed ring. With this modification, at least one high-speed ring is saved and two additional low-speed ADMs are used. So the total cost is decreased by

$$2.5 - 2 = 0.5$$

which again is impossible as $f = (f_1, \cdots, f_n)$ is already optimal. ■

As a corollary of Lemma 12.6, in any canonical optimal grooming, any high-speed ring can carry traffic from at most three nodes.

The next lemma states that, at any node, when a traffic demand from a node is at most one, it should be always put in a low-speed ring; and when a traffic demand is more than three, it should be always put in a high-speed ring.

LEMMA 12.7 *Let $f = (f_1, \cdots, f_n)$ be any optimal traffic partition. Then for any $1 \leq i \leq n$, if $r_i \leq 1$, $f_i = r_i$; and if $r_i > 3$, $f_i = 0$.*

Proof. The first part follows directly from Lemma 12.6. Now we assume that $r_i > 3$ and $f_i > 0$. From Lemma 12.4 and Lemma 12.6, $0 < f_i \leq 1$, and thus $r_i - f_i > 2$. The $r_i - f_i$ amount of traffic from node i must share some traffic from other nodes, for otherwise we can put all traffic from node i in the high-speed ring and decreases the cost by at least one. From Lemma 12.6 if there is some traffic, from any node, carried in a high-speed ring, its amount is more than one. Thus the $r_i - f_i$ amount of traffic from node i share one high-speed ring with some amount, denoted by x_i, of traffic from exactly one node, say j. Note that

$$1 < x_i \leq 4 - r_i + f_i.$$

So we consider the following modification to a canonical optimal solution. We replace the f_i amount of traffic from node i in some low-speed ring by the f_i amount of traffic from node j. This may save one low-speed ADM. We then place the $x_i - f_i$ in a dedicated low-speed ring as

$$x_i - f_i \leq 4 - r_i < 1.$$

This adds two low-speed ADMs. Finally, we place all traffic from node i in the high-speed ring originally carrying the $r_i - f_i$ amount of traffic from node i and x_i amount of traffic from node j. This saves one high-speed ADM. Thus after the modification, the total ADM cost is decreased by at least $2.5 - 2 = 0.5$, which contradicts to the optimality of $f = (f_1, \cdots, f_n)$. ■

The above lemma implies that if $r_i \leq 1$ for any node $1 \leq i \leq n$, then all traffic must be carried in low-speed rings. In particular, if the traffic is uniform with amount r, the total ADM cost is $F(1, r, n)$. If $r_i > 3$ for any node $1 \leq i \leq n$, then all traffic must be carried in high-speed rings. As in the canonical grooming, the traffic demand from any node must be carried in a dedicated high-speed ring. Thus $2n$ high-speed ADMs are needed with cost $5n$ in total. A remark is such cost only accounts for the working ring, if we consider the protection as well, the total cost should then be doubled.

5.2 ALL TRAFFIC DEMANDS ARE AT MOST TWO

In the next lemma, we show that when the traffic demand from each node is at most two, then there is an optimal traffic partition in which none of them is carried in both low-speed rings and high-speed rings.

LEMMA 12.8 *If $r_i \leq 2$ for all $1 \leq i \leq n$, then there is an optimal traffic partition f with $S(f) = \emptyset$.*

Proof. We prove it by contradiction. Let $f = (f_1, \cdots, f_n)$ be any optimal traffic partition with the smallest $|S(f)|$. Let $i \in S(f)$ and consider any canonical optimal grooming. From Lemma 12.6, $0 < f_i \leq 1$ and $r_i - f_i > 1$. Thus in any canonical optimal grooming, the traffic from node i is carried in exactly one low-speed ring and exactly one high-speed ring. We concentrate on the high-speed ring carrying the $r_i - f_i$ amount of traffic from node i. It can carry traffic from at most three nodes. First of all, it must also carry some traffic from other nodes, for otherwise we can fill it with all traffic from node i and decreases the cost by at least one. Secondly, it is impossible that this high-speed ring carries the traffic from only two nodes, for otherwise we can put all traffic from these two nodes in this high-speed ring, which can also save at least one low-speed ADM. Thus this high-speed ring must carry traffic from exactly three nodes. We denote the other two nodes other than node i by j and k. We show that $j, k \in U(f)$. Suppose to the contrary. We modify the placement of the traffic from these three nodes as follows. We use the high-speed ring to carry the whole traffic from node i and the whole traffic from node j and nothing else. We add at most two new dedicated low-speed rings to carry the traffic from node k. We save one high-speed ADM and add at most two more low-speed ADMs. Thus the modification decreases the total cost by at least 0.5, which contradicts to the optimality of f. Therefore both j and k are in $U(f)$, that is all traffic from node j and node k are carried in the high-speed ring. As $r_i - f_i > 1$,

$$r_j + r_k \leq 4 - (r_i - f_i) < 4 - 1 = 3.$$

So we can modify the placement of the traffic from nodes i, j and k as follows. We place all the traffic from node i and nothing else in the high-speed ring, and use at most three new low-speed rings to carry all traffic from nodes j and k. Then two high-speed ADMs are saved, and at most five low-speed ADMs are added. The resulting solution has the same cost as f but it contains one less nodes whose traffic are carried in both low-speed rings and high-speed rings. This contradicts to that $|S(f)|$ is the smallest. Thus the lemma is true. ■

The next lemma states that when the traffic demand from each node is at most $\frac{3}{2}$, then we can put all traffic in the low-speed rings.

LEMMA 12.9 *If $r_i \leq \frac{3}{2}$ for all $1 \leq i \leq n$, then the traffic partition $f = (f_1, \cdots, f_n)$ where $f_i = r_i$ for all $1 \leq i \leq n$ is optimal.*

Proof. We prove it by contradiction. Let $f = (f_1, \cdots, f_n)$ be any optimal traffic partition with $f_i = 0$ or r_i for all $1 \leq i \leq n$ and the smallest number of zero entries. Consider any canonical optimal grooming. As any high-speed ring carries traffic from at most three nodes. We consider the following three cases.

If a high-speed ring carries traffic from only one node, we can use at most two new low-speed rings to carry all traffic from this node. This modification saves two high-speed ADMs and uses at most four low-speed ADMs. Thus the cost is decreased by 0.5, which contradicts to the optimality of $f = (f_1, \cdots, f_n)$.

If a high-speed ring carries traffic from two nodes, we can use at most three new low-speed rings to carry all traffic from these nodes. This modification saves three high-speed ADMs and uses at most seven low-speed ADMs. Thus the cost is decreased by 0.5, which also contradicts to the optimality of $f = (f_1, \cdots, f_n)$.

If a high-speed ring carries traffic from three nodes, we use at most four new low-speed rings to carry all traffic in this high-speed ring. This modification saves four high-speed ADMs and uses at most ten low-speed ADMs. The resulting solution has the same cost as f , but the number of zero entries is decreased by three, which contradicts to the selection of f. Therefore, the lemma is true. ■

The above lemma implies if the traffic is uniform with demand $r \leq \frac{3}{2}$, the minimum cost of ADMs is $F(1, r, n)$.

We now consider the traffic with demands more than $\frac{3}{2}$ but at most two.

LEMMA 12.10 *Suppose that $\frac{3}{2} < r_i \leq 2$ for all $1 \leq i \leq n$. If n is even, then the traffic partition $f = (f_1, \cdots, f_n)$ where $f_i = 0$ for all $1 \leq i \leq n$ is optimal. If n is odd, then for any $1 \leq j \leq n$ the traffic partition $f = (f_1, \cdots, f_n)$ where $f_i = 0$ for $i \neq j$ and $f_j = r_j$ is optimal.*

Proof. We also prove it by contradiction that there is an optimal traffic partition $f = (f_1, \cdots, f_n)$ with $f_i = 0$ or r_i for all $1 \leq i \leq n$ and at most one non-zero entries. Let $f = (f_1, \cdots, f_n)$ be any optimal traffic partition with $f_i = 0$ or r_i for all $1 \leq i \leq n$ and the smallest number of non-zero entries. Assume that $f_i = r_i$ and $f_j = r_j$. Consider any canonical optimal grooming. There are two low-speed rings devoted to node i and two low-speed rings devoted to node j. We relocate the traffic from node i and node j to one new high-speed ring. This modification saves 8 low-speed ADMs and uses 3 high-speed ADMs. The total cost is decreased by 0.5. This contradicts to the optimality of $f = (f_1, \cdots, f_n)$.

Now let $f = (f_1, \cdots, f_n)$ be any optimal traffic partition with $f_i = 0$ or r_i for all $1 \leq i \leq n$ and at most one non-zero entries. Note that in any canonical optimal grooming, each high-speed ring must carry traffic from two nodes, for otherwise we can move it to two low-speed rings and the cost would be decreased by 1. Thus if n is even, $f_i = 0$ for all $1 \leq i \leq n$, and if n is odd, there is exactly one $1 \leq i \leq n$ with $f_i = r_i$. ■

From the above lemma, if $\frac{3}{2} < r_i \leq 2$ for all $1 \leq i \leq n$ we can provide optimal grooming as follows. If n is even, then all traffic is carried in high-speed rings, and each high-speed ring carries the whole traffic from two nodes. It requires totally $1.5n$ high-speed ADMs with total cost

$$1.5n \cdot 2.5 = 3.75n.$$

If n is odd, then the traffic from one node is carried in two low-speed rings to carry the whole traffic from a node, and the traffic from all other nodes are carried in the high-speed rings, with each ring dedicated to a pair of nodes. Thus 4 low-speed ADMs and $1.5\,(n-1)$ high-speed ADMs are used. So the total ADM cost is

$$4 + 1.5\,(n-1) \cdot 2.5 = 3.75n + 1.5.$$

5.3 ALL TRAFFIC DEMANDS ARE MORE THAN TWO

In general, each high-speed ring can carry traffic from at most three nodes. The next lemma states that if all traffic demands are more than two, then in any canonical optimal grooming no high-speed ring can carry traffic from three nodes.

LEMMA 12.11 *If $r_i > 2$ for all node i, then in any canonical optimal grooming each high-speed ring carries traffic from at most two nodes.*

Proof. We prove it by contradiction. Consider a canonical optimal grooming with traffic partition $f = (f_1, \cdots, f_n)$. Assume that three nodes i, j and k appear in a high-speed ring. Then $i, j, k \in S(f)$ for otherwise

$$(r_i - f_i) + (r_j - f_j) + (r_k - f_k) > 2 + 1 + 1 = 4.$$

As

$$(r_i - f_i) + (r_j - f_j) + (r_k - f_k) \leq 4,$$

we have

$$f_i + f_j + f_k \geq r_i + r_j + r_k - 4 > 2.$$

As $f_k \leq 1$, $f_i + f_j > 1$, so are $f_i + f_k$ and $f_j + f_k$. This means that all the three nodes must appear in three distinct low-speed rings. Assume these three rings carry x_i, x_j and x_k amount of the traffic from other nodes respectively. Then we have

$$x_i + x_j + x_k \leq 3 - (f_i + f_j + f_k) < 1.$$

Note that

$$r_i + r_j + r_k \leq f_i + f_j + f_k + 4 \leq 7,$$

As $r_k > 2$, $r_i + r_j < 5$, so are $r_i + r_k$ and $r_j + r_k$.

Now we relocate the traffic carried in these three low-speed rings and the high-speed ring as follows. We place the whole traffic from node i in the high-speed ring, place the whole traffic from node j and $4-r_j$ amount of traffic from node k in a new high-speed ring, and place $r_j + r_k - 4$ amount of traffic from node k in a low-speed ring as

$$0 < r_j + r_k - 4 < 1.$$

The x_i, x_j and x_k amount of the traffic from other nodes are carried exclusively in another low-speed ring. After the relocation, we save three low-speed ADMs and add one high-speed ADM. So the total cost is decreased by 0.5, which is a contradiction. ∎

The following lemma states that if all traffic demands are greater than two, we can concentrate on those canonical grooming in which exactly one node in each high-speed ring has its whole traffic carried in this high-speed ring.

LEMMA 12.12 *If $r_i > 2$ for all $1 \leq i \leq n$, then there is a canonical optimal grooming in which exactly one node in each high-speed ring has its whole traffic carried in this high-speed ring.*

Proof. We prove it by contradiction. Consider a canonical optimal grooming with traffic partition $f = (f_1, \cdots, f_n)$ with $f_i \leq 2$ for all $1 \leq i \leq n$. From Lemma 12.6, $f_i \leq 1$ for all $1 \leq i \leq n$. Thus for all $1 \leq i \leq n$,

$$r_i - f_i > 2 - 1 = 1.$$

If a high-speed carries traffic from only one node, then it must carry the whole traffic from that node. Now we consider a high-speed ring which carries traffic from two nodes $i, j \in S(f)$.

We relocate the traffic from node i and node j as follows. The high-speed ring carries r_i amount of traffic from node i, and $4 - r_i$ amount of traffic from node j. We replace the original f_i amount of traffic from node i in a low-speed ring by f_i amount of traffic from node j. The cost of the result grooming is not increased. We repeat such procedure for all high-speed rings which each carry traffic from two nodes that are both in $S(f)$. In the end, we come up with a grooming in which each high-speed ring carries the whole traffic from at least one node. Finally we use a canonical grooming to place all traffic carried in low-speed rings. Then the resulting grooming satisfies the requirement given in the lemma. ∎

When all traffic demands are greater than $\frac{5}{2}$, the following lemma gives an optimal traffic partition.

LEMMA 12.13 *If $r_i > \frac{5}{2}$ for all $1 \leq i \leq n$, then the traffic partition $f = (f_1, \cdots, f_n)$ where $f_i = 0$ for all $1 \leq i \leq n$ is optimal.*

Proof. We consider a canonical optimal grooming with the traffic partition $f = (f_1, \cdots, f_n)$ in which each high-speed ring carries the whole traffic from at least one node. Assume that $f_i > 0$ for some $1 \leq i \leq n$. From Lemma 12.6, $f_i \leq 1$. Furthermore, the high-speed ring where node i appears must carry the whole traffic from another node, say j, and no other traffic. As

$$(r_i - f_i) + r_j \leq 4,$$

we have

$$f_i \geq r_i + r_j - 4 > 1.$$

This contradicts to $f_i \leq 1$. ■

The above lemma suggests that if all traffic demands are more than $\frac{5}{2}$, we should carry all traffic in high-speed rings. In this optimal traffic partition, the canonical grooming is unique and each high-speed ring carries exclusively the whole traffic from only one node. Thus the minimal total ADM cost (in the working ring) is $5n$.

Finally we consider the traffic with demands at most $\frac{5}{2}$ but more than two. The next lemma states that if all traffic demands are at most $\frac{5}{2}$, then in any optimal grooming there is at most one high-speed ring which carries exclusively the whole traffic from exactly one node.

LEMMA 12.14 *If $r_i \leq \frac{5}{2}$ for all $1 \leq i \leq n$, then in any optimal grooming at most one high-speed ring carries exclusively the whole traffic from exactly one node.*

Proof. We prove it by contradiction. Consider an optimal grooming with traffic partition $f = (f_1, \cdots, f_n)$ in which there are two high-speed ring dedicated to node i and node j repulsively. We relocate the traffic from node i and node j as follows. We place r_i amount of traffic from node i, and $\min\{4 - r_i, r_j\}$ amount of traffic from node j on one high-speed ring, and if $r_i + r_j > 4$ we place $r_i + r_j - 4$ amount of traffic from node j on one low-speed ring. This modification saves one high-speed ADM and adds at most two low-speed ADMs. The cost is decreased by at least 0.5, which is a contradiction. ∎

From Lemma 12.11, 12.12 and 12.14, if n is even and $2 < r_i \leq \frac{5}{2}$ for all $1 \leq i \leq n$, then there is a canonical optimal grooming in which half nodes have their traffic carried in high-speed rings and the half node have their traffic carried in both high-speed rings and low-speed ring, and each high-speed is *fully* filled with the whole traffic from one node in the first half and a portion of traffic from a node in the second half. if n is odd and $2 < r_i \leq \frac{5}{2}$ for all $1 \leq i \leq n$, then there is a canonical optimal grooming in which the traffic from one node is carried exclusively in a high-speed ring and the traffic from other nodes are carried in the same way as the number of nodes is even. However, how to select the set of nodes to be carried wholly in high-speed rings and how to form node pairs to appear in high-speed rings remains open. But if the traffic is uniform, these two questions can be easily solved. We can select any $\lceil \frac{n}{2} \rceil$ nodes to be carried wholly in high-speed rings, and the pairing between those nodes and the remaining nodes can be selected arbitrarily. Thus, for uniform traffic with demand $2 < r \leq \frac{5}{2}$, the total ADM cost in the working ring is

$$3.25n + F\left(1, 2r - 4, \frac{n}{2}\right)$$

if n is even, and is

$$\begin{aligned} & 5 + 3.25\,(n-1) + F\left(1, 2r - 4, \frac{n-1}{2}\right) \\ = {} & 1.75 + 3.25n + F\left(1, 2r - 4, \frac{n-1}{2}\right) \end{aligned}$$

if n is odd.

6. SUMMARY

In this chapter, we first consider the optimal traffic grooming in single-hub WDM self-healing rings with single line-speed. For uniform traffic demands, we have presented the optimal grooming algorithm and the analytic formula for minimal ADM cost. For non-uniform traffic demands, efficient approximation algorithm is developed and its performance is also analyzed.

The second part of the this chapter deals with the optimal traffic grooming in single-hub WDM self-healing rings with two line-speeds. For uniform traffic demands, we have provided optimal traffic partition and grooming, which is summarized in Table 12.1. For non-uniform traffic demands, optimal or suboptimal solutions have been developed depending on the range of all demands. If all demands are at most 1.5, then all of them are carried in low-speed rings. If all traffic demands are greater than 1.5 but less than two, then with even n, all of them are carried in high-speed rings and the total ADM cost is $3.75n$; with odd n, all of them except an arbitrary one are carried in high-speed rings and the total ADM cost is $3.75n + 1.5$. These costs remain the same as long as all demands are greater than 1.5 but less than two. If all traffic demands are greater than 2.5, all of them are carried in high-speed rings and the total ADM cost is $5n$. Such cost also remain the same as long as all demands are greater than 2.5. When all traffic demands are greater than two but less than 2.5, the optimal solution is a little complicated. We first pair up the n nodes. If n is odd, some node is stand-alone and its whole traffic is carried in a high-speed ring. For each pair of nodes i and j, we use a high-speed ring to carry the whole traffic from node i and the remaining capacity is used to carry the traffic from node j.

Range of all r's	$(f_1, f_2, \cdots, f_n)$
$(0, 1.5]$	$f_i = r, \forall i$
$(1.5, 2]$,$n = 2k$	$f_i = 0, \forall i$
$(1.5, 2]$,$n = 2k + 1$	$f_i = 0, \forall i \neq j; f_j = r$
$(2, 2.5]$	$f_{2i-1} = 0, f_{2i} = 2r - 4$
$(2.5, 4]$	$f_i = 0, \forall i$

Table 12.1. Optimal grooming of uniform traffic demands in WDM self-healing ring with two line-speeds.

When the WDM self-healing rings have more than two line-speeds, the techniques and properties developed in this chapter can be extended to provide general optimal or suboptimal groomings. The details are omitted here.

Appendix A
Regular Interconnection Topologies

1. HYPERCUBE

The n-dimensional hypercube Q_n, or n-cube in short, has $N = 2^n$ nodes which are labeled by n-bit binary numbers. The nodal degree of n-cube is n. For each node $0 \leq a \leq N-1$, its outgoing links are

$$a \rightarrow a \oplus 2^i, 0 \leq i \leq n-1$$

and its incoming links are

$$a \oplus 2^i \rightarrow a, 0 \leq i \leq n-1$$

where the operator $\oplus$ is the parity operator (bit-wise exclusive or). The hypercube presents several attractive properties, such as simple self-routing, logarithm diameter and high fault-tolerance.

2. GENERALIZED DE BRUIJN DIGRAPH

The generalized de Bruijn digraph is the generalization of de Bruijn digraph [32] by I. Imase and M. Itoh [80] and S.M. Reddy, D.K. Pradhan and J.G. Kuhl [131], independently. A generalized de Bruijn digraph $D(N,p)$ has N vertices and each vertex has p outgoing links and p incoming links. For a vertex $0 \leq a \leq N-1$, its outgoing links are

$$a \rightarrow (ap+i) \bmod N, 0 \leq i \leq p-1$$

and its i-th incoming link is

$$\left\lfloor \frac{jN+a}{p} \right\rfloor \rightarrow a, 0 \leq j \leq p-1$$

Fig. A.1 shows a generalized de Bruijn digraph $D(8,4)$.

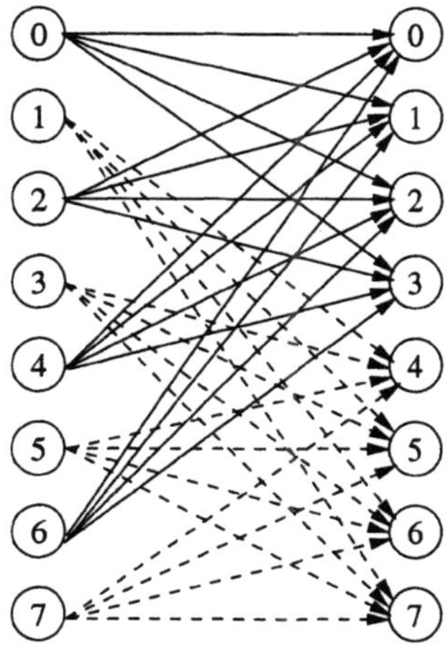

Figure A.1. The generalized de Bruijn digraph $D(8,4)$

When $N = p^k$ for some $k > 0$, then $D(N,p)$ is isomorphic to the de Bruijn graph with the same size and degree. The generalized de Bruijn graph $D(N,p)$ has a lot of attractive properties. First as hypercubes, it has logarithm diameter and good connectivity. $D(N,p)$ has diameter at most $\lceil \log_p N \rceil$ and is $(p-1)$-connected [82]. Secondly, it also has some unique properties that hypercube does not have. In the generalized de Bruijn digraph, the network size and the degree of each vertex are independent, and the network can have any number of size. Therefore, he generalized de Bruijn digraph have a flexibility in scaling the network size. This is one of the main motivations that we propose the generalized de Bruijn digraph as regular communication pattern.

3. GENERALIZED KAUTZ DIGRAPH

The generalized Kautz digraph is also called Imase-Itoh digraph. It is the generalization of Kautz digraph [89] by I. Imase and M. Itoh [81]. A generalized Kautz digraph $K(N,p)$ has N vertices and each vertex has p outgoing links and p incoming links. For an vertex $0 \le a \le N-1$, its outgoing links are

$$a \to ((N-a-1)p+i) \bmod N, 0 \le i \le p-1$$

and its incoming links are

$$N - 1 - \left\lfloor \frac{jN + a}{p} \right\rfloor \to a, 0 \le j \le p - 1$$

It was shown in [55] that when $N = p^t + p^{t-1}$ for some $t > 0$, then $K(N, p)$ is isomorphic to the Kautz digraph with same size and degree. The generalized Kautz digraph $K(N, p)$ has a lot of attractive properties. It has diameter at most $\lceil \log_p N \rceil$. Moreover, Imase and Itoh [81] proved that if $N = p^s + p^{s+t}$ for some s and some odd t, then $K(N, p)$ has diameter $\lceil \log_p n \rceil - 1$. $K(N, p)$ also has a good connectivity. M. Imase, T. Soneoka and K. Okada [82] is at least $p - 1$-connected.

4. STAR GRAPH

The star graph is a member of a class of graphs called *Cayley graphs*. This class of graph uses a group-theoretic approach as a basis for defining graphs. Let G be a finite group and S a set of generators for G. The generator set S is required to be closed under inversion, i.e., for any $s \in S$, s^{-1} is also in S. The *Cayley graph of G with generating set S*, denoted by $Cay(S : G)$, is defined as follows.

1. Each element of G is a vertex of $Cay(S : G)$.
2. For x and y in G, there is a link between x and y if and only if $x \cdot s = y$ for some $s \in S$.

The star graph is defined through the permutation group. Let P_n be the set consisting of all permutations on n symbols $\{1, 2, \cdots, n\}$. A permutation $\sigma \in P_n$ is represented by $\sigma(1)\sigma(2)\cdots\sigma(n)$. For example, the permutation $\sigma = 2143$ represents that $\sigma(1) = 2, \sigma(2) = 1, \sigma(3) = 4$ and $\sigma(4) = 3$. The permutation $1234\cdots n$ is called the identity permutation. A *transposition* is a permutation which interchanges only two symbols. In this paper, we are interested in the following $n - 1$ transpositions which transpose the symbol 1 and any other symbols:

$$\begin{aligned} s_2 &= 213\cdots n, \\ s_3 &= 321\cdots n, \\ &\vdots \\ s_n &= n23\cdots 1. \end{aligned}$$

Traditionally, the product of two permutations has two different definitions. The two definitions differ at the order in which the multiplication is taken.

1. The multiplication is taken from left to right. We use "$\cdot$" to denote this multiplication operation. The product of any two permutations α and β is given by

$$\alpha \cdot \beta(i) = \beta(\alpha(i))$$

for any $1 \leq i \leq n$.

2. The multiplication is taken from right to left. We use "$\circ$" to denote this multiplication operation. The product of any two permutations α and β is given by

$$\alpha \circ \beta(i) = \alpha(\beta(i))$$

for any $1 \leq i \leq n$.

It's well-known that under both multiplications,

1. P_n is a group;
2. s_k is the inverse of itself for any $2 \leq k \leq n$;
3. the set $S = \{s_2, s_3, \cdots, s_n\}$ is a generator set of P_n.

An n-dimensional star graph S_n, also to as n-star, is the Cayley graph $Cay(S : G)$ where G is P_n, and S consists of $n-1$ transpositions $\{s_2, s_2, \cdots, s_n\}$. The topology of S_n depends upon the multiplication operation of the permutation group. Figure A.2 illustrates the 3-star and 4-star if the multiplication is taken from left to right. Figure A.3 illustrates the 3-star and 4-star if the multiplication is taken from right to left.

The star graphs have a lot of elegant properties. All star graphs are vertex and edge symmetric [5]. The diameter of the n-star is $\lfloor \frac{3(n-1)}{2} \rfloor$ [6].

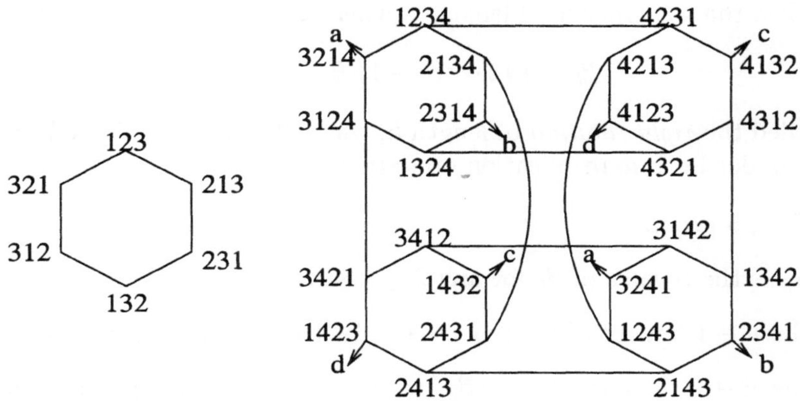

Figure A.2. The topology of 3-star and 4-star of the model that the multiplication in the permutation group is taken from left to right.

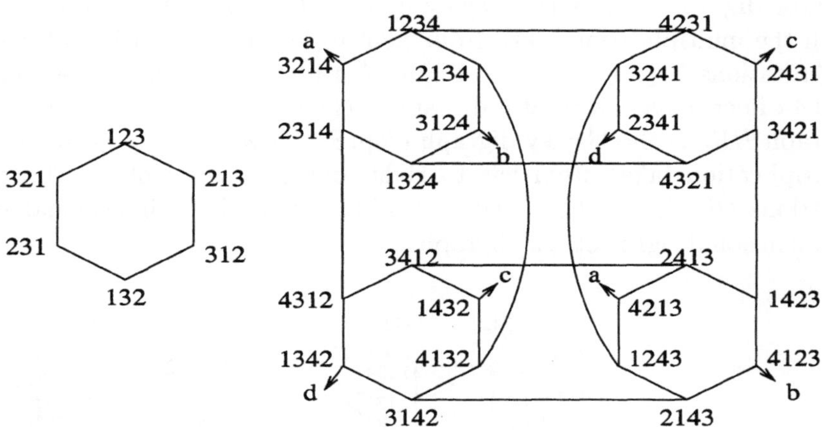

Figure A.3. The topology of 3-star and 4-star of the model that the multiplication in the permutation group is taken from right to left.

5. ROTATOR DIGRAPH

The rotator digraph is another member of Cayley digraphs. The rotator digraph is also defined through the permutation group. The permutation

$$\alpha_k = 23 \cdots k1(k+1) \cdots n$$

is called the *left rotation* of length k, where $2 \leq k \leq n$. The permutation

$$\beta_k = k12\cdots(k-1)(k+1)\cdots n$$

is called the *right rotation* of length k, where $2 \leq k \leq n$. It's well-known that under both multiplication operations

1. P_n is a group;
2. α_k is the reverse of β_k for any $2 \leq k \leq n$;
3. the $n-1$ left rotations $\{\alpha_2, \alpha_3, \cdots, \alpha_n\}$ is a generator set for P_n;
4. the $n-1$ right rotations $\{\beta_2, \beta_3, \cdots, \beta_n\}$ is also a generator set for P_n.

There are two models of rotator digraphs. The n-dimensional left rotator digraph LR_n is the Cayley digraph $Cay(S:G)$ where G is P_n with the multiplication taken from right to left and S consists of $n-1$ left rotations $\{\alpha_2, \alpha_3, \cdots, \alpha_n\}$. Figure A.4 illustrates the 3-dimensional and 4-dimensional left rotator digraphs. The n-dimensional right rotator digraph RR_n is the Cayley digraph $Cay(S:G)$ where G is P_n with the multiplication taken from left to right and S consists of $n-1$ right rotations $\{\beta_2, \beta_3, \cdots, \beta_n\}$. Figure A.5 illustrates the 3-dimensional and 4-dimensional right rotator digraphs.

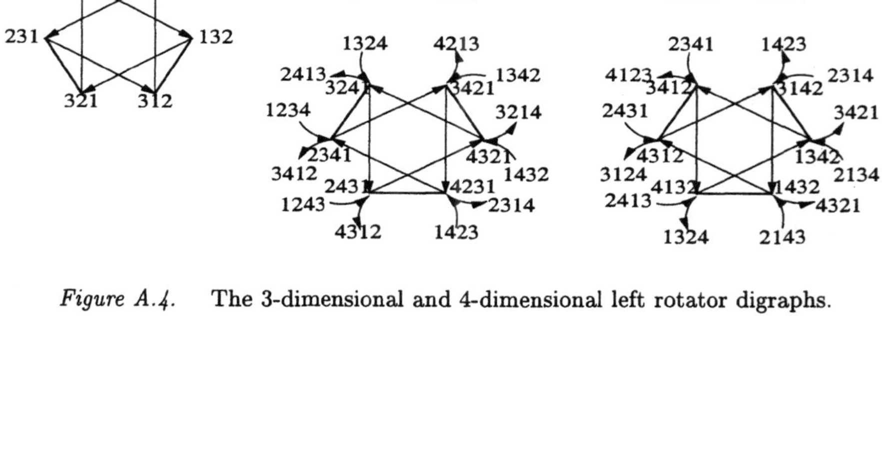

Figure A.4. The 3-dimensional and 4-dimensional left rotator digraphs.

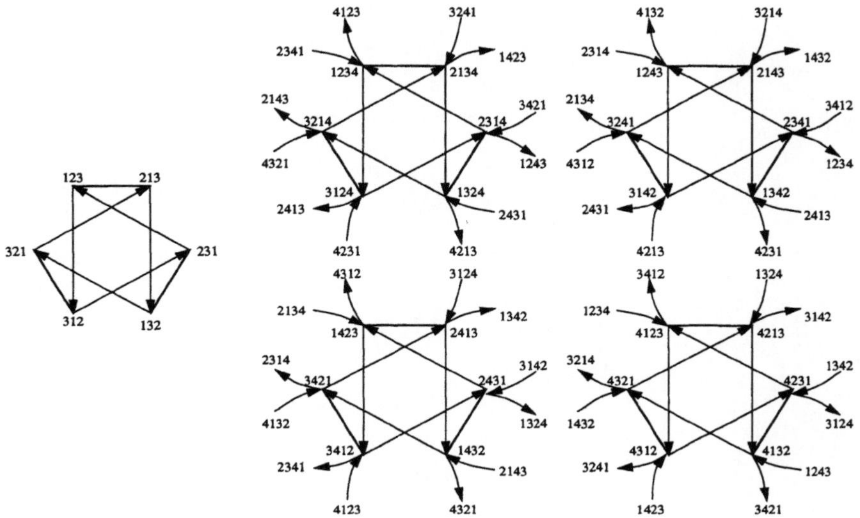

Figure A.5. The 3-dimensional and 4-dimensional right rotator digraphs.

The rotator digraphs have a lot of attractive properties. All rotator digraphs are vertex and edge symmetric. In [43], it is proved that the diameter of the n-rotator is $n-1$, and has a simple and optimal routing algorithm. They share the fault handling capacity of hierarchical Cayley graphs, including the Star and Pancake graphs and the binary hypercubes, with good performance possible because of the small diameter and average routing distance.

References

[1] Kevin M. Able, "Optical Fiber Standards Evolve", *Lightwave Xtra!*, February 1998.

[2] A.S. Acampora, M.J. Karol, M.G. Hluchyj, "Terabit Lightwave Networks: The Multihop Approach", *AT&T Technical Journal*, vol. 66, no. 6, Nov/Dec 1987, pp. 21-34.

[3] Adams et al., "Nonlinearities in semiconductor laser amplifiers", *Opt. and Quantum Elect.*, vol. 27, 1995, pp. 113.

[4] A. Aggarwal, A. Bar-Noy, D. Coppersmith, R. Ramaswani, B. Schieber, and M. Sudan, "Efficient Routing and Scheduling Algorithms for Optical Networks", *Proceedings of the 5th Annual ACM-SAM Symposium on Discrete Algorithms*, pp. 412–423, 1994.

[5] S.B. Akers, B. Krishnamurthy, "A Group-theoretic Model for Symmetric Interconnection Networks", *IEEE Transaction on Computers*, Vol. 38, pp. 555-566, 1989.

[6] S.B. Akers, D. Harel, B. Krishnamurthy, "The Star Graph: An Attractive Alternative to The n-cube", *Proc. Int. Conf. Parallel Processing*, pp. 393-400, 1987.

[7] Daniel Y. Al-Salameh, Mohammad T. Fatehi, William J. Gartner, Stan Lumish, Bruce L. Nelson, and Kamal K. Raychaudhuri, "Optical Networking", *Bell Labs Technical Journal*, January–March 1998, pp. 39-61.

[8] S. Alexander, et. al., "A precompetitive Consortium on Wide-Band All Optical Networks", *Journal of Lightwave Technology*, vol. 11, pp. 714-733, No. 5/6, May/June 1993.

[9] Rod C. Alferness, Paul A. Bonenfant, Curtis J. Newton, Kevin A. Sparks, and Eve L. Varma, "A Practical Vision for Optical Transport Networking", Bell Labs Tech. J., Vol. 4, No. 1, Jan.–Mar. 1999, pp. 1–16.

[10] K. A. Aly, " Reconfigurable WDM Interconnection of Large-scale Multiprocessor Systems", Ph.D dissertation, Department of Electrical And Computer Engineering, SUNY at Buffalo, July 1993.

[11] K. A. Aly, P. W. Dowd, " A class of Scalable Optical Interconnection Networks through Discrete Broadcast-select Multi-domain WDM", *Proc. IEEE INFOCOM'94*, (Torando, Canda), June 1994.

[12] V. Arya, D. W. Sherrer, A. Wang, R. O. Claus, and M. Jones, "Temperature compensation scheme for refractive index grating–based optical fiber devices", *Proceedings of the SPIE*, vol. 2594, pp. 52–59, 1996.

[13] Y. Aumann and Y. Rabani, "Improved Bounds for All Optical Routing", *Proceedings of the Sixth Annual ACM-SIAM Symposium on Discrete Algorithms*, pp. 567-576, San Francisco, California, pp. 22–24 January 1995.

[14] B. Awerbuch, Y. Azar, A. Fiat, S. Leonardi, and A. Rosen, "Online Competitive Algorithms for Call Admission in Optical Network", *Proceedings of the 4th Annual European Symposium on Algorithms*, Lecture Notes in Computer Science 1136, pp. 431–444, Springer-Verlag, 1996.

[15] R. Ballart, Y.C. Ching, "SONET: Now It's the Standard Optical Network", *IEEE Communications Magazine*, pp. 8-15, March 1989.

[16] J. Babcock. SONET: A Practical Perspective. In Business Communications Review, Sept 1990: 59–63.

[17] B. Baker, "Approximation algorithms for NP-complete problems on planar graphs," *Proc. 24th IEEE Symp. on Foundations of Computer Science*, 1983.

[18] R. Bar-Yehuda, S. Even, "On approximating a vertex cover for planar graphs," *Proc. 14th ACM Symp. on Theory of Computing*, 1982.

[19] R.A. Barry and P.A. Humblet, "Bounds on the Number of Wavelengths Needed in WDM Networks", *LEOS'92 Summer Tropical Mtg. Digest*, pp. 114–127, 1992.

[20] R.A. Barry and P.A. Humblet, "On the Number of Wavelengths and Switches in All Optical Networks", *IEEE Transactions on Communications*, pp. 583–591, Vol. 42, No. 2/3/4, February/March/April, 1994.

[21] Y. Bartal, A. Fiat, and S. Leonardi, "Lower Bounds for On-Line Graph Problems with Application to On-Line Circuit and Optical Routing", *Proceedings of the 28th ACM Symposium on Theory of Computing*, pp. 531–540, 1996.

[22] Y. Bartal and S. Leonardi, "On-Line Routing in All-Optical Networks", *Proceedings of the 24th International Colloquium on Automata, Languages and Programming*, LNCS 1256, pp. 516–526. Springer-Verlag, 1997.

[23] J.-C. Bermond, L. Gargano, S. Perennes, A. Rescigno, and U. Vaccaro, "Efficient collective communication in optical networks", *Proceedings of the 23th International Colloquium on Automata, Languages and Programming (ICALP '96)*, July 1996.

[24] B. K. Bhattacharya and D. Kaller, "An $O(m + n \log n)$ algorithm for the maximum-clique problem in circular-arc graphs", *Journal of Algorithms*, 25(2): 336-358, November, 1997.

[25] L.N. Bhuyan, D.P. Agrawal, "Generalized Hypercube and Hypercube Structures for a Computer Network", *IEEE Transaction on Computers*, vol. C-33, pp. 323-333, April 1984.

[26] C.A. Brackett, "On the Capacity of Multiwavelength Optical-Star Packet Switches", *IEEE Lightwave Magazine*, May 1991, pp. 33-37.

[27] Blache, G., Karpinski, M., and Wirtgen, J. (1998), "On approximation intractability of the bandwidth problem", *Technical Report TR98-014, ECCC*, ftp://ftp.eccc.uni-trier.de/pub/eccc/reports/1998/TR98-014/index.html

[28] C. A. Brackett, " Dense wavelength division multiplexing networks: Principles and applications", *IEEE Journal on Selected Areas in Communications*, vol. 8, pp. 948–964, Aug. 1990.

[29] C. A. Brackett et al., "A scalable multiwavelength multihop optical network: a proposal for research on all-optical networks", *IEEE/OSA Journal of Lightwave Technology*, vol. 11, pp. 736–753, May/June 1993.

[30] S. Bregni, "A historical perspective on telecommunications network synchronization", IEEE Communications Magazine Volume: 36(6), June 1998, pp 158 –166.

[31] D. Brooks and S. Ruschin, "Integrated electrooptic multielectrode tunable filter", *J. Lightwave Tech.*, vol. 13, no. 7, July 1995, pp. 1508-13.

[32] N.G. de Bruijn, "A Combinatorial Problem", *Nederl, Akad. Wetensh. Proc.* 49 (1946), pp. 758-764.

[33] W.E. Burr, "The FDDI Optical Data Link", *IEEE Communications Magazine*, Vol.24, No.5, May 1986.

[34] Yair Caro, and Raphael Yuster, "Covering Graphs: The Covering Problem Solved", *Journal of Combinatorial Theory*, Series A 83, 273 282 (1998).

[35] Tamra Carpenter, Steven Cosares, and Iraj Saniee, "Demand Routing and Slotting on Ring Networks", *DIMACS Technical Report* 97-02, 1997.

[36] M.S. Chen, N.R. Dono, R. Ramaswami, "A Media Access-Protocol for Packet-Switched Wavelength Division Multiaccess Metropolitan Networks", *IEEE Journal on Selected Areas in Communications*, vol. 8, no. 6, Aug 1990, pp. 1048-1057.

[37] I. Chlamtac, A. Ganz, "Toward Alternative High-Speed Network Concepts: The SWIFT Architecture", *IEEE Trans. on Communications*, vol. 38, no. 4, Apr 1990, pp. 431-439.

[38] H. Choi, H.-A. Choi and M. Azizoglu, "On the All-to-All Broadcast Problem in Optical Networks," *Proc. Infocom'97*, Kobe, Japan, April 1997.

[39] V. Chvátal, "A Greedy Heuristic for the Set-Covering Problem", *Mathematics of Operations Research*, Vol. 4, No. 3, pp. 233–235, 1979.

[40] E. G. Coffman Jr. , M. R., Garey, and D. S. Johnson, "Approximation algorithms for bin-packing - a survey", in *Approximation algorithms for NP-hard problems*, PWS Publishing Company, Boston, pp. 46-93, 1997.

[41] C.J. Colbourn, and A. Rosa, *Triple Systems*, Oxford University Press, 1999.

[42] C. Colbourn, and P.-J. Wan, "Minimizing Drop Cost for SONET/WDM Networks with $\frac{1}{8}$ Wavelength Requirements", submitted for preparation.

[43] P.F. Corbett, "Rotator Graphs: AN Efficient Topology for Point-to-Point Multiprocessor Networks", *IEEE Transactions on Parallel And Distributed Systems*, Vol. 3, No. 5, Sep 1992. pp-622-626.

[44] G. Cornuejols. M. Fisher, G. Nemhauser, "Location of Bank Accounts to Optimize Float", *Management Science*, vol. 23, pp. 789–810, 1977.

[45] S. Cosares and I. Saniee, "An optimization problem related to balancing loads on SONET rings", *Telecommunications Systems*, 3:165-181, 1994.

[46] T. Cox, F. Dix, C. Hemrick, J. McRoberts, "SMDS: The Beginning of WAN Superhighways", *Data Communications*, April 1991.

[47] T.Q. Dam, K.A. Williams, D.H.C. Du, "A Media-Access Protocol for Time and Wavelength Division Multiplexed Passive Star Networks", Technical Report 91-63, Computer Science Dept., University of Minnesota.

[48] X. Deng, G. Li, W. Zang and Y. Zhou, "2-Approximation algorithm for path coloring on trees of rings", *Eleventh Annual International Symposium on Algorithms And Computation*, December, 2000.

[49] E.A. Dinits, A.V. Karzanov, and M.L. Lomonosov, "On the structure of a family of minimal weighted cuts in a graph", in *Studies in Discrete Mathematics*, pp. 290-306, 1976.

[50] P. Dowd, "Wavelength Division Multiple Access Channel Hypercube Processor Interconnection", *IEEE Transactions on Computers*, vol. 41, no. 10, Oct 1992, pp. 1223-1241.

[51] R. Downey, M. Fellows, "Parametrized Computational Feasibility," in *Feasible Mathematics II*, P. Clote and J. Remmel, eds., Birkhauser, 1994.

[52] C. Dragone, "Efficient N x N Star Coupler Based on Fourier Optics", *Electronics Letters*, vol. 24, no. 15, Jul 1988, pp. 942-944.

[53] C. Dragone, C. A. Edwards, and R. C. Kistler, "Integrated optics $N \times N$ multiplexer on silicon", IEEE Photon. Technol. Lett., vol. 4, pp. 896–899, 1991.

[54] J. Drake, "A review of the four major SONET/SDH rings", IEEE International Conference on Communications, 1993, Volume 2, pp 878 -884.

[55] D.-Z. Du, F. Cao, F. Hsu, "de Bruijn Digraph, Kautz Digraph, And Their Generalizations", *Combinatorial Network Theory*, D.-Z. Du, Hsu D.F. eds, Kluwer Academic Publishers, 1995, pp. 65-105.

[56] Georgios Ellinas, Krishna Bala, G.K. Chang, "Scalability of a Novel Wavelength Assignment Algorithm for WDM Shared Protection Rings", *OFC'98*, San Jose, CA, February, 1998.

[57] Patrick Emery, Carol Cowley, "Not all optical fibers were created equal", *Lightwave Xtra!*, July 1998.

[58] R.D. Feldman, E.E. Harstead, S. Jiang, T.H. Wood, and M. Zirngibl, "An evaluation of architectures incorporating wavelength division multiplexing for broad-band fiber access", Journal of Lightwave Technology, Volume: 169, Sept. 1998, pp 1546 –1559.

[59] T. Erlbach, K. Jansen, "Scheduling of Virtual Connections in Fast Networks", *Proceedings of the 4th Parallel Systems and Algorithms Workshop (PASA'96)*, World Scientific Publishing, 1997, pp. 13–32.

[60] T. Erlbach, K. Jansen, "Call Scheduling in Trees, Rings and Meshes", *Proceedings of the 30th Hawaii International Conference on System Sciences (HICSS-30)*, IEEE Computer Society Press, 1997, pp. 221–222.

[61] M. Fukui, K. Oda, H. Toba, K. Okamoto, and M. Ishii, "10 channel 10Gbit/s WDM add/ drop multiplexing/transmission experiment over 240 km of dispersion-shifted fiber employing unequally-spaced arrayed-waveguide-grating ADM filter with fold-back configuration", *Electron. Lett.*, vol. 31, no. 20, pp. 1757-1759, Sept. 1995.

[62] M. R. Garey and D. S. Johnson. Computers and Intractability: *A Guide to the Theory of NP-Completeness.* W.H. Freeman and Co., San Francisco, CA, 1979.

[63] M. Garey, D. Johnson, G. Miller, and C. Papadimitriou, The complexity of coloring circular arc graphs and chords, *SIAM Journal of Discrete Math*, vol. 1, no. 2, pp. 216-227, 1980.

[64] N. Garg, "Multicommodity Flows and Approximation Algorithms", Ph.D thesis, Department of Computer Science and Engineering, Indian Institute of Technology, Delhi, April 1994.

[65] M. Gerla, L. Fratta, " Tree Structured Fiber Optics MANs", *IEEE Journal on Selected Areas in Communications*, vol. 6, pp. 934-943, July 1988.

[66] O. Gerstel, G. Sasaki, and R. Ramaswami, "Dynamic channel assignment for WDM optical networks with little or no wavelength conversion", *Proc. Allerton* 1996, Monticello, IL, Oct. 1996, pp. 32–43.

[67] O. Gerstel, P. Lin, and G. Sasaki, "Wavelength assignment in a WDM ring to minimize cost of embedded SONET rings", *Proc. IEEE INFOCOM'98* (San Francisco, CA), Vol. 1, pp. 94 –101.

[68] O. Gerstel, P. Lin, and G. Sasaki, "Combined WDM and SONET Network Design", *Proc. IEEE INFOCOM'99* (New York, NY), Vol. 2, pp. 734 –743.

[69] R.L. Graham, M. Grötschel, and L. Lóvasz, *Handbook of Combinatorics*, MIT Press, 1995.

[70] P.E. Green, Jr., *Fiber Optic Networks*, Prentice Hall, 1993.

[71] I. Haque, W. Kremer, and K. Raychauduri, Self-Healing Rings in a synchronous environment, *SONET/SDH: a sourcebook of synchronous networking*, Eds. C.A. Siller and M. Shafi, IEEE Press, New York, pp. 131-139, 1996.

[72] J. Haralambides, F. Makedon, and B. Monien, "Bandwidth minimization: an approximation algorithm for caterpillars", *Math. Systems Theory* 24, 169-177.

[73] J. Hecht, *Understanding Lasers: An Entry-Level Guide*, New York: IEEE Press, 1992.

[74] J. Hecht, *Understanding Lasers: An Entry-Level Guide to Fiber Optics*, vol. 2. Indianapolis, IN: Sams Publishing, 1993

[75] P.S. Henry, "Lightwave Primer", *IEEE Journal of Quantum Electronics*, vol. QE-21, pp. 1862-1879., December, 1985.

[76] H. Herrmann, K. Schafer, and C. Schmidt, "Low loss tunable integrated acoustooptical wavelength filter in $LiNbO_3$ with strong sidelobe suppression, *IEEE PTL*, vol. 10, Jan. 1998", pp. 120-22.

[77] M.G. Hluchyj, M.J. Karol, "Shuffle Net: An Application of Generalized Perfect Shuffles to Multihop Lightwave Networks", *Journal of Lightwave Technology*, vol. 9, no. 10, Oct 91, pp. 1386-1396.

[78] K.J. Hood et al., "Optical Distribution Systems for Television Studio Applications", *Journal of Lightwave Technology*, vol. 11, pp. 680-687, No. 5/6, May/June 1993.

[79] K. Hwang, F.A. Briggs, "Computer architecture and parallel processing", *New York : McGraw-Hill*, c1984.

[80] M. Imase, M. Itoh, "Design to Minimize a Diameter on Building Block Network", *IEEE Trans. on Computers*, C-30, 1981, pp. 439-443.

[81] M. Imase, M. Itoh, "A Design for Directed Graph with Minimal Diameter", *IEEE Trans. on Computers*, C-32, 1983, pp. 782-784.

[82] M. Imase, T. Soneoka, K. Okada, "Connectivity of Regular Directed Graphs with Small Diameter", *IEEE Trans. on Computers*, C-34, 1985, pp. 267-273.

[83] J. Iness, S. Banerjee, B. Mukherjee, "GEMNET: A Generalized, Shuffle-Exchange-Based, Regular, Scalable, Modular, Multihop, WDM Lightwave Network", *IEEE/ACM Transactions on Networking*, vol. 3, No. 4, pp. 470-476, August 1995.

[84] A. Iocco, H. G. Limberger, and R. P. Salathe, "Bragg grating fast tunable filter", *Elect. Lett.*, vol. 33, no. 25, Dec. 1997, pp. 2147-48.

[85] O. Ishida, H. Takahashi, S. Suzuki, and Y. Innoue, "Multichannel frequency-selective switch employing an arrayed-waveguide grating multiplexer with fold-back optical paths", *IEEE Photon. Technol. Lett.*, vol. 6, pp. 1219-1221, Oct. 1994.

[86] Joergensen et al., "All-Optical Wavelength Conversion at Bit Rates Above 10Gb/s Using Semiconductor Optical Amplifiers", *IEEE J. Sel. Topics Quantum Elect.*, vol. 3, no. 5, 1997, pp. 1168-80.

[87] J.R. Jump, "YACSIM Reference Manual", Department of Electrical And Computer Engineering, Rice University, 1.2 ed., August 1992.

[88] V. Kann "Maximum bounded H-matching is MAX SNP-complete", *Inform. Process. Lett.* 49, 309-318.

[89] W.H. Kautz, "Bounds on directed (d, k) graphs", *Theory of Cellular Logic Networks And Machines*, AFCRL-68-0668 Final Report, 1968, pp. 20-28.

[90] I.A. Karapetian, "On coloring of arc graphs", *Dokladi of the Academy of Science of the Armenian Soviet Socialist Republic*, 70(5): 306-311, 1980. (In Russian)

[91] R.M. Karp, "Reducibility among combinatorial problems", *Complexity of Computer Computations*, R.E. Miller, J.W. Thatcher, Eds., New York: Plenum Press, 1972. pp. 85–103.

[92] M. Karpinski, and J. Wirtgen, "On approximation hardness of the bandwidth problem", *Technical Report TR 97-041, ECCC*, ftp://ftp.eccc.uni-trier.de/pub/eccc/reports/1997/TR97-041/index.html

[93] M. Karpinski, J. Wirtgen, and A. Zelikovsky, "An approximation algorithm for the bandwidth problem on dense graphs", *Technical Report TR97-017, ECCC*, ftp://ftp.eccc.uni-trier.de/pub/eccc/reports/1997/TR97-004/index.html

[94] S. Khanna, "A polynomial-time approximation scheme for the SONET ring loading problem", *Bell Laboratories Tech. J.*, Vol. 2, No. 2, Spring 1997.

[95] H. A. Kierstead and W. T. Trotter, "An Extremal Problem in Recursive Combinatorics", *Congressus Numerantium*, 33:143-153, 1981.

[96] J. Kleinberg and E. Tardos, "Disjoint Paths in Densely Embedded Graphs", *Proceedings of the 36th Annual IEEE Symposium on Foundations of Computer Science*, pp. 52–61, 1995.

[97] J. Kleinberg, A. Kumar, Wavelength Conversion in Optical Networks, Proc. 10th ACM-SIAM Symposium on Discrete Algorithms, 1999.

[98] T. Kloks, D. Kratsch, and H. Müller, "Approximating the bandwidth for asteroidal triple-free graphs", *Proc. 3rd Ann. European Symp. on Algorithms*, Lecture Notes in Comput. Sci. 979, Springer-Verlag, 434-447.

[99] M.E. Kramer, J. van Leeuwen, "The complexity of wire routing and finding the minimum area layouts for arbitrary VLSI circuits", *Advances in Computing Research 2: VLSI Theory*, F.P. Preparata, Ed., London: JAI Press, 1984. pp. 129–146.

[100] V. Kumar, E. Schwabe, "Improved access to optical bandwidth," *Proc. 8th ACM–SIAM Symp. on Discrete Algorithms*, 1997.

[101] Wing C. Kwong, Guu-Chang Yang, "An Algebraic Approach to the Unequal-Spaced Channel-Allocation Problem in WDM Lightwave Systems", *IEEE TRANSACTIONS ON COMMUNICATIONS*, VOL. 45, NO. 3, MARCH 1997, pp. 352-359.

[102] B. Li , A. Ganz, "Virtual Topologies for WDM Star LANs - The Regular Structures Approach", *IEEE INFOCOM'92*, pp. 1-8, May 1992.

[103] Y. P. Li, and C. H. Henry, "Silica-based optical integrated circuits", IEE Proceedings on Optoelectronics, Volume 143, Issue 5 , Oct. 1996, Page(s): 263 -280.

[104] R.A. Linker, "Frequency Division Multiplexed Optical Networks Using Heterodyne Detection",*IEEE Network Magazine*, von. 3, no. 2, Mar 1989, pp. 13-20.

[105] J. Y. Liu and K. M. Johnson, "Analog smectic C^* ferroelectric liquid crystal Fabry Perot optical tunable filter", *IEEE PTL*, vol. 7, no. 11, Nov. 1995, pp. 1309-11.

[106] L.-W. Liu, X.-Y. Li, P.-J. Wan, and O. Frieder, "Wavelength Assignment in WDM Rings to Minimize SONET ADMs", *Proc. INFOCOM 2000*, (Tel-Aviv, Israel), Vol. 2, pp. 1020 -1025.

[107] Y. Liu, "Dispersion-shifted large-effective-area fiber for amplified high-capacity long-distance systems", *Proceedings of the Optical Fiber Communication Conference*, Dallas, TX (Feb. 16-21, 1997).

[108] John Lively, "Dealing with the critical problem of chromatic dispersion", *Lightwave Xtra!*, September 1998.

[109] L. Lovasz, *Combinatorial Problems and Exercises*, North Holland, 1979.

[110] M. O'Mahony, et. al., "The Design of European Optical Network", *Journal of Lightwave Technology*, vol. 11, pp. 817-828, No. 5/6, May/June 1993.

[111] N.F. Maxemchuck, "Routing in Manhattan Street Network", *IEEE Trans. Commun.,*, vol. COM-35, pp. 503-512, May 1987.

[112] N.F. Maxemchuck, "Regular Mesh Topologies in Local and Metropolitan Area Networks", *AT&T Tech. Journal*, vol. 64, pp. 1659-1686, September 1985.

[113] Kenneth A. McGreer, "Arrayed Waveguide Gratings for Wavelength Routing", *IEEE Communications Magazine*, December 1998, pp. 62-68.

[114] M. Mihail, C. Kaklamanis, S. Rao, "Efficient access to optical bandwidth," *Proc. 36th IEEE Symp. on Foundations of Computer Science*, 1995, pp. 548-557.

[115] E. Modiano and A. Chiu, "Traffic Grooming Algorithms for Minimizing Electronic Multiplexing Costs in Unidirectional SONET/WDM Networks", *CISS '98*, Princeton, NJ, March, 1998.

[116] B. Mukherjee, "WDM-based Local Lightwave Networks-Part I: Single-hop Systems", *IEEE Network*, vol. 6, pp. 12-27, May 1992.

[117] B. Mukherjee, "WDM-based Local Lightwave Networks-Part II: Multihop Systems", *IEEE Network*, vol. 6, pp. 20-32, July 1992.

[118] B. Mukherjee, *Optical Communication Networks*, New York: McGraw-Hill, July 1997.

[119] Derek Nesset, Tony Kelly, and Dominique Marcenac, "All-Optical Wavelength Conversion Using SOA Nonlinearities", *IEEE Communications Magazine*, December 1998, pp. 56-61.

[120] C. Nomikos, S. Zachos, "Coloring Maximum Number of Paths in Graphs", *Workshop on Algorithmic Aspects of Communication*, Bologna, July 11-12, 1997.

[121] R.K. Pankaj, "Architecture for Linear Lightwave Networks", Ph.D thesis, MIT, 1992.

[122] R.K. Pankaj, R. G. Gallager, "Wavelength Requirements of All-Optical Networks", *IEEE/ACM Trans. on Networking* 3 (1995), pp. 269–280.

[123] J.H. Patel , "Performance of Processor-Memory Interconnection for Multiprocessors", *IEEE Trans. on Computers*, vol. C-30, no. 10, Oct 1981, pp. 771-780.

[124] A. Pavan, S. Bhattacharya, D.H.C. Du, "Reverse Channel Augmented Multihop Lightwave Networks", *IEEE Infocom-93 Proceedings.*

[125] A. Pavan, P.-J. Wan, S.-R. Tong, D.H.C. Du: A New Multihop Lightwave Networks Based on Generalized de Bruijin Graphs, *Proceedings of the 21st IEEE Annual Conference on Local Computer Networks*, pp. 498-057, 1996.

[126] A.W. Pollack, "Speed Record: One Trillion Bits a Second", D2, *The New York Times*, March 1, 1996.

[127] C. Qiao, Y. Mei, M. Yoo, X, Zhang, Polymorphic Control for Cost-Effective Design of Optical Networks", in *Multichannel Optical Networks: Theory and Practice*, eds by P.-J. Wan, D.-Z. Du, P.M. Pardalos, Ameriacan Mathematical Press, 1998.

[128] Y. Rabani, "Path Coloring on the Mesh", *Proc. of 37th Annual Symposium on Foundations of Computer Science (FOCS)* (Oct. 1996).

[129] P. Raghavan and U. Upfal, "Efficient Routing in All Optical Networks", *Proceedings of the 26th Annual ACM Symposium on Theory of Computing*, pp. 133–143, 1994.

[130] R. Ramaswami, "Multiwavelength Lightwave Networks for Computer Communication", *IEEE Communication Magazine*, vol. 31, pp. 78-88, February 1993.

[131] S.M. Reddy, D.K. Pradhan, J.G. Kuhl, "Directed Graphs with Minimal Diameter And Maximal Connectivity", *School of Engineering Oakland Univ. Tech. Rep.*, 1980.

[132] G.R. Ritchie, "SONET Lays the Roadbed for Broadband Networks", Networking Management 1990.

[133] John P. Ryan, "WDM: North American Deployment Trends", *IEEE Communications Magazine*, February 1998, pp. 40-44.

[134] Y. Saad, M.H. Schultz, "Topological Properties of Hypercubes", *IEEE Transaction on Computers*, vol. C-37, pp. 867-872, January 1988.

[135] Dan Sadot and Efraim Boimovich, "Tunable Optical Filters for Dense WDM Networks", *IEEE Communications Magazine*, December 1998, pp. 50-55.

[136] I. Saniee, "Optimal routing in self-healing communications networks", *International Transactions in Operations Research*, 3, pp. 187-195, 1996.

[137] A. Schrijver, P.D. Seymour, and P. Winkler. "The ring loading problem", *SIAM J. Disc. Math.*, Vol. 11, No. 1, pp. 1-14, February 1998.

[138] Michael Shimazu, "Emerging 1310-nm amplifiers reopen a legacy window", *Lightwave Xtra!*, August 1998.

[139] J. Simmons, E. Golstein, and A. Saleh, "Quantifying the Benefit of Wavelength Add-Drop in WDM Rings with Distance-Independent and Dependent Traffic", *Journal of Lightwave Technology*, Vol. 17, No. 1, January 1999, pp. 48-57.

[140] K. Sivarajan, R. Ramaswami, "Multihop Lightwave Networks Based on de Bruijin Graphs", *INFOCOM 90*, vol. 2, 1990, pp. 1001-1011.

[141] M. K. Smit, "New focusing and dispersive planar component based on an optical phased array", Electron. Lett., vol. 24, pp. 385–386, 1988.S.

[142] M. K. Smit, and C.V. Dam, "PHASAR-based WDM-devices: Principles, design and applications", IEEE Journal on Selected Topics in Quantum Electronics, Volume 2, Issue 2, June 1996, Page(s): 236 -250.

[143] A. Srivastava et al., "32 x 10 Gb/s wdm transmission over 640 km using broadband, gain-flattened erbium-doped silica fiber amplifiers", *Optical Fiber Communication Conference*, Dallas, TX (Feb. 16-21, 1997), post-deadline paper.

[144] H.S. Stone , "Parallel Processing with the Perfect Shuffle", *IEEE Trans. on Computers*, vol. C-20, Feb 1971, pp. 153-161.

[145] J. Stone and L. W. Stulz, "Pigtailed high finesse tunable FP interferometer with large medium, and small FSR", *Elect. Lett.*, no. 23, 1987, pp. 781-83.

[146] S. Suzuki, A. Himeno, Y. Tachikawa, and Y. Yamada, "Multichannel optical wavelength selective switch with arrayed-waveguide grating multiplexer", Electron. Lett., vol. 30, no. 13, pp. 1091–1092, June 1994.

[147] H. Takahashi, Y. Hibino, and I. Nishi, "Polarization-insensitive arrayed waveguide grating wavelength multiplexer on silicon", Opt. Lett., vol. 3, pp. 491–501, 1992.

[148] Y. Tachikawa, Y. Inoue, M. Ishii, and T. Nozawa, "Arrayed-waveguide grating multiplexer with loop-back optical paths and its applications", J. Lightwave Technol., vol. 14, pp. 977–984, June 1996.

[149] Y. Tachikawa and K. Okamoto, "32 wavelength tunable arrayed-waveguide grating laser based on special input/output arrangement", Electron. Lett., vol. 31, no. 19, pp. 1665–1666, Sept. 1995.

[150] Sheau-Ru Tong, "Efficient Designs for High Speed Network Architectures", *Ph.D. Thesis, University of Minnesota*, Feb. 1994.

[151] A. Tucker, "Coloring a family of circular arcs", *SIAM J. Appl. Math.*, 29(3):493-502, 1975.

[152] R. Vachani, A. Shulman, P. Kubat, J. Ward, "Multicommodity Flows in Ring Networks", *INFORMS Journal on Computing*, 8, pp. 235-242, 1996.

[153] B. H. Verbeek et al., "Integrated four channel Mach-Zehnder multi/demultiplexer fabricated with phosphorous doped SiO_2 waveguides on Si", *J. Lightwave Tech.*, vol. 6, June 1988, pp. 1011-15.

[154] P.-J. Wan: TWDM Multichannel Lightwave Hypercube Networks, *Theoretical Computer Science*, (194)1-2 (1998) pp. 123-136.

[155] P.-J. Wan: Optimal Conflict-free Channel Assignment for an Optical Cluster Interconnection Network Based on Rotator digraphs, *Theoretical Computer Science*, 1998.

[156] P.-J. Wan: Near-Optimal Conflict-free Channel Assignment for an Optical Cluster Hypercube Interconnection Network, *Journal of Combinatorial Optimization* 1, 179-186 (1997).

[157] P.-J. Wan: TWDM Multihop Lightwave Networks Based on Rotator Digraphs, *Proceedings of the Sixth IEEE International Conference on Computer and Communication Networks*, pp. 154-157, 1997.

[158] P.-J. Wan: Conflict-Free Channel Assignment for an Optical Cluster Interconnection Network Based on Star Graphs, accepted by *INFORMATICA*, 1997, also appearing in *Proceeding of the Ninth IASTED International Conference on Parallel and Distributed Computing and Systems*, 1997.

[159] P.-J. Wan: TWDM Lightwave Networks Based on Kautz Digraphs, *Proceedings of the Fifth IEEE International Conference on Computer and Communication Networks*, pp. 196-199, 1996.

[160] P.-J. Wan, L.-W. Liu, "Maximal Throughput in Wavelength-Routed Optical Networks, *DIMACS Workshop on Optical Networks,* April 1998.

[161] P.-J. Wan, L.-W. Liu, and O. Frieder, "Optimal Placement of Wavelength Converters in Trees and Trees of Rings", IEEE ICCCN'99.

[162] P.-J. Wan, L.-W. Liu, and O. Frieder, " Wavelength Assignment to Minimize Requirement on Tunable Range of Optical Transceivers in WDM Networks", IEEE ICCCN'99.

[163] P.-J. Wan, L.-W. Liu, and O. Frieder, "Grooming of Arbitrary Traffic in SONET/WDM Rings", IEEE Globecom'99.

[164] P.-J. Wan, A. Pavan: TWDM Media Access Protocol for Single-hop Lightwave Network, *Proceedings of the 21st IEEE Annual Conference on Local Computer Networks*, pp. 486-490, 1996.

[165] G. T. Wilfong., "Minimizing wavelengths in an all-optical ring network", *Proceedings of 7th International Symposium on Algorithms and Computation*, pages 346-355, 1996.

[166] P. Wilfong, P. Winkler, "Ring Routing and Wavelength Translation", *Proc. 9th ACM-SIAM Symp. on Discrete Algorithms*, 1998.

[167] K.A. Williams, D.H.C. Du, "Time and Wavelength Division Multiplexed Architectures for Optical Passive Stars Networks", *Technical Report 92-44, Computer Science Dept., University of Minnesota.*

[168] J. M. Wiesenfeld, "Gain Dynamics and Associated Nonlinearities in Semiconductor Optical Amplifiers", *Int'. J. High Speed Elect. and Sys.*, vol. 7, no. 1, 1996, pp. 179-222.

[169] L. Wooten et al., "Rapidly tunable narrowband wavelength filter using $LiNbO_3$ unbalanced Mach-Zehnder interferometers", *J. Lightwave Tech.*, vol. 14, no. 11, Nov. 1996, pp. 2530-36.

[170] Dan Yang, "Amplifiers at 1590 nm double the DWDM bandwidth", *Lightwave Xtra!*, September 1998.

[171] M. Yannakakis and F. Gavril,"The Maximum k-Colorable Subgraph Problem for Chordal Graphs", *Information Processing Letters*, vol. 24, pp. 133–137, 1987.

[172] A. C. Yao, "New Algorithms in Bin Packing", technical report, Computer Science Department, Stanford, 1978.

[173] C. E. Zah, F. J. Favire, B. Pathak, R. Bhat, C. Caneau, P. S. D. Lin, A. M. Zirngibl, C. H. Joyner, and L. W. Stulz, "WDM receiver by monolithic integration of an optical preamplifier, waveguide grating router and photodiode array", Electron. Lett., vol. 31, no. 7, pp. 581–582, Mar. 1995.

[174] S. Q. Zheng, "Maximum independent sets of circular-arc graphs: Simplified algorithm and proofs", *Networks*, 1(28): 15-19, 1996.

Index

MIX
Papier aus verantwortungsvollen Quellen
Paper from responsible sources
FSC® C105338

If you have any concerns about our products,
you can contact us on
ProductSafety@springernature.com

In case Publisher is established outside the EU,
the EU authorized representative is:
Springer Nature Customer Service Center GmbH
Europaplatz 3, 69115 Heidelberg, Germany

Printed by Libri Plureos GmbH
in Hamburg, Germany